Reconnecting People with Nature through Agriculture

Reconnecting People with Nature through Agriculture

Editors

José Luis Vicente-Vicente
Cristina Quintas-Soriano
María D. López-Rodríguez

MDPI • Basel • Beijing • Wuhan • Barcelona • Belgrade • Manchester • Tokyo • Cluj • Tianjin

Editors
José Luis Vicente-Vicente
Leibniz Centre for Agricultural Landscape Research (ZALF)
Germany

Cristina Quintas-Soriano
University of Almeria
Spain

María D. López-Rodríguez
Universitat Oberta de Catalunya
Spain

Editorial Office
MDPI
St. Alban-Anlage 66
4052 Basel, Switzerland

This is a reprint of articles from the Special Issue published online in the open access journal *Agriculture* (ISSN 2077-0472) (available at: https://www.mdpi.com/journal/agriculture/special_issues/People_Nature_Agriculture).

For citation purposes, cite each article independently as indicated on the article page online and as indicated below:

LastName, A.A.; LastName, B.B.; LastName, C.C. Article Title. *Journal Name* **Year**, *Volume Number*, Page Range.

ISBN 978-3-0365-3855-6 (Hbk)
ISBN 978-3-0365-3856-3 (PDF)

Cover image courtesy of Dr. José Luis Vicente-Vicente

Contents

About the Editors

José Luis Vicente-Vicente (PhD) is a postdoctoral environmental scientist at the Leibniz Centre for Agricultural Landscape Research (ZALF), currently focused on assessing the biophysical impacts of the implementation of agri-food initiatives in alternative food networks within the EU FoodSHIFT2030 project. He assesses how agroecology can play a key role to the transformation of the agri-food systems. In order to achieve this, he is also working on how foodshed and potential food self-sufficiency analyses can contribute to the transformation. He also works on climate change mitigation and adaptation in agriculture, participating in different research on this topic and being an expert reviewer in some IPCC reports.

Cristina Quintas-Soriano (PhD) is a Marie Curie researcher at the University of Almería (Spain). Currently, she is the principal investigator of the SCALABLE project on "Social-ecological pathways and gender perspectives for future conservation of biocultural mountain agroecosystems". Her interests cover the impacts of land abandonment on rural social–ecological systems and how participatory approaches and the incorporation of gender lenses can reveal sustainable options for managing rural areas.

Maria D. López-Rodríguez (PhD) is a postdoctoral researcher at the University of Almería and associated researcher at the Universitad Oberta de Catalunya (Spain). Her main area of research is focused on conducting inter- and trans-disciplinary research to bridge the links between science and other parts of society in order to advance the governance of social–ecological systems in different cultural, social, and institutional frameworks. Her motivation is to conduct research so that scientific knowledge can be a driver of societal learning and policy changes to advance towards sustainability.

agriculture

MDPI

Editorial

A Transformative (r)Evolution of the Research on Agriculture through Fostering Human-Nature Connectedness—A Special Issue Editorial

José Luis Vicente-Vicente [1,*], Cristina Quintas-Soriano [2,3] and María D. López-Rodríguez [2,4]

1 Leibniz Centre for Agricultural Landscape Research (ZALF), Eberswalder Str. 84, 15374 Müncheberg, Germany
2 Andalusian Centre for the Assessment and Monitoring of Global Change (CAESCG), Department of Biology and Geology, University of Almeria, La Cañada de San Urbano, 04120 Almeria, Spain; cqs572@ual.es (C.Q.-S.); mariad.lopezrod@ual.es or mlopezrodrigu@uoc.edu (M.D.L.-R.)
3 FRACTAL Collective, San Remigio 2, 28022 Madrid, Spain
4 Internet Interdisciplinary Institute (IN3), Universitat Oberta de Catalunya (UOC), 08860 Castelldefels, Spain
* Correspondence: vicente@zalf.de or jlvicentevi@gmail.com

Citation: Vicente-Vicente, J.L.; Quintas-Soriano, C.; López-Rodríguez, M.D. A Transformative (r)Evolution of the Research on Agriculture through Fostering Human-Nature Connectedness—A Special Issue Editorial. *Agriculture* **2022**, *12*, 522. https://doi.org/10.3390/agriculture12040522

Received: 29 March 2022
Accepted: 6 April 2022
Published: 7 April 2022

More and more people live in cities. In recent decades, this, combined with rural abandonment, has resulted in increased land ownership concentration and land grabbing [1–4], with an increase in agricultural intensification [5,6]. This process is leading to an increasingly polarized landscape between abandonment of traditional farming activities and highly intensive agriculture lands. Rural land abandonment is motivated mainly by socio-cultural factors, such as population aging and migration patterns from rural to urban areas [7]. Land abandonment has been described as a complex process with implications at ecological and socio-cultural levels [8]. Primarily, it can support ecological restoration, increase carbon storage or improve habitat quality. However, at social and cultural levels, it can endanger local ecological knowledge, cultural heritage, local identity and can negatively impact rural livelihoods through the loss of agricultural and forest products. On the other hand, highly intense agricultural farming systems are formed by large monocrops, which are extremely simplified systems, very often combined with the application of high rates of pesticides, the plantation of genetically modified species, and the removal of all kinds of wild biological diversity. A similar process has been observed in terms of livestock, with an increase in intensification in farming systems and the appearance of highly intense facilities (i.e., factory farms) [9], to the detriment of the extensive farming systems, which are less economically profitable but have a stronger link to the territory and integration within the available natural resources [10]. This has resulted in trade-offs with different ecosystem services [5,11–14] due to the prioritization of provisioning services (such as food) at the detriment of other supporting, regulating and cultural services. In addition, agricultural intensification is currently threatening the maintenance of traditional indigenous and peasant farming, whose practices have been proven to be beneficial for building up resilient agroecosystems that sustain both ecosystems and societal well-being [15]. This has led, ultimately, to the loss of the connection of people with nature [16,17].

The loss of human–nature connectedness in Western and urbanized societies has one of its paradigmatic examples in the commodification of food, which takes place in a context of an increasingly complex [18,19] and highly vulnerable [20–22] globalized food system. Therefore, it is clear that a transformation of the agri-food system is urgently needed [23].

In this SI, we have collected eleven studies assessing, using a diversity of approaches, how human–nature connectedness can be recovered through agriculture. Many of them are focused on the application of a systemic approach, by considering a set of sustainable agricultural practices, whereas some are focused on studying what management practices can be applied in agricultural systems in order to reconnect people with nature. One article addresses principles of good governance to create inclusive and integrative processes that support healthy communities and resilient ecosystems, whereas another one is focused on the consumer's side in order to foster societal transformation.

Integrating Natural and Social Considerations for the Transformation of Agri-Food Systems

To transform agri-food systems, alternative frameworks, approaches and methodologies are needed. A long-standing framework such agroecology [24] seems to be one of the most suitable methods to include both the ecological and the social dimensions of agri-food systems [25,26] and eventually to contribute to agri-food system change [27]. In this SI, different contributions include agroecology as a framework for the transition toward more sustainable agri-food systems (e.g., [28–31]). However, so far, in the Global North, the majority of the proposed frameworks have been based on applying incremental or reformative practices (e.g., sustainable intensification, sustainable agriculture, conservation agriculture, or integrated pest management) [32], whereas agroecology has been more frequently applied in countries of the Global South [33].

Nevertheless, embracing a new framework is not sufficient to transform the agri-food system. This has to be accompanied by the combination of a set of social research methods and approaches that lead to transformative change in the agricultural systems for sustainability. Thus, in this SI, alternative approaches have been employed, such as conservation biology [34], nature's contribution to people [35], ecosystem services [36,37] and multi-actor authorship approaches [29], which go beyond the biophysical analysis by considering socio-cultural aspects under a more systemic perspective. Most of these approaches imply a broadening in the scale of the study while incorporating knowledge, perceptions, preferences and values from local stakeholders. While typically, agricultural studies are focused at the plot level, embracing these new approaches leads to a broadening of the scale, especially to farm or landscape scales [29,31,35,37–39], and to a lesser extent to regional or foodshed [40] or to country scales [34].

These approaches incorporate methodologies from social sciences that are not usually included in agricultural studies. Thus, many of the studies used questionnaires as their main data source, although acquisition was different depending on the purposes and specificities of the study. For instance, Schwartz et al. [36] used GIS software to develop a participatory mapping exercise, Murillo-López et al. [35] used semi-structured interviews to collect qualitative data, whereas Chen et al. [39] and Gugerell et al. [38] developed questionnaires and interviews for quantitative analysis to evaluate farmers' ecological cognition and different proximities, respectively. However, other studies required mixed complex methods incorporating agroecology and involving multi-actor, agricultural knowledge, and innovation systems [29,37]. Besides these innovative approaches, there are others not included in this SI that can also be used or combined with those mentioned. For instance, very often, agroecology is combined with the use of participatory research methods, such as participatory action research [41], and on the other hand, over the last few years, citizen-science methods have become an emerging topic in sustainable agriculture [42].

Therefore, the application of systemic frameworks comprising a set of approaches and methods covering the socio-ecological dimensions of agriculture would address the study of human–nature connectedness in agricultural systems. However, this would imply the inclusion of methods that are more commonly used in social sciences and ecology. Therefore, we think that future studies on agriculture should focus on a more effective integration of social considerations.

Transdisciplinarity as a Driver of Policy Changes and Sustainability Learning in Agri-Food Systems

In addition to the above-mentioned academic efforts, there is growing acceptance that sustainability transitions also require transdisciplinary work schemes to encourage changes in institutional practices and individual behaviors to progress towards the implementation of a sustainable agri-food system (e.g., [43]). Transdisciplinary work schemes aim to address sustainability challenges by integrating knowledge from various scientific and societal bodies of knowledge [44] through co-learning and knowledge co-production processes [45].

In doing so, scientists from many disciplines (e.g., ecology, agriculture, sociology, and anthropology) and non-scientists (e.g., decision makers, technicians, farmers, and local communities) work together to find evidence-based solutions to deal with policy needs and societal concerns in the field of sustainability. The conceptual basis of transdisciplinary science recognizes that research questions and solutions are framed in policy and societal contexts to provide realistic and context-specific pathways to help the policy community and social actors progress towards a sustainable future [46]. On this basis, transdisciplinary work schemes create methodological tools through which scientists can adopt a more active role and produce mission-oriented research and innovation to help agricultural systems transit towards sustainability [47]. Therefore, they can be considered a means for scientific knowledge to be a driver of societal learning and policy changes while facilitating a culture of shared responsibility for sustainability among the public, academia, private sectors, and consumers to advance towards sustainability [48]. Even though transdisciplinary research has been gradually increasing in multiple areas of sustainable governance such as water, forests, and even agriculture [49,50], more studies from local to global scales are needed to foster people's awareness of our dependence on nature and to collectively support a real transformative (r)evolution of the agri-food system towards sustainability. In this SI, some studies contribute to progress in this direction by providing successful multi-actor initiatives that show how collaborative work at different scales can generate positive impacts to reorientate agri-food systems towards sustainability [30,31].

Shifting the Paradigm in the Research on Agriculture

The need for a reconnection of people with nature implies a shift in the paradigm that places the agricultural system within nature and its ecological boundaries. For that purpose, sustainable production and consumption should be the starting point to develop agricultural and societal transformation pathways (Figure 1). As the articles presented in this SI show, pathways toward sustainability can be based on scientific evidence acquired through the application of systemic frameworks, including new socio-ecological approaches and socio-cultural methodologies. In addition, inclusive governance approaches will be needed in order to provide suitable conditions for developing sustainable transformative transitions (Figure 1).

We have identified some specific aspects that, in our opinion, should be changed in agricultural research as part of this paradigm shift:

- From "crops" to "agroecosystems". Reducing the consideration of agricultural systems to just crops implies treating food only as a commodity, and therefore, to something that can be easily displaced far from its place of origin. On the contrary, when looking at crops within agroecosystems, it is acknowledged that food—a provisioning service—is produced through complex socio-ecological relationships, and therefore, its production has an impact on many different ecosystem services (e.g., soil fertility, biodiversity, climate regulation, and culture). Thereby, recognizing agri-food systems as coupled human–environment systems would support enhanced sustainability outcomes from agroecosystems.
- From the "plot" to the "landscape" vision of a "farm". The scale of a plot does not include the many ecological relationships that are taking place at the landscape level. A farm should be managed as a whole, as a habitat, instead of being perceived as a set of plots that can be managed separately. Future research should be focused on the farm level, in order to capture different socio-ecological impacts. In addition, this scale may make the multiple human–nature connections that these areas provide visible.
- From "agriculture" to "(agri) food systems". Agriculture is just one part of the whole agri-food system. By tackling only one side of the system we are avoiding the consumption side as well as the socio-cultural aspects of food production. In order to achieve sustainability, it will be mandatory to consider the multiple dimensions of agri-food systems, as well as the telecouplings associated with them.

- From "food as a commodity" to a sustainable food system of "landscape products". The study of agri-food systems is more than just studying the interactions along the supply chain or the food network. It should also imply the study of the social–ecological conditions and the non-chain actors in the areas of production. In this sense, the incorporation of landscape products as a way to consider food from multiple (ecological, social, and economic) dimensions will promote more resilient social–ecological systems [51].
- From the "social", "ecological" or "agronomic" perspectives to the "transdisciplinarity" vision. Agri-food systems should be addressed by people with different epistemologies, backgrounds and perspectives, tackling it as a complex system with multiple ramifications and interlinkages, which requires the alignment of scientific advances, policy needs, and societal concerns at different scales to transit collectively towards sustainability.
- From the "top-down/hierarchical and sectoral decision-making", to the "inclusive and integrative governance". So far, traditional governance approaches based on top-down models and sectoral policies across different scales have not necessarily resulted in positive outcomes for sustainability. The active participation of stakeholders and local communities in governance systems is increasingly recognized as crucial to strengthening the links between governmental and non-governmental institutions to: (1) facilitate social learning processes that encourage building understanding and trust in the sustainability framework, and (2) promote the policy community and the rest of society to act as agents of change to advance together towards sustainability. Moreover, the articulation of institutional efforts at multiple scales and sectors is pivotal in developing coherent policies and actions that support biodiversity conservation and human well-being. By implementing inclusive and integrative governance approaches it would be possible to develop the policy and social changes needed to implement sustainable transformative transitions in agri-food systems.

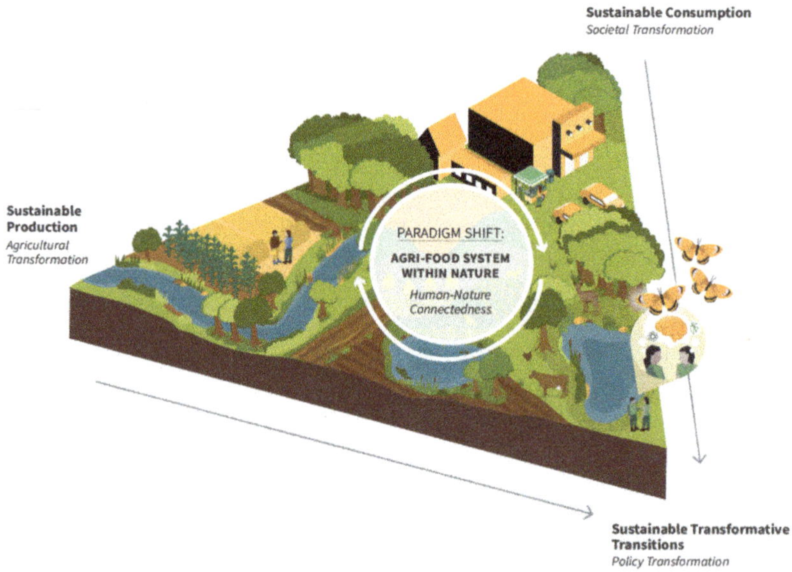

Figure 1. A shift in the paradigm is needed to reconnect people with nature. The new paradigm uses sustainable production and consumption as the starting points to develop a transformative transition to an agri-food system within nature. To achieve that, a policy transformation based on inclusive and integrative governance approaches is needed, enabling conditions for fostering human-nature connectedness, and eventually, for developing sustainable transitions.

Our planet is in a state of emergency. Agri-food systems can worsen the problem (e.g., by emitting greenhouse gasses or contributing to the decrease in biodiversity) or can be part of the solution, by mitigating (e.g., soil organic carbon sequestration, reducing the emission of greenhouse gasses or fostering biodiversity) and adapting (e.g., resilient agroecosystems and food networks) to climate change. There is wide consensus that the current globalized agri-food system is contributing to worsening the social–ecological planetary emergency, and that transformative solutions are urgently needed. We encourage researchers in agriculture to adopt the aforementioned paradigm shift and to develop transformative studies so that we can all contribute to creating agri-food systems within the planetary boundaries.

Author Contributions: Conceptualization, J.L.V.-V., C.Q.-S. and M.D.L.-R.; investigation, J.L.V.-V., C.Q.-S. and M.D.L.-R..; writing—original draft preparation, J.L.V.-V., C.Q.-S. and M.D.L.-R.; writing—review and editing, J.L.V.-V., C.Q.-S. and M.D.L.-R.; visualization, J.L.V.-V.; supervision, J.L.V.-V. All authors have read and agreed to the published version of the manuscript.

Acknowledgments: We thank Olivia S. Meyer for supporting us with the graphic design and Annette Piorr for providing insights for the elaboration of this text. We also thank all the authors that have contributed to this Special Issue.

Conflicts of Interest: The authors declare no conflict of interest.

References

1. Bunkus, R.; Theesfeld, I. Land Grabbing in Europe? Socio-Cultural Externalities of Large-Scale Land Acquisitions in East Germany. *Land* **2018**, *7*, 98. [CrossRef]
2. Estel, S.; Kuemmerle, T.; Alcántara, C.; Levers, C.; Prishchepov, A.; Hostert, P. Mapping Farmland Abandonment and Recultivation across Europe Using MODIS NDVI Time Series. *Remote Sens. Environ.* **2015**, *163*, 312–325. [CrossRef]
3. Palšová, L.; Bandlerová, A.; Machničová, Z. Land Concentration and Land Grabbing Processes—Evidence from Slovakia. *Land* **2021**, *10*, 873. [CrossRef]
4. Van der Zanden, E.H.; Verburg, P.H.; Schulp, C.J.E.; Verkerk, P.J. Trade-Offs of European Agricultural Abandonment. *Land Use Policy* **2017**, *62*, 290–301. [CrossRef]
5. Clough, Y.; Kirchweger, S.; Kantelhardt, J. Field Sizes and the Future of Farmland Biodiversity in European Landscapes. *Conserv. Lett.* **2020**, *13*, e12752. [CrossRef]
6. Bruno, D.; Sorando, R.; Álvarez-Farizo, B.; Castellano, C.; Céspedes, V.; Gallardo, B.; Jiménez, J.J.; López, M.V.; López-Flores, R.; Moret-Fernández, D.; et al. Depopulation Impacts on Ecosystem Services in Mediterranean Rural Areas. *Ecosyst. Serv.* **2021**, *52*, 101369. [CrossRef]
7. Rey Benayas, J.M. Abandonment of Agricultural Land: An Overview of Drivers and Consequences. *CAB Rev.* **2007**, *2*, 1–14. [CrossRef]
8. Quintas-Soriano, C.; Buerkert, A.; Plieninger, T. Effects of Land Abandonment on Nature Contributions to People and Good Quality of Life Components in the Mediterranean Region: A Review. *Land Use Policy* **2022**, *116*, 106053. [CrossRef]
9. Sanchez Carpio, D.; Stanka, B. *The Urgent Case to Stop Factory Farms in Europe*; Friends of the Earh Europe and Food & Water Action Europe: Brussels, Belgium, 2020; p. 36.
10. Bertolozzi-Caredio, D.; Garrido, A.; Soriano, B.; Bardaji, I. Implications of Alternative Farm Management Patterns to Promote Resilience in Extensive Sheep Farming. *J. Rural Stud.* **2021**, *86*, 633–644. [CrossRef]
11. Raven, P.H.; Wagner, D.L. Agricultural Intensification and Climate Change Are Rapidly Decreasing Insect Biodiversity. *Proc. Natl. Acad. Sci. USA* **2021**, *118*, e2002548117. [CrossRef]
12. Eggleton, P. The State of the World's Insects. *Annu. Rev. Environ. Resour.* **2020**, *45*, 61–82. [CrossRef]
13. Dreoni, I.; Matthews, Z.; Schaafsma, M. The Impacts of Soy Production on Multi-Dimensional Well-Being and Ecosystem Services: A Systematic Review. *J. Clean. Prod.* **2022**, *335*, 130182. [CrossRef]
14. Machovina, B.; Feeley, K.J.; Ripple, W.J. Biodiversity Conservation: The Key Is Reducing Meat Consumption. *Sci. Total Environ.* **2015**, *536*, 419–431. [CrossRef]
15. FAO. *Indigenous Peoples' Food Systems: Insights on Sustainability and Resilience from the Front Line of Climate Change*; FAO, Alliance of Bioversity International, and CIAT: Rome, Italy, 2021.
16. Pacheco-Romero, M.; Kuemmerle, T.; Levers, C.; Alcaraz-Segura, D.; Cabello, J. Integrating Inductive and Deductive Analysis to Identify and Characterize Archetypical Social-Ecological Systems and Their Changes. *Landsc. Urban Plan.* **2021**, *215*, 104199. [CrossRef]
17. Riechers, M.; Pătru-Dușe, I.A.; Balázsi, Á. Leverage Points to Foster Human–Nature Connectedness in Cultural Landscapes. *Ambio* **2021**, *50*, 1670–1680. [CrossRef]
18. Puma, M.J. Resilience of the Global Food System. *Nat. Sustain.* **2019**, *2*, 260–261. [CrossRef]

19. Puma, M.J.; Bose, S.; Chon, S.Y.; Cook, B.I. Assessing the Evolving Fragility of the Global Food System. *Environ. Res. Lett.* **2015**, *10*, 024007. [CrossRef]
20. Savary, S.; Akter, S.; Almekinders, C.; Harris, J.; Korsten, L.; Rötter, R.; Waddington, S.; Watson, D. Mapping Disruption and Resilience Mechanisms in Food Systems. *Food Sec.* **2020**, *12*, 695–717. [CrossRef]
21. Bailey, R.; Wellesley, L. *Chokepoints and Vulnerabilities in Global Food Trade*; Chatham House, the Royal Institute of International Affairs: London, UK, 2017; ISBN 978-1-78413-230-9.
22. Kummu, M.; Kinnunen, P.; Lehikoinen, E.; Porkka, M.; Queiroz, C.; Röös, E.; Troell, M.; Weil, C. Interplay of Trade and Food System Resilience: Gains on Supply Diversity over Time at the Cost of Trade Independency. *Glob. Food Secur.* **2020**, *24*, 100360. [CrossRef]
23. Webb, P.; Benton, T.G.; Beddington, J.; Flynn, D.; Kelly, N.M.; Thomas, S.M. The Urgency of Food System Transformation Is Now Irrefutable. *Nat. Food* **2020**, *1*, 584–585. [CrossRef]
24. Altieri, M.A.; Farrell, J.G.; Hecht, S.B.; Liebman, M.; Magdoff, F.; Murphy, B.; Norgaard, R.B.; Sikor, T.O. *Agroecology: The Science of Sustainable Agriculture*, 2nd ed.; CRC Press: Boca Raton, FL, USA, 2019; ISBN 978-0-429-49546-5.
25. FAO. *The 10 Elements of Agroecology: Guiding the Transition to Sustainable Food and Agricultural Systems*; FAO: Italy, Rome, 2018.
26. HLPE HLPE Report #14—Agroecological and Other Innovative Approaches for Sustainable Agriculture and Food Systems That Enhance Food Security and Nutrition. 2019, p. 163. Available online: https://www.fao.org/family-farming/detail/en/c/1263887/ (accessed on 28 March 2022).
27. Gliessman, S. Transforming Food Systems with Agroecology. *Agroecol. Sustain. Food Syst.* **2016**, *40*, 187–189. [CrossRef]
28. Mehrabi, S.; Perez-Mesa, J.C.; Giagnocavo, C. The Role of Consumer-Citizens and Connectedness to Nature in the Sustainable Transition to Agroecological Food Systems: The Mediation of Innovative Business Models and a Multi-Level Perspective. *Agriculture* **2022**, *12*, 203. [CrossRef]
29. Giagnocavo, C.; de Cara-García, M.; González, M.; Juan, M.; Marín-Guirao, J.I.; Mehrabi, S.; Rodríguez, E.; van der Blom, J.; Crisol-Martínez, E. Reconnecting Farmers with Nature through Agroecological Transitions: Interacting Niches and Experimentation and the Role of Agricultural Knowledge and Innovation Systems. *Agriculture* **2022**, *12*, 137. [CrossRef]
30. Wilkes, J. Reconnecting with Nature through Good Governance: Inclusive Policy across Scales. *Agriculture* **2022**, *12*, 382. [CrossRef]
31. González-Rosado, M.; Parras-Alcántara, L.; Aguilera-Huertas, J.; Lozano-García, B. Building an Agroecological Process towards Agricultural Sustainability: A Case Study from Southern Spain. *Agriculture* **2021**, *11*, 1024. [CrossRef]
32. Pretty, J. Intensification for Redesigned and Sustainable Agricultural Systems. *Science* **2018**, *362*, eaav0294. [CrossRef]
33. Altieri, M.A.; Nicholls, C.I. Agroecology: A Brief Account of Its Origins and Currents of Thought in Latin America. *Agroecol. Sustain. Food Syst.* **2017**, *41*, 231–237. [CrossRef]
34. Molina-Pardo, J.L.; Rodríguez-Caballero, E.; Cueto, M.; Barranco, P.; Sánchez-Robles, M.; Laguía-Allué, A.; Giménez-Luque, E. Effects of Agricultural Use on Endangered Plant Taxa in Spain. *Agriculture* **2021**, *11*, 1097. [CrossRef]
35. Murillo-López, B.E.; Castro, A.J.; Feijoo-Martínez, A. Nature's Contributions to People Shape Sense of Place in the Coffee Cultural Landscape of Colombia. *Agriculture* **2022**, *12*, 457. [CrossRef]
36. Schwartz, C.; Shaaban, M.; Bellingrath-Kimura, S.D.; Piorr, A. Participatory Mapping of Demand for Ecosystem Services in Agricultural Landscapes. *Agriculture* **2021**, *11*, 1193. [CrossRef]
37. Acebes, P.; Iglesias-González, Z.; Muñoz-Galvez, F.J. Do Traditional Livestock Systems Fit into Contemporary Landscapes? Integrating Social Perceptions and Values on Landscape Change. *Agriculture* **2021**, *11*, 1107. [CrossRef]
38. Gugerell, C.; Sato, T.; Hvitsand, C.; Toriyama, D.; Suzuki, N.; Penker, M. Know the Farmer That Feeds You: A Cross-Country Analysis of Spatial-Relational Proximities and the Attractiveness of Community Supported Agriculture. *Agriculture* **2021**, *11*, 1006. [CrossRef]
39. Chen, Z.; Sarkar, A.; Hasan, A.K.; Li, X.; Xia, X. Evaluation of Farmers' Ecological Cognition in Responses to Specialty Orchard Fruit Planting Behavior: Evidence in Shaanxi and Ningxia, China. *Agriculture* **2021**, *11*, 1056. [CrossRef]
40. Mouléry, M.; Sanz Sanz, E.; Debolini, M.; Napoléone, C.; Josselin, D.; Mabire, L.; Vicente-Vicente, J.L. Self-Sufficiency Assessment: Defining the Foodshed Spatial Signature of Supply Chains for Beef in Avignon, France. *Agriculture* **2022**, *12*, 419. [CrossRef]
41. López-García, D.; Cuéllar-Padilla, M.; de Azevedo Olival, A.; Laranjeira, N.P.; Méndez, V.E.; Peredo y Parada, S.; Barbosa, C.A.; Barrera Salas, C.; Caswell, M.; Cohen, R.; et al. Building Agroecology with People. Challenges of Participatory Methods to Deepen on the Agroecological Transition in Different Contexts. *J. Rural Stud.* **2021**, *83*, 257–267. [CrossRef]
42. Ebitu, L.; Avery, H.; Mourad, K.A.; Enyetu, J. Citizen Science for Sustainable Agriculture–A Systematic Literature Review. *Land Use Policy* **2021**, *103*, 105326. [CrossRef]
43. Lewandowski, I. *Bioeconomy: Shaping the Transition to a Sustainable, Biobased Economy*; Springer: Berlin/Heidelberg, Germany, 2017; ISBN 978-3-319-68152-8.
44. Lang, D.J.; Wiek, A.; Bergmann, M.; Stauffacher, M.; Martens, P.; Moll, P.; Swilling, M.; Thomas, C.J. Transdisciplinary Research in Sustainability Science: Practice, Principles, and Challenges. *Sustain. Sci.* **2012**, *7*, 25–43. [CrossRef]
45. Angelstam, P.; Andersson, K.; Annerstedt, M.; Axelsson, R.; Elbakidze, M.; Garrido Rodriquez, P.; Grahn, P.; Jönsson, K.I.; Pedersen, S.; Schlyter, P.; et al. Solving Problems in Social-Ecological Systems: Definition, Practice and Barriers of Transdisciplinary Research. *Ambio* **2013**, *42*, 254–265. [CrossRef]

46. Spangenberg, J.H. Sustainability Science: A Review, an Analysis and Some Empirical Lessons. *Environ. Conserv.* **2011**, *38*, 275–287. [CrossRef]
47. Egea, F.J.; López-Rodríguez, M.D.; Oña-Burgos, P.; Castro, A.J.; Glass, C.R. Bioeconomy as a Transforming Driver of Intensive Greenhouse Horticulture in SE Spain. *New Biotechnol.* **2021**, *61*, 50–56. [CrossRef]
48. López-Rodríguez, M.D. Science—Policy Interfaces in the Area of Environmental Governance: Empirical Tests for Its Methodological Development. PhD Thesis, University of Almería, Almería, Spain, 2016.
49. Schneider, F.; Fry, P.; Ledermann, T.; Rist, S. Social Learning Processes in Swiss Soil Protection—The 'From Farmer–To Farmer' Project. *Hum. Ecol.* **2009**, *37*, 475–489. [CrossRef]
50. Castro, A.J.; López-Rodríguez, M.; Giagnocavo, C.; Giménez, M.; Céspedes, L.; La Calle, A.; Gallardo, M.; Pumares, P.; Cabello, J.; Rodríguez, E.; et al. Six Collective Challenges for Sustainability of Almería Greenhouse Horticulture. *Int. J. Environ. Res. Public Health* **2019**, *16*, 4097. [CrossRef]
51. Garcia-Martin, M.; Torralba, M.; Quintas-Soriano, C.; Kahl, J.; Plieninger, T. Linking Food Systems and Landscape Sustainability in the Mediterranean Region. *Landsc. Ecol.* **2021**, *36*, 2259–2275. [CrossRef]

agriculture

MDPI

Article

Know the Farmer That Feeds You: A Cross-Country Analysis of Spatial-Relational Proximities and the Attractiveness of Community Supported Agriculture

Christina Gugerell [1,*], Takeshi Sato [2], Christine Hvitsand [3], Daichi Toriyama [2], Nobuhiro Suzuki [2] and Marianne Penker [1]

1 Department of Economics and Social Sciences, Institute for Sustainable Economic Development, University of Natural Resources and Life Sciences, Vienna, 1180 Vienna, Austria; marianne.penker@boku.ac.at
2 Department of Global Agricultural Sciences, University of Tokyo, Tokyo 113-8654, Japan; atakeshi@mail.ecc.u-tokyo.ac.jp (T.S.); toriyama-daichi@g.ecc.u-tokyo.ac.jp (D.T.); asuzukiz@mail.ecc.u-tokyo.ac.jp (N.S.)
3 Faculty of Biosciences, Norwegian University of Life Sciences, 1432 Ås, Norway; christine.hvitsand@nmbu.no
* Correspondence: christina.gugerell@boku.ac.at; Tel.: +436-641-308-606

Citation: Gugerell, C.; Sato, T.; Hvitsand, C.; Toriyama, D.; Suzuki, N.; Penker, M. Know the Farmer That Feeds You: A Cross-Country Analysis of Spatial-Relational Proximities and the Attractiveness of Community Supported Agriculture. *Agriculture* **2021**, *11*, 1006. https://doi.org/10.3390/agriculture11101006

Academic Editors: José Luis Vicente-Vicente, Cristina Quintas-Soriano and María D. López-Rodríguez

Received: 16 August 2021
Accepted: 7 October 2021
Published: 14 October 2021

Abstract: While food production and consumption processes worldwide are characterized by geographical and social distance, alternative food networks aim to reconnect producers and consumers. Our study proposes a framework to distinguish multiple dimensions of proximity in the context of Community Supported Agriculture (a type of alternative food network) and to quantitatively evaluate them. In a principal component analysis, we aggregated various detailed proximity items from a multinational survey using principal component analysis and examined their relationship with the attractiveness of Community Supported Agriculture in a multiple regression analysis. Our findings highlight the importance of relational proximity and thus of increasing trust, collaboration, and the sharing of values and knowledge within and across organizations in the food system. Rather than focusing on spatial proximity, increasing relational proximity might support alternative food networks, such as Community Supported Agriculture.

Keywords: community supported agriculture; alternative food networks; spatial proximity; relational proximity; cross-national case study

1. Introduction

The current agricultural and food industry is based on labor division and connects companies in different regions, countries, and sometimes also continents [1]. As a result, production and consumption processes often take place at a great geographical and social distance [2]. Alternative food networks (AFNs) aim to overcome this distance by anchoring food in its socio-ecological context and thus promote direct producer-consumer relationships [3,4]. AFNs therefore pose an alternative to the mainstream, industrial food system [3,5]. Community Supported Agriculture (CSA) is a food production and distribution model in which farming responsibilities, risks, and rewards are shared between farmers and consumers [6]. They have mainly been established in or close to urban areas, where people are more spatially separated and alienated from food production than people in rural areas [7,8].

Previous research on AFNs has pointed especially to the importance of spatial dynamics and the essential role of place in building alternative food systems [9,10], as well as the socio-cultural embeddedness of food in local relations of food provision [3,11]. Therefore, scholars highlighted the positive effects of local contexts on social ties and innovation processes [3]. However, food system actors are interconnected due to various spatial–relational configurations [12]. Close producer–consumer relations may also be performed "at a distance" [13–15]. To provide an attractive alternative to conventional food provision, AFNs aim to rebuild production and consumption processes [5].

In this context, we find it purposeful to utilize the term "proximity" and Boschma's [16] differentiation between relational (i.e., social, cognitive, institutional, and organizational) proximity and spatial (i.e., geographical) proximity dimensions [16,17]. Using the proximity concept could be one way to expand our knowledge of what makes AFNs, such as CSAs, attractive, and to better understand what constitutes attractive relationships between CSA members (i.e., consumers and producers) and between CSA members and society in general [16,18].

While CSA literature [19,20] highlights implications of geographical proximity, to our knowledge, only one study related Boschma's [16] broader perspective on proximity dimensions to CSAs [21]. However, in an ever-evolving body of knowledge, critical questions on various spatial–relational configurations associated with AFNs are being debated [15,22–25]. With respect to CSAs, this includes motivations to join the CSA scheme [26–30], challenges CSAs face in retaining members [31–34], the institutionalization of CSA principles [35] and up-scaling processes [36,37], as well as the extent to which CSAs succeed in creating an alternative to conventional practices in the market [3]. Furthermore, the appeal of CSAs has been investigated in previous studies [38–41]. Interrelating the latter to the different dimensions of spatial and relational proximity configurations promises new insights for better understanding the role of spatial–relational proximity for the attractiveness of CSA and other AFNs. Thus, we also hope to gain some insights into what factors should be used to promote AFNs—a knowledge gap that has been attributed to their recentness [25].

More generally, we want to contribute to relational rural sociology. In theory, human-to-human relations and relations between humans and their bio-physical context (farm, land, infrastructure) are well debated (for an overview, see [42]). However, the relational perspective still poses various methodological challenges, such as shifting the analytical attention from nodes, objects, and subjects to their relations [42]. Taking the example of CSA, we want to demonstrate that proximity theory can help to operationalize geographical, social, cognitive, institutional, and organizational relations of CSA members with their social and bio-physical contexts using a quantitative multi-variate analysis and thus complement Actor–Network Theory, providing graphical or visceral methods that help to empirically analyze human-to-human, human–technology, or human–nature relations [42].

Our literature analysis revealed that there are hardly any studies quantitatively differentiating between spatial–relational proximity dimensions and their role in AFN attractiveness. Taking the example of CSA, an AFN implemented in different parts of the world, this study examines the interrelation of spatial–relational proximity with CSA's attractiveness. CSA attractiveness has been investigated in several studies, but, to our knowledge, not yet regarding different proximity dimensions. More generally, the measurement of organizational attraction dates back to early research, such as Vroom [43], who measured the attractiveness of different organizations to potential job seekers using a single item. A few years later, Singh [44] applied information integration theory to organization choice using a single item that assessed the likelihood of accepting a job with the company. We assume that organizational attractiveness can also help to understand the membership in non-profit organizations, such as CSAs. Recent studies have analyzed member satisfaction within CSAs [38–41]. In the literature, CSA attractiveness and satisfaction have been measured with single items, so there is no multi-item attractiveness scale yet.

The empirical analysis is based on data from several countries. We selected Austria, Japan, and Norway for this cross-national case study, as their national CSA movements have developed differently. However, the organization of CSA movements in these countries is similar (see Section 3 for further justification of study sites).

By interviewing CSA members in different (peri-)urban contexts, we aim to understand better the relevance of proximity dimensions for the attractiveness of the CSA model. We distinguish between spatial and relational proximity among CSA members (CSA-internal proximity) as well as between CSA members and CSA-external actors, structures, and resources (CSA-external proximity). The central research question of our study is: How are spatial and relational proximity within and outside CSAs related to the attractiveness of CSAs in (peri-)urban contexts? Based on proximity and the CSA literature (see Section 2), we hypothesize that there is a positive correlation between all dimensions of social proximity and attractiveness, except for institutional and organizational proximity to external actors (as members may seek to distance themselves from dominant food organizations and deviate from prevailing rules and standards).

This paper is structured as follows. First, we briefly review proximity literature and present assumptions about proximity and CSAs (Section 2). We then describe our research design and data collection process in Section 3. In Section 4, we create proximity variables using principal components analysis. In a multiple linear regression, we analyze the interrelation between these proximity variables and CSA attractiveness. Section 5 discusses the results and the limitations of the study. Finally, in Section 6, we conclude the paper by highlighting its empirical and methodological contributions.

2. Theoretical Background on Proximity and Operationalization for CSA

Theoretical definitions of proximity dimensions have been proposed by scholars [16,45,46] aiming to understand the coordination of economic activities. Boschma [16] differentiated between five dimensions of proximity: geographical proximity (i.e., spatial proximity), as well as social, institutional, cognitive, and organizational proximity. The latter four can be subsumed under the umbrella of relational proximity (i.e., non-spatial proximity), because they conceptually overlap (i.e., they are intangible dimensions based on affinity and similarity) and often coexist in practice [47]. The five proximity dimensions were later adapted to the field of sustainability innovation [48]. The sustainability of AFNs, such as CSAs, has been addressed in previous studies [49–51]. The CSA concept represents an alternative, sustainability-oriented model of food provision that addresses social justice, community, and environmental sustainability. Thus, we conceptualize CSA as a social innovation [52,53]. While previous scholars have examined proximity dimensions with a focus on innovation [16], this paper analyzes the exploratory value of proximity dimensions for CSA attractiveness. Since proximity dimensions have not previously been operationalized for analyzing CSA attractiveness, we ground our assumptions on a broader base in the literature on proximity and CSA.

Scholars associate geographical proximity with physical distance between actors [16,48] and local availability of natural resources [48]. Cognitive proximity is understood as a base of knowledge, competence, and expectation shared between actors. Knowledge and expectations that lead to the emergence of innovations need to be shared to create a mutual understanding between actors [16,48]. Social proximity is defined by trust-building activities between actors. Mutual trust based on friendship, kinship, and mutual experience is a prerequisite for collaborations before knowledge or resources are deployed between actors [16,48]. Institutional proximity refers to the similarity of contextual rules, norms, and values, e.g., the similarities of actors to external institutions, such as prevailing rules and regulations within a system (i.e., the rules and regulations by which actors play) [16,48,54]. Finally, organizational proximity refers to the extent to which relationships are shared among actors in a formal, organizational arrangement, including the degree of autonomy and control under which actors can experiment and share knowledge [16,48]. The different proximity dimensions may support, complement, or replace each other [55,56]. Thus, the occurrence of relational proximity could replace the need for geographical proximity as a precondition for experimentation and learning. Furthermore, social proximity complemented by cognitive proximity can support the transmission of "value-laden information" between actors without the need to enforce external standards [22]. However, previous studies point to the positive effects of proximity while neglecting the potential impediments that arise from it [48]. Thus, geographic proximity might constrain organizations in accessing land and resources and in competing with other local actors. While institutional proximity of alternative (e.g., social) innovations to prevailing food system structures could promote effective cross-level learning and coordination, being too rule-bound could hinder experimentation [48]. The greater the trust relationships within or between actors, the less organizational control is required by or between actors. However, tendencies toward excessive trust between actors can also be detrimental to their collaboration [48].

Due to their complementary, substitutive, and supporting nature, the analytically clearly delineated proximity dimensions can be quite messy in real-life and therefore difficult to measure empirically. Therefore, we opted for an explorative approach (see Section 3.1). Based on previous definitions by scholars [16,48,54] and interpretations of proximity dimensions in the context of CSAs [21], we operationalized social, cognitive, institutional, organizational, and geographical proximity:

- Operationalization of geographical proximity: The spatial distance among CSA members (i.e., their access to the CSA farm) (CSA-internal) and the local availability of resources and structures for the CSA farm (e.g., farmland, urban area, infrastructure) (CSA-external) [16,21,48].
- Operationalization of cognitive proximity: The degree to which CSA members empathize with CSA ideas and thus share knowledge, competence, and expectations with respect to CSAs (CSA-internal), and, as CSA-external actors, the degree of interest in and understanding of the CSA model (CSA-external) [16,21,48].
- Operationalization of social proximity: The degree of connections among CSA members (i.e., their trust in each other) (CSA-internal) and societal acceptance (i.e., attitudes) between CSA members and CSA-external actors (CSA-external) [16,21,48].
- Operationalization of institutional proximity: The extent to which CSA rules, norms, and values are shared among CSA members (CSA-internal), and the similarities of the CSA institutions to external, prevailing food system institutions (i.e., production and market mechanisms of dominant food system actors) (CSA-external) [16,21,48,54].
- Operationalization of organizational proximity: The degree to which the CSA members are connected to other CSA members (CSA-internal) and CSA-external actors (CSA-external) in a formal, organizational arrangement [16,21,48].

Figure 1 illustrates the operationalization of spatial and relational proximity dimensions in the context of CSAs. The figure differentiates between CSA-internal proximity (i.e., arrows illustrating proximity among CSA members) and CSA-external proximity (i.e., arrows illustrating proximity between CSA members and CSA-external actors, structures, and resources).

Operationalizing the proximity dimensions for the CSA context and a literature review on CSAs in Austria, Japan, Norway, and beyond helped to make assumptions about how the different proximity dimensions might affect CSAs and their attractiveness in these countries. This review also helped to tailor the statements and questions for the cross-national contexts (see Section 3.1).

- Geographical proximity: In general, CSAs seem to face a trade-off between the locational advantages of rural and urban areas. While CSAs target affordable access to biophysically suitable farmland that is predominantly located in rural areas, a CSA which has a location in or near a city with mainly urban CSA consumers represents a locational advantage (e.g., access to public transportation, infrastructure, networking opportunities) [21]. Thus, by being close to rural and urban areas, a CSA could stimulate a mutual understanding (i.e., cognitive proximity) between people in rural and urban areas (see next point) [30].
- Cognitive proximity: CSA members in Austria share knowledge, competence, and expectations of CSA ideas (e.g., pricing based on self-assessment) with each other, and therefore predominantly connect with individuals already connected to the CSA community (i.e., members of other CSA initiatives) [21]. CSA members' empathy for CSA ideas promotes their endorsement of the CSA [57]. However, Austrian CSA members raised the concern that CSA ideas might be too difficult to under-

stand for actors outside the CSA [21]. With the expansion of mainstream organic food marketing channels in Japan, the interest in CSAs among CSA-external actors is decreasing [58,59]. Thus, in terms of cognitive proximity, Japanese teikei might lack the ability to adapt to the expectations of today's consumers [21]. In contrast, the growing demand for locally and organically produced food and a trend toward urban gardening in Norway might explain the growing interest of Norwegians in CSA and the rapid growth of CSAs in Norway [30,60–62].

- Social proximity: Personal contact with food system actors can increase trust or distrust in the system [63]. CSAs aim to create social proximity among their members by connecting them through network relationships, organizing meetings and events, and participatory decision making [21,30,57,60]. CSA members in Austria highlighted that trust-building activities among CSA members and with society are important for the CSA. Though they have built strong connections with other local CSA actors, relations with other (dominant) food system actors are rare, as stated by CSA members [21]. In Japan, building trusting relationships with actors outside their (teikei) community might be even more challenging due to a more collectivist pattern [64]. While trust within established and stable relationships (such as the teikei community) might be higher than in individualistic societies (i.e., Norway and Austria), it has been observed that Japanese tend to distrust actors outside these relationships [65].
- Institutional proximity: Several studies indicate that Austrian, Japanese, and Norwegian CSA members try to avoid institutionalizing the CSA but rather aim to disrupt conventional food provision practices, rules, norms, and values [21,35,59,66]. They aim to contrast the mainstream and seek an alternative form of food provision [67,68], characterized by typical CSA features (e.g., small-scale operation, short value chains, transparent food provision, social and ecological sustainability) [18,25,60]. Austrian and Norwegian CSAs emerged in response to the conventionalization of the organic food market (i.e., a process in which the organic food market increasingly takes on the characteristics/institutions of mainstream industrial agriculture), and thus CSA members tend to criticize the dominant structures of the food system [21,60,69,70]. In contrast, CSAs emerged in Japan before the Japanese organic food market became conventional, in response to the negative effects of chemically intensive and mechanized agriculture. However, the expansion and institutionalization (i.e., the introduction of a certification system and other government policies to adapt to the dominant structures of the conventional food system) of the organic market since the 1980s, as well as the introduction of a certification system for organic food, were largely responsible for the decline of CSAs in Japan [59].
- Organizational proximity: Due to the shared organizational arrangement, organizational proximity among members of the original teikei type (i.e., OF–OC teikei scheme) and European CSA organizations is high. However, formal collaboration between CSAs and other (dominant) food system actors seems to be less relevant for Austrian and Japanese CSA members [21,59]. In contrast, Norwegian CSAs receive financial and technical support as well as advisory services. The association Organic Norway, the Agricultural Extension Service, the Norwegian Agriculture Agency, and several county governors have been particularly supportive of CSAs, promoting them, and playing an important role in the development of CSAs in Norway [60,71,72]. Although closer links to non-CSA actors, such as government and public institutions, could generate additional resources for CSAs, they may also lead to a loss of independence [73].

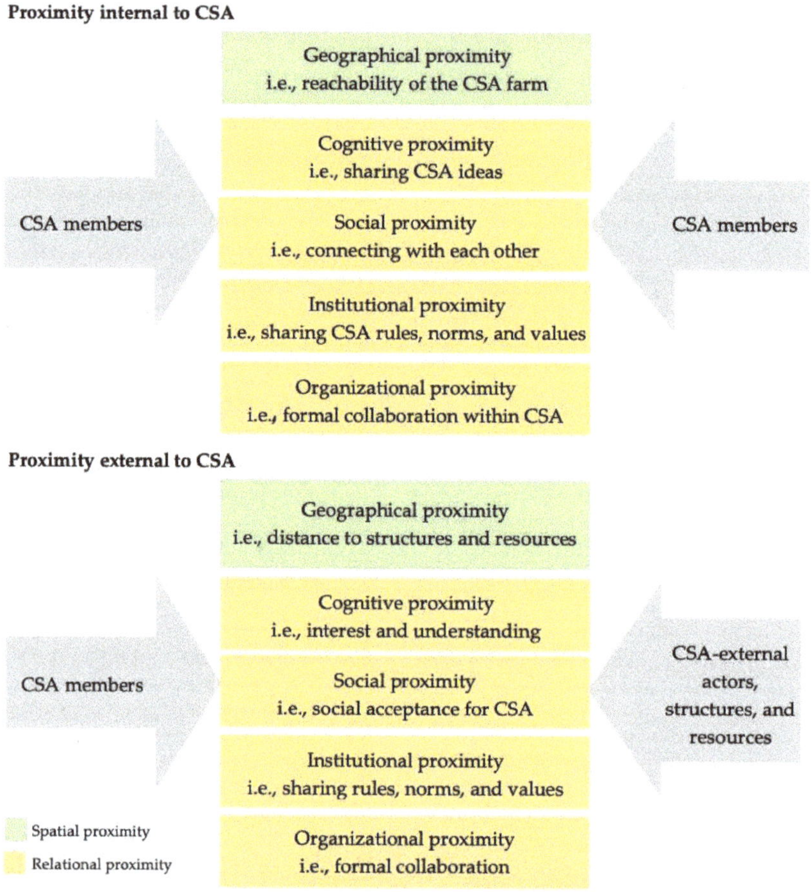

Figure 1. Operationalized spatial–relational proximity dimensions for the CSA (Community Supported Agriculture) context.

Finally, demographic factors could also be related to the attractiveness of a CSA. They might partly explain the development of CSAs in Japan, Austria, and Norway. To maintain the essence of CSA, CSA members are strongly encouraged to actively participate in various activities [58], regardless of their age and gender [5,74]. However, while young people might be less interested in joining a CSA, the physical support expected by CSAs (e.g., work in the fields) can be particularly challenging for older people [58,59]. Furthermore, it can be difficult to work full time and participate in a CSA [58]. Most CSA members are women [57,58]. In Japan, housewives have historically been the driving force behind CSAs, but as more women pursue a career, membership is declining [58].

3. Data and Methodology

This paper analytically differentiates between various proximity dimensions in the context of CSA and examines how these dimensions relate to CSA attractiveness. This section explains the research design used, including site selection, the design of the quantitative analysis, the creation of proximity variables, and their interrelation with CSA attractiveness.

3.1. Site Selection

We applied the proximity framework in three very different national contexts. Drawing on a literature review, Internet research, and informal talks with CSA coordinators, we selected six CSAs in Austria, Japan, and Norway because they share organizational similarities, even though CSA development paths differ in these countries.

CSA has its origins in various countries. One of them is Japan, where the CSA movement, also known as teikei, originated in 1971 [75]. In Japan, there are different types of teikei schemes, ranging from associations with 20–30 households and a single farm to hundreds or thousands of households and multiple farmers [75]. Most of today's teikei systems trade agricultural products to individual consumers who are not organized (e.g., farmers delivering vegetable boxes to consumers). Hence, they require little or no consumer participation (e.g., for agricultural and delivery labor) [75–77]. The original form of teikei, consisting of a group of organized farmers and consumers (OF–OC teikei scheme), experienced rapid growth until the 1980s (there were about 238 teikeis in 1990) [58,77]. Since then, it has gradually lost popularity, especially among younger families [58], and in 2019, there were only about ten active OF–OC teikei schemes [78]. This study focuses only on the OF–OC teikei scheme, as its formal arrangement is similar to the CSA schemes in Austria and Norway.

Austria experienced an increase in CSAs in the first years after the introduction of CSA in 2011. However, CSA in Austria developed late and slowly compared to other countries in Europe and beyond [6,67,79]. In 2020, there were approximately 30 CSA organizations in Austria [66]. The development of CSAs in Austria has been stagnating in recent years [21].

In contrast, the popularity of CSAs in Norway, a non-EU country, has grown rapidly since their initial introduction in 2006 [60]. It is expected that their popularity will continue to grow [62]. In Norway, the number of registered CSAs reached 92 in 2020 [72].

Because CSA arrangements vary across initiatives [6], we selected six similar CSA cases (two per country) for our study. The six selected CSAs have a similar formal structure (i.e., organizational proximity), in that product prices are collectively negotiated and there is an emphasis on the year-round commitment of members.

3.2. Setting up the Quantitative Analysis

For data collection, we designed a cross-national survey on proximity related to CSA attractiveness in Austria, Japan, and Norway. We collected data from CSA members, including farm owners/managers via online and in-person questionnaires. Based on the literature presented in Section 2, the proximity dimensions were operationalized for the CSA questionnaire. The common questionnaire first introduced the objectives of the cross-national study. The first questions addressed CSA-internal relational proximity among CSA members. Furthermore, the questionnaire included questions on CSA-internal geographical proximity (i.e., CSA members' accessibility to the CSA farm) and the geographical proximity of the CSA farm to external structures and resources (i.e., suitable farmland, urban areas, services, network structures, and other community activities). In order to gather information on CSA-external relational proximity, respondents were asked about broader societal contexts of the CSA, such as attitudes, interest, and the level of support by CSA-external actors. The questionnaire included other parts for different research objectives not presented here (see Supplementary Materials). In cases where respondents did not hold information, they could skip questions about CSA collaboration with other food system actors and questions about the policy context that influences the CSA. For these two topics, we relied on the answers of respondents who indicated that they were in a leading position within a CSA ($n = 14$) (as demonstrated by the number and types of activities as well as the working hours for the CSA stated in the questionnaires) to avoid guessing and to ensure the validity of the answers. The questionnaires concluded with demographic questions about the respondent. We translated the questionnaires into German, Japanese, and Norwegian and distributed them to members of six CSAs (two CSAs per country, each in a different city) (Table 1).

Table 1. Selected CSAs and number of respondents in Austria, Japan and Norway.

Country	(Peri-)urban Areas	CSA Members	Surveys (n = 209)	Organizational Similarities
Austria	Vienna Graz	About 300 About 100	51 27	Collective price negotiation;
Norway	Sandefjord Porsgrunn	About 140 About 120	39 49	Year-round commitment of members;
Japan	Tokyo Tsukuba	About 40 About 40	25 18	Participative decision-making processes

Regarding the total number of CSA members, CSA coordinators indicated a lack of data, as the number is constantly changing. In addition, one or more family member/s often split one harvest share (i.e., the amount of produce dedicated to one CSA member), but the exact number is missing. So, we cannot assess the representativeness of the sample. However, this should not be an issue as we do not aim to provide representative insights into the CSA model, but to analyze the relationships between proximity and attractiveness. Data collection resulted in a total of 209 questionnaires (after excluding 19 surveys with too many missing values and/or outliers) that were analyzed using principal component analysis, and 208 questionnaires that were included in the regression modeling (only respondents whose gender was indicated). IBM SPSS Statistics 24 software (IBM, Armonk, NY, USA) supported both principal component analysis and regression modeling. Table 2 illustrates the demographic characteristics of CSA members who responded to our survey.

Table 2. Demographic characteristics of respondents.

Variable	Category	Austria (in %)	Japan (in %)	Norway (in %)
Country		37.3	20.6	42.1
Gender	Female	65.4	74.4	81.4
	Male	34.6	25.6	17.4
	Diverse	0.0	0.0	1.2
Age	>24 years	6.5	0.0	0.0
	25–44 years	50.6	25.6	19.8
	45–64 years	33.8	37.2	45.3
	>65 years	9.1	37.2	34.9
Work condition	Working full-time	25.3	9.3	37.6
	Working part-time	24.0	14.0	9.4
	Being self-employed	14.7	20.9	15.3
	Being not employed (studying, retirement, parental leave, unemployment)	28.0	41.9	36.5
	Other	8.0	14.0	1.2

3.3. Creating Proximity Variables

To create the variables for our model, we measured the spatial–relational proximity items on six-point scales with equally distanced intervals (interval scale of 1 (not significant/disagree/not given/not attractive) to 6 (very significant, completely agree/absolutely given/very attractive). Proximity variables measured with more than a single item on graphically equally distanced 6-point scales were treated as continuous data. Thus, we measured proximity variables with more than a single item and ensured graphically equal distances between response patterns in the survey design [80]. Similar to Rossi and Woods [41] and Galt [38], who measured satisfaction with CSA on a single-item scale, we measured CSA attractiveness on a six-point scale based on the question: "To what extent is CSA attractive to you?

The operationalization of spatial and relational proximity dimensions for the CSA context provided the basis for developing the proximity statements. Table 3 presents all operationalized proximity items in our survey. We asked about the importance of the proximity items to CSA participation (i.e., CSA-internal relational proximity), for the extent to which proximity items were present by participant (i.e., CSA-internal and -external geographical proximity), and for participants' agreement with proximity items (i.e., relational proximity to CSA-external actors). We used an explorative principal component analysis to weight, reduce, and linearly combine the operationalized proximity items (i.e., items describing the overlapping, complementary, and partially substitutive proximity dimensions in the context of CSA presented above). Principal component analysis allowed us to create a small number of synthetic variables (i.e., principal components reflecting different proximity dimensions) from a large number of operationalized proximity items and to test whether the structure of the principal components could be related to latent proximity dimensions similar to those described in the literature [16,21,48]. The resulting variables (i.e., principal components) then served as explanatory variables for the multiple linear regression [81].

The survey also captured perceptions about proximity among CSA members. These proximities refer to linkages within the same CSA to assess social, cognitive, institutional, and geographical proximity among CSA members. For internal linkages, we asked CSA members about the significance of several items for their participation in a CSA: connection with the local CSA community and farmer (i.e., social proximity); empathy with the CSA idea of risk sharing and ensuring a secure income for local farmers (i.e., cognitive proximity); independence from the regular food market and its prices, thus supporting a new food market; and traceability and transparency of food production (i.e., institutional proximity). In addition, we asked CSA members about the accessibility of the CSA farm from their homes by car, bike, or on foot, as well as by public transportation) (i.e., geographical proximity).

We also operationalized the proximity of CSA members to actors, structures, and resources outside of CSAs. Thus, the survey included questions on perceptions of the social, cognitive, institutional, and organizational proximity of CSA members to CSA-external actors, as well as the geographical proximity of CSA members to the urban areas, infrastructure, and agricultural land. Hence, we asked CSA members to assess how they perceive CSA-external actors' attitudes toward the CSA (i.e., social proximity), how understandable the CSA model is to CSA-external actors, and how they perceive the public interest in the CSA (i.e., cognitive proximity). Because members characterized the CSA preferably by institutional distance from the dominant structures of the food system [21,57], we also asked about external institutional linkages. Thus, we asked CSA members about their agreement with CSA's institutional orientation on independence from dominant product and market mechanisms of the food system to avoid institutional proximity to the latter. Furthermore, we asked members in a leading position within the CSA about the degree and type of support they received from CSA-external actors (i.e., organizational proximity). Finally, CSA members were asked about the availability of infrastructure and social activities near their CSA farm, access to suitable land for agricultural production, and the proximity of their CSA farm to an urban area (i.e., geographical proximity).

3.4. Interrelating Proximity to CSA Attractiveness

To analyze the interrelation between proximity variables and CSA attractiveness, we applied both a binary logistic model (which divides the responses on CSA attractiveness into two groups: "very attractive" and "less attractive") and a multiple linear regression (which measures CSA attractiveness on a 6-point interval scale based on equal distances between response patterns in the survey). The two analyses showed basically the same outcome, indicating the robustness of the results. Although an ordered logit model might be more appropriate in terms of model assumptions, linear regression also has some advantages. Therefore, we chose to present the linear regression results here because they can be interpreted more intuitively. In addition, as users of the results, CSA members are more familiar with linear regression results. Finally, the simpler model is equally well suited for presenting the results.

Table 3. Operationalized items of spatial–relational proximity dimensions.

CSA-Internal Proximity	Operationalized Proximity Items as Presented in the Questionnaire	Mean	Standard Deviation
Social proximity among CSA members	Significance of connecting with the CSA community	4.53	1.360
	Significance of direct connection with the CSA farmer	4.83	1.227
Cognitive proximity among CSA members	Significance of empathy for CSA ideas of risk sharing and ensuring a secure income for local farmers	5.23	1.145
Institutional proximity among CSA members	Significance of traceability of food and transparency of production	5.48	0.818
	Significance of becoming more independent from the regular agricultural market and its prices	4.95	1.298
	Significance to support the development of a new and more sustainable agricultural market	5.63	0.758
Geographical proximity among CSA members	Extent of connection to CSA farm via road network for driving	5.48	0.871
	Extent of connection to CSA farm via road network for biking/walking	4.93	1.308
	Extent of connection of public transport system to the CSA farm	3.90	1.659
CSA-external proximity	**Operationalized proximity item in survey**	**Mean**	**Standard deviation**
Social proximity between members and CSA-external actors	Agreement that attitudes of the CSA are in general positive	4.26	1.300
Cognitive proximity between CSA-external actors and CSA members	Agreement that local interest in CSA is increasing in recent years	4.25	1.552
	Agreement that CSA model is easy to understand for CSA-external actors	3.28	1.557
	Agreement that media often reports about CSAs *	2.03	1.202
Organizational proximity between CSA-external actors and CSA members	Agreement to support/impediment by CSA-external actors (e.g., by governmental organizations, agricultural associations, food businesses, farmers, other CSAs, NGOs, private actors) **		
	Agreement that the CSA should cooperate with dominant actors and organizations of the food system and encourage them to become more sustainable *	3.34	1.797
Institutional proximity between CSA-external actors and CSA members	Agreement that the CSA should stay independent and small-scale, to be an alternative to the production and market mechanisms of the dominant actors of the food system *	4.57	1.846
	Agreement that the CSA should not adapt to the production and market mechanisms of the dominant actors of the food system, to grow faster and gain power *	5.10 recoded	1.207
Geographical proximity between CSA farm and urban area, infrastructure, and agricultural land	Extent of suitability of land and climate for agricultural production	5.33	0.829
	Extent of proximity of the CSA farm to the city *	4.58	1.340
	Extent of services nearby the CSA farm	3.16	1.646
	Extent of other community activities nearby the CSA farm	3.28	1.575
	Extent of networking opportunities nearby the CSA farm	3.19	1.446

* Items have been excluded before conducting the principal component analysis, as all correlations to other items were ≤ 0.3 (two-tailed Pearson correlation) ** Items have been excluded before conducting the principal component analysis, as only members in a leading position within the CSA responded. Results are not presented in the table but are qualitatively described in Section 4.2.

Multiple linear regression shows the correlation between CSA attractiveness (i.e., the dependent variable) and the latent proximity dimensions identified in the principal component analysis (i.e., the explanatory variables) (see Section 4.1). Furthermore, we added dummy-coded categorical variables to the regression to examine the extent to which demographic variables might explain CSA attractiveness. We selected country, age, gender, and work situation based on the demographic variables highlighted in the CSA literature (see Section 2). We also collected data on the geographical distance (measured as the linear distance in kilometers based on zip codes) of the location of CSA members and the CSA farm and distance in minutes needed to access the farm. Since these variables did not show correlations with the attractiveness variable, we did not include them in the regression. Before running the multiple linear regression, we checked the data for linearity, multicollinearity, and homoscedasticity [81].

4. Results

We created five latent proximity variables that served as explanatory variables for the multiple linear regression to explain CSA attractiveness [81]. The results of the principal component analysis and the reliability analysis are shown in Table 4.

Table 4. Results of the principal component analysis and the reliability analysis (n = 209).

Factor Loadings ▼ Principal Components ►	1	2	3	4	5
Principal component 1: Social–cognitive proximity among CSA members					
Connection with CSA farmer(s) (CSA-internal social proximity)	0.845				
Connection with CSA community (CSA-internal social proximity)	0.682				
Empathy for CSA ideas (CSA-internal cognitive proximity)	0.675				
Principal component 2:CSA farm'sgeographic proximity toCSA members and land					
Road for biking/walking (CSA-internal geographical proximity)		0.797			
Road for driving (CSA-internal geographical proximity)		0.724			
Suitability of land (CSA-external geographical proximity)		0.679			
Public transport (CSA-internal geographical proximity)		0.552			
Principal component 3: CSA farm's geographic proximity to external structures and resources					
Community activities nearby (CSA-external geographical proximity)			0.793		
Services nearby (CSA-external geographical proximity)			0.748		
Networking nearby (CSA-external geographical proximity)			0.687		
Principal component 4: CSA-external social–cognitive proximity					
Positive attitudes about CSA (CSA-external social proximity)				0.742	
Local interest in CSA (CSA-external cognitive proximity)				0.720	
Understanding CSA model (CSA-external cognitive proximity)				0.624	
Principal component 5: Institutional proximity among CSA members					
Support of the new food market (CSA-internal proximity)					0.842
Independence from the regular market (CSA-internal proximity)					0.578
Traceability and transparency (CSA-internal proximity)					0.540
Eigenvalue	2.068	2.019	1.887	1.766	1.617
% of Variance	12.928	12.620	11.791	11.039	10.106
Cumulative %	12.928	25.548	37.340	48.379	58.485
Cronbach's Alpha	0.696	0.646	0.723	0.636	0.546

Note: Extraction method: principal component analysis (Bartlett's test of Sphericity: Significance: 0.000 (i.e., highly significant); Kaiser-Meyer Olkin Measure of Sampling Adequacy: 0.651 (i.e., relatively low but sufficient for our study, should be greater than 0.5 as a bare minimum); Residuals: there are 57 (47.0%) non-redundant residuals with absolute values greater than 0.05 (i.e., albeit the residuals with 47% of >0.05 are relatively high, they are below the 50% threshold) Rotation method: Varimax with Kaiser Normalization. Only factor loadings over 0.5 are shown. Rotation converged in 5 iterations [81]).

Table 4 shows that analysis results in five principal components with an Eigenvalue greater than 1 [82]. In total, these principal components explain 55.616% of the variance. All factor loadings of the five principal components are above the acceptable limit of 0.5 [81]. Principal components 1–4 are internally consistent, as the values of Cronbach's alpha (i.e., a measure of internal consistency that indicates the extent to which all items in a test measure describe the same concept or construct) are in the range of 0.636 and 0.723, which are satisfactory values for exploratory research [83,84]. In contrast to the other principal components, Cronbach's alpha of principal component 5 is low, with a value of 0.546. Because this value is still respectable for social science studies [84], we included principal component 5 in the regression. The resulting factors in the rotated component matrix correspond to five different proximity dimensions:

- Principal component 1 groups CSA-internal social and cognitive proximities among CSA members. We labelled this factor *social–cognitive proximity among CSA members.*
- Principal component 2 includes variables describing *CSA farm's geographic proximity to CSA members and land* (hence the name of this component). The variables illustrate the location conflict between the proximity to CSA members, mainly located in the city, and suitable land for cultivation by the CSA farm.
- Principal component 3 also contains geographic variables that ask about *the CSA farm's geographic proximity to external structures and resources* (i.e., the name of this component), such as infrastructures and nearby services.
- Principal component 4 captures the CSA-external social and cognitive relations between the CSA members and CSA-external actors. We have referred to principal component 4 as *CSA-external social–cognitive proximity.*
- Principal component 5 contains variables on CSA members' institutional proximity. Therefore, we termed principal component 5 *institutional proximity among CSA members.*

4.1. Interrelating Proximity to CSA Attractiveness

Multiple linear regression allowed us to explain the value of CSA attractiveness (i.e., the dependent variable) with the latent proximity variables (i.e., the explanatory variables) and demographic data (Table 5).

Table 5. Results of the multiple linear regression (n = 208).

No.	Variables	B $_1$	Standard Error $_2$	β $_3$	SIGNIFICANCE $_4$
	Constant	5.574	0.160		0.000
1	Principal component 1	0.248	0.052	0.330	0.000
2	Principal component 2	0.031	0.057	0.041	0.587
3	Principal component 3	−0.050	0.053	−0.066	0.350
4	Principal component 4	0.200	0.062	0.264	0.002
5	Principal component 5	0.115	0.053	0.144	0.032
6	Country: Japan	0.039	0.174	0.021	0.823
7	Country: Norway	0.108	0.139	0.070	0.436
8	Age: <24	−1.038	0.371	−0.193	0.006
9	Age: 25–44	−0.065	0.124	−0.040	0.601
10	Age: >65	−0.047	0.153	−0.027	0.758
11	Gender: Male	−0.251	0.118	−0.145	0.035
12	Employment: Full-time	−0.086	0.151	−0.050	0.572
13	Employment: Part-time	0.104	0.167	0.050	0.533
14	Employment: Self-employed	−0.098	0.165	−0.048	0.552
15	Employment: Other	−0.014	0.227	−0.004	0.952

Dependent variable: CSA attractiveness; in bold when $p < 0.05$. Reference variables: Age: 45–64; Country: Austria, Gender: Female; Work situation: Not employed (i.e., studying, retired, on parental leave, unemployed). (1): The B-values refer to the relationship between CSA attractiveness and each predictor. A positive value indicates a positive relationship between the predictor and the dependent variable, whereas a negative coefficient represents a negative relationship [81]. (2): The standard error associated with each B value indicates how these values vary in different samples [81]. (3): Beta values (β) are standardized versions of the B values. They are measured in standard deviation units and are directly comparable (as they do not depend on the units of measure of the variables). Thus, they provide better insight into the importance of a predictor in the model [81]. (4): If the t-test associated with a B-value is significant (if the significance value is less than 0.05), then the predictor contributes significantly to the model. The smaller the significance value, the greater the contribution of the predictor [81].

Our results show a statistically significant fit of the data, as indicated by an *F*-test statistic of 3.953 (i.e., the F-test looks at whether using the regression model predicts the values of the dependent variable significantly better than using the mean of the dependent variable. If the improvement from fitting the regression model is much greater than the imprecision within the model, then the *F*-value is greater than 1 [81]) and a *p*-value below the 0.05 level. The model explains 24.8% of the variance in CSA attractiveness [81]. Principal component 1 (i.e., social–cognitive proximity among CSA members) and principal component 4 (i.e., CSA-external social–cognitive proximity) are positively related to CSA attractiveness ($p < 5\%$). The standardized beta value for principal component 1 ($\beta = 0.330$) indicates that social–cognitive proximity among CSA members shows the strongest interrelation with the attractiveness rating, followed by principal component 4 ($\beta = 0.264$) (i.e., CSA-external social–cognitive proximity). Furthermore, our results suggest that principal component 5 ($\beta = 0.144$) (i.e., institutional proximity among CSA members) is also positively related to CSA attractiveness ($p < 0.05$). Finally, principal component 2 (i.e., CSA farm's geographical proximity to members and land), and principal component 3 (i.e., CSA farm's geographical proximity to external structures) are not significantly related to the respondents' attractiveness ratings.

Compared to their reference group, the regression coefficients of two dummy variables in the multiple linear regression proved to be statistically significant: first, CSA members aged under 24 years ($\beta = -0.193$) consider CSAs less attractive than the reference group of CSA members aged between 45 and 64 years; second, male CSA members ($\beta = -0.145$) consider CSAs less attractive than their female counterparts.

4.2. Descriptive Analysis of Country-Specific Results on Institutional and Organizational Proximity

The regression does not indicate a country effect. However, we also wanted to take a closer look at institutional and organizational proximity variables. Although these variables were collected in the survey, they were excluded from the analysis due to a lack of correlations or respondents (see proximity items highlighted with * and ** in Table 3). For institutional proximity between CSA-external actors and CSA members, participants rated their agreement to adapt their CSA to, and independence from, production and market mechanisms of the dominant food system actors. Table 6 shows that CSA members agreed (Ø = 4.57) and disagreed (Ø = 1.70) with CSA's independence from production and market mechanisms of the dominant actors. A cross-country comparison reveals that CSA members in all three countries disagreed with the CSA's adaption to dominant food system structures. However, while Austrian and Norwegian CSA members agree with CSA's independence from dominant food system structures, Japanese CSA members slightly disagree with the latter (Ø = 3.19).

Table 6. Institutional proximity to dominant food system structures ($n = 209$).

	CSA Independence from Dominant Structures		**CSA Adaption to Dominant Structures**	
	Mean	**Standard Deviation**	**Mean**	**Standard Deviation**
Total ($n = 209$)	4.57	1.864	1.70	1.282
Austria	5.54	0.878	1.65	1.215
Japan	3.19	2.239	1.81	1.500
Norway	4.40	1.797	1.68	1.282

In terms of organizational proximity, CSA members in all three countries did not fully agree (Ø = 3.34, $n = 209$) that CSAs should work with dominant food system actors to encourage them to become more sustainable. Furthermore, members who hold leadership positions within their CSAs ($n = 14$) rated the level of support and hindrance from other organizations in the food system to reveal their organizational proximity to the CSA.

Norwegian CSA members perceived financial support from local, federal, and provincial governments (e.g., by Innovation Norway and county governors) during the establishment phase, but also desired support thereafter. The Norwegian CSA network, organized by the association Organic Norway (formerly OIKOS), has supported CSAs with networking opportunities and has increased their visibility. Furthermore, the Norwegian Agricultural Extension Service provides training and advice to organic farmers, including CSAs.

In contrast, Japanese and Austrian CSA members perceive the local, federal and provincial government, as well as organic associations, as rather unsupportive. Although they receive farm subsidies from the government (like any other farm), there is no specific financial support for the CSA scheme. Austrian CSA members point to the support from other CSAs, private individuals, farmers, and farmer markets in the form of financial support, space and infrastructure, networking opportunities, and advice. Japanese CSA members mentioned that they have been mainly supported by private individuals and a CSA study group in terms of visibility, networking, infrastructure, and machinery.

5. Discussion

In our exploratory analysis, we operationalized spatial–relational proximity dimensions for a multivariate analysis of CSA attractiveness. We differentiate not only between geographical, social, organizational, institutional, and cognitive proximity, but also between CSA-internal relations among members and CSA-external relations between members and external actors, as well as structures and resources. In the first step of our analysis, we used principal component analysis to create five latent proximity variables for CSA.

Principal components 2 and 3 (i.e., items loading on CSA geographical proximity) and 5 (items loading on institutional proximity) indicate latent variables corresponding to the proximity dimensions differentiated in the literature. In principle component 2, we have items describing geographical proximity to other members (internal) and land (which we labeled as external geographical proximity). However, the respondents seem to distinguish less between the human–bio-physical divide and more between what they perceive as part of the CSA, which for them includes members and farmland. In retrospect, this makes a lot of sense. Social–cognitive principal components 1 and 4 combine two proximity variables that have been analytically differentiated in the literature [16,21,48]. On the one hand, this result might confirm the supportive, complementary, or substitutive nature of proximity dimensions [55,56]. The dimensions that are clearly differentiated analytically might be messily interwoven in real life. On the other hand, the complementarity of social and cognitive proximity dimensions might be due to inadequate operationalization in survey items.

Multiple linear regression (as well as binary logistic regression) showed differences in the interrelations of latent proximity variables with members' CSA attractiveness ratings in Austria, Japan, and Norway. As hypothesized, relational proximities (i.e., social, cognitive, and institutional proximity) significantly predict CSA attractiveness in our model. Surprisingly and contrary to our hypothesis, however, this was not the case for the two geographical proximity variables. Social–cognitive proximity among CSA members (i.e., principal component 1) shows the strongest interrelation with member attractiveness ratings in the model. Thus, connection to other CSA members and farmer(s), as well as the sharing of CSA ideas, seem to be closely related to members' perceptions of CSA attractiveness. Furthermore, CSA-external social–cognitive proximity (i.e., principal component 4) shows the second highest correlation with CSA attractiveness in the model. Thus, CSA attractiveness might increase with a growing understanding of a rising interest in and a positive attitude toward the CSA and its concept in society. Our results confirm the importance of trust-building interactions within and outside the CSA [21]. Additionally, we confirm that empathy for the CSA model (i.e., cognitive proximity) promotes approval of the CSA, which was also addressed by Samoggia et al. [57].

Institutional proximity: Previous studies [18,21,60,67] emphasized that CSA institutions (i.e., rules, norms, values) contrast with the dominant institutions of the food system.

Therefore, in this study, we assumed that institutional proximity among CSA members reflects their shared values and identity based on being different from dominant food system structures. However, the related component 5 (i.e., institutional proximity among CSA members) shows low reliability with a Cronbach's alpha of 0.546. Future analyses are needed with other or more items to increase the reliability of an institutional proximity scale [81]. Multiple linear regression suggests that institutional proximity among CSA members (i.e., principal component 5) might be positively related to CSA attractiveness. Thus, the latter increases as CSA members strive for more independence from the regular food market and the establishment of a new one, as well as for traceable and transparent food (production).

Descriptive analysis shows that respondents criticized prevailing rules, norms, and values in the food system, wanted to change the latter, and aimed to avoid institutionalization of the CSA scheme, which is consistent with the findings of previous studies [21,60]. Most respondents in the three countries studies agreed that CSA schemes should rather avoid an adaption to the dominant institutions of the food system. In other words, they do not want to conform to the latter. Following Coenen et al. [48], alternative (e.g., social) innovations (such as CSA), could be limited in their freedom and experimentation if they were oriented towards dominant institutions. Thus, too much institutional proximity to CSA-external (dominant) food system actors could have a negative impact on CSA attractiveness, as our study shows. However, the institutional distance of CSAs from dominant structures might also hinder cross-level learning, collaboration, and coordination between CSAs and dominant food system actors.

Organizational proximity: In Austria and Japan, political support for CSAs seems to be low. Austrian and Japanese CSA members stated that there has been support, if any, from other alternative innovations or private actors. In contrast, Norwegian CSA members pointed to various supporting measures for their CSAs from government organizations and interest groups, which Devik [71] and Hvitsand [30] had already pointed out. This might explain why organizational proximity of the CSA to dominant food system actors is perceived as relatively low, especially by Austrian and Japanese respondents (as described in Section 4.2). CSA members slightly disagree that their CSA should collaborate with dominant actors to encourage them to become more sustainable. CSA members might lack trust toward dominant food system actors (i.e., lack of social proximity) [21] and may be afraid of too much dependence and organizational control by the latter [48,73].

Geographical proximity: The regression demonstrated that the principal components related to geographical proximity (i.e., principal components 2 and 3) do not predict CSA attractiveness. Thus, the latter is neither significantly increased by the accessibility to members of a CSA farm from their homes nor by CSA farms' access to suitable farmland, the urban area, infrastructure, and social activities nearby. Linear distance (kilometers) and travel time variables from respondents' homes to the CSA farm did not correlate with the attractiveness ratings. This result might be different if we had also included non-members in our sample or members who live far away. The CSA membership of our respondents might result from a self-selection process that is strongly influenced by geographical proximity. On the other hand, the distance between members and the CSA farm is less relevant for CSA models in which members do not pick up the food at the farm but at one of several collection sites near the CSA members. In this case, distance to food collection points is more important than distance to the farm. Therefore, our results do not necessarily indicate that geographical proximity is irrelevant to sustainable food systems. However, our model suggests that relational proximity might be more relevant to CSA attractiveness than spatial proximity (i.e., geographical proximity). Although the overall goal of CSAs is to connect producers and consumers [3,5], which might be easier in spatially proximate situations, the latter might also be achieved "at a distance" [5,14,15]. Therefore, the focus of CSAs on relational proximity could reduce or even partially replace the importance of spatial (i.e., geographical) proximity [22].

Demographic variables: The generally low proportion of young members in our sample, especially in the Japanese and Norwegian subsamples, is in line with the Japanese literature [55,56]. The regression also shows that CSA attractiveness is significantly lower for the youngest age group (age: <24) compared to the reference group (age: 45–64). Furthermore, we found that most CSA members in all three country subsamples are females, as already highlighted by previous scholars [57,58]. Consequently, the regression demonstrated that male CSA members consider CSAs less attractive than female respondents. Finally, neither respondent nationality nor work situation showed a significant interrelation with attractiveness ratings. A limitation of our analysis is that we could not include comparable economic data (such as household income) that have been identified as relevant in other studies [39]. Furthermore, the survey was conducted only with Austrian, Japanese, and Norwegian CSA members (and not with former members or non-members) of six CSAs in three different countries. This limitation of our study points to the importance of studying CSAs in different countries and with nonmembers.

Finally, the development of CSAs has been stagnating in Austria and even declining in Japan. In Norway, on the other hand, the number of CSA farms has been steadily increasing, partly due to the supportive attitude of public bodies and various agricultural organizations, especially the association Organic Norway, towards CSAs.

6. Conclusions

Since AFNs (such as CSAs) have only recently come into existence, there still is a lack of knowledge about which factors should be used to promote them [25]. This article shows that the notion of proximity can help operationalize geographical, socio-cognitive, organizational, and institutional relations as explanatory variables in a linear regression model of CSA attractiveness. Multivariate analysis of empirical data from six CSA groups in Norway, Japan, and Austria highlights the importance of social–cognitive and institutional proximity to CSA attractiveness and thus, the relevance of increased trust, collaboration, shared knowledge, and shared values within and across organizations in the food system. Rather than focusing on geographical proximity, supporting social–cognitive and institutional relations within the CSA and beyond might support CSAs' attractiveness. The lack of a country effect suggests that the findings might be robust across socio-cultural and political contexts.

Future research could address this study's possible limitations of operationalization (i.e., the complementarity of social and cognitive proximity; the low reliability of principal component 5), and limitations of our sample (i.e., no inclusion of non-CSA members and economic data of respondents).

In our study, items for geographical, social, cognitive, institutional, and organizational dimensions of proximity were operationalized and tested. They cover network-internal and -external relations, human-to-human relations, and the relations of AFN members to their bio-physical context of land or infrastructure. We hope that our small methodological contribution will be useful for future structured AFN surveys and the advancement of diverse methods in relational rural sociology.

Supplementary Materials: The following are available online at https://www.mdpi.com/article/agriculture11101006/s1.

Author Contributions: Conceptualization, methodology, data collection and analysis, writing—original draft, preparation, visualization, C.G.; conceptualization, methodology, data analysis, writing—review and editing, T.S.; conceptualization, methodology, data collection, writing—review and editing, C.H.; conceptualization, data collection and analysis, D.T.; conceptualization, methodology, N.S.; conceptualization, methodology, writing—review and editing, supervision, project administration, and funding acquisition, M.P. All authors have read and agreed to the published version of the manuscript.

Funding: This work was supported by the Vienna Science and Technology Fund (WWTF) Vienna, Austria [grant number ESR17042].

Institutional Review Board Statement: Considering the UNESCO Recommendation for Science and Scientific Researchers, this international comparative study followed social science ethical standards: transparency on study purpose, informed consent by CSA leaders and by individual respondents, privacy and anonymity, care in methods selection and analysis, no vulnerable groups involved. As BOKU established its Ethics Commission after data collection, this study was not subject to a formalized ethical review and approval.

Data Availability Statement: The data that support the findings of this study are available on request from the corresponding author. The data are not publicly available due to privacy restrictions.

Acknowledgments: The authors thank all survey participants for their enthusiastic participation. Without them, this research would not have been possible. We are particularly grateful for Christina Roder's editing support.

Conflicts of Interest: The authors declare no conflict of interest.

References

1. Ermann, U.; Langthaler, E.; Penker, M.; Schermer, M. *Agro-Food Studies: Eine Einführung*; UTB Böhlau Verlag: Vienna, Austria, 2018; p. 260.
2. Krausmann, F.; Langthaler, E. Food regimes and their trade links: A socio-ecological perspective. *Ecol. Econ.* **2019**, *160*, 87–95. [CrossRef]
3. Hinrichs, C.C. Embeddedness and local food systems: Notes on two types of direct agricultural market. *J. Rural. Stud.* **2000**, *16*, 295–303. [CrossRef]
4. Penker, M. Mapping and measuring the ecological embeddedness of food supply chains. *Geoforum* **2006**, *37*, 368–379. [CrossRef]
5. Renting, H.; Marsden, T.; Banks, J. Understanding alternative food networks: Exploring the role of short food supply chains in rural development. *Environ. Plan. A Econ. Space* **2003**, *35*, 393–411. [CrossRef]
6. Weckenbrock, P.; Volz, P.; Parot, J.; Cressot, N. Introduction to Community Supported Agriculture in Europe. In *Overview of Community Supported Agriculture in Europe*; European CSA Research Group: Aubagne, France, 2016; pp. 8–10.
7. Jossart-Marcelli, P.; Bosco, F.J. Alternative food projects, localization and neoliberal urban development. *Métropoles* **2014**, *15*. [CrossRef]
8. Watson, D.J. Working the fields: The organization of labor in community supported agriculture. *Organization* **2020**, *27*, 291–313. [CrossRef]
9. Brunori, G.; Bartolini, F. Local agri-food systems in a global world: Market, social and environmental challenges. *Eur. Rev. Agric. Econ.* **2013**, *40*, 408–411. [CrossRef]
10. Darnhofer, I.; Gibbon, D.; Dedieu, B. *Farming Systems Research into the 21st Century: The New Dynamic*; Springer: Dordrecht, The Netherlands, 2012.
11. Schermer, M. From "Food from Nowhere" to "Food from Here:" Changing producer—Consumer relations in Austria. *Agric. Hum. Values* **2015**, *32*, 121–132. [CrossRef]
12. DuPuis, E.M.; Goodman, D. Should we go "home" to eat? Towards a reflexive politics of localism. *J. Rural. Stud.* **2005**, *21*, 359–371. [CrossRef]
13. Milestad, R.; Westberg, L.; Geber, U.; Björklund, J. Enhancing adaptive capacity in food systems: Learning at farmers' markets in Sweden. *Ecol. Soc.* **2010**, *15*, 29–46. [CrossRef]
14. Kneafsy, M.; Venn, L.; Schmutz, U.; Trenchard, L.; Eyden-Wood, T.; Bos, E.; Sutton, G.; Blackett, M. Short Food Supply Chains and Local Food Systems in the EU. 2013. Available online: https://www.europarl.europa.eu/thinktank/en/document.html?reference=EPRS_BRI(2016)586650 (accessed on 18 January 2021). [CrossRef]
15. Watts, D.C.H.; Ilbery, B.; Maye, D. Making reconnections in agro-food geography: Alternative systems of food provision. *Prog. Hum. Geogr.* **2005**, *29*, 22–40. [CrossRef]
16. Boschma, R. Proximity and innovation: A critical assessment. *Reg. Stud.* **2005**, *39*, 61–74. [CrossRef]
17. Aubry, C.; Kebir, L. Shortening food supply chains: A means for maintaining agriculture close to urban areas? The case of the French metropolitan area of Paris. *Food Policy* **2013**, *41*, 85–93. [CrossRef]
18. Abrahams, C.N. Globally useful conceptions of alternative food networks in the developing south: The case of Johannesburg's urban food supply system. In *Alternative Food Geographies: Representation and Practice*; Maye, D., Holloway, L., Kneafsey, M., Eds.; Emerald: Bingley, UK, 2007; pp. 95–114.
19. Sitaker, M.; McGuirt, J.T.; Wang, W.; Kolodinsky, J.; Seguin, R.A. Spatial considerations for implementing two direct-to-consumer food models in two states. *Sustainability* **2019**, *11*, 2081. [CrossRef]
20. Struś, M.; Kalisik-Medelska, M.; Nadolny, M.; Kachniarz, M.; Raftowicz, M. Community-supported agriculture as a perspective model for the development of small agricultural holding in the region. *Sustainability* **2020**, *12*, 2656. [CrossRef]
21. Gugerell, C.; Penker, M. Change Agents' Perspectives on Spatial–Relational Proximities and Urban Food Niches. *Sustainability* **2020**, *12*, 2333. [CrossRef]

22. Dubois, A. Translocal practices and proximities in short quality food chains at the periphery: The case of North Swedish farmers. *Agric. Hum. Values* **2019**, *236*, 763–778. [CrossRef]
23. Edelmann, H.; Quiñones-Ruiz, X.F.; Penker, M. Analytic Framework to Determine Proximity in Relationship Coffee Models. *Sociol. Rural.* **2019**, *60*, 458–481. [CrossRef] [PubMed]
24. Kebir, L.; Torre, A. Geographical proximity and new short supply food chains. In *Creative Industries and Innovation in Europe, Concepts, Measures, and Comparative Case Studies*; Lazzeretti, L., Ed.; Routledge: New York, NY, USA, 2013; p. 328.
25. De Bernardi, P.; Bertello, A.; Venuti, F.; Foscolo, E. How to avoid the tragedy of alternative food networks (AFNs)? The impact of social capital and transparency on AFN performance. *Br. Food J.* **2020**, *122*, 2171–2186. [CrossRef]
26. Galt, R.; O'Sullivan, L.; Beckett, J.; Hiner, C.C. *Community Supported Agriculture (CSA) in and around California's Central Valley: Farm and Farmer Characteristics, Farm-Member Relationships, Economic Viability and Emerging Issues*; University of California: Oakland, CA, USA, 2011.
27. Bougheraraa, D.; Grolleaub, G.; Mzoughic, N. Buy local, pollute less: What drives households to join a community supported farm? *Ecol. Econ.* **2009**, *68*, 1488–1495. [CrossRef]
28. Brehm, J.M.; Eisenhauer, B.W. Motivations for participating in community-supported agriculture and their relationship with community attachment and social capital. *J. Rural. Soc. Sci.* **2008**, *23*, 5.
29. Cox, R.; Holloway, L.; Venn, L.; Dowler, L.; Hein, J.R.; Kneafsey, M.; Tuomainen, H. Common ground? Motivations for participation in a community-supported agriculture scheme. *Local Environ.* **2008**, *13*, 203–218. [CrossRef]
30. Hvitsand, C. *Organic Spearhead—The Role of Community Supported Agriculture in Enhancing Bio Economy, and Increased Knowledge about and Consumption of Organic Food*; Title Translated from Norwegian; Telemark Research Institute: Bø, Norway, 2014.
31. Kane, D.; Lohr, L. *The Dangers of Space Turnips and Blind Dates: Bridging the Gap Between CSA Shareholders' Expectations and Reality*; CSA Farm Network: Stillwater, NY, USA, 1998.
32. Cone, C.; Myhre, A. Community-supported Agriculture: A Sustainable Alternative to Industrial Agriculture? *Hum. Organ.* **2000**, *59*, 187–199. [CrossRef]
33. Galt, R.E.; Bradley, K.; Christensen, L.O.; Munden-Dixon, K. The (un)making of "CSA people": Member retention and the customization paradox in Community Supported Agriculture (CSA) in California. *J. Rural Stud.* **2019**, *65*, 172–185. [CrossRef]
34. Witzling, L.; Shaw, B.R.; Strader, C.; Sedlak, C.; Jones, E. The role of community: CSA member retention. *Br. Food J.* **2020**, *122*, 2289–2302. [CrossRef]
35. Vitari, C.; Whittingham, E. Tackling Conventional Agriculture: The Institutionalization of Community Supported Agriculture's (CSA) Principles. In Proceedings of the Research & Degrowth Conference, Malmö, Sweden, 21–25 August 2018.
36. Balázs, B.; Pataki, G.; Lazányi, O. Prospects for the future: Community supported agriculture in Hungary. *Futures* **2016**, *83*, 100–111. [CrossRef]
37. Nost, E. Scaling-up local foods: Commodity practice in community supported agriculture (CSA). *J. Rural Stud.* **2014**, *34*, 152–160. [CrossRef]
38. Galt, R.E.; Bradley, K.; Christiensen, L.; Van Soelen Kim, J.; Lobo, R. Eroding the Community in Community Supported Agriculture (CSA): Competition's Effects in Alternative Food Networks in California. *Sociol. Rural.* **2016**, *56*, 491–512. [CrossRef]
39. Galt, R.E.; Bradley, K.; Christiensen, L.; Fake, C.; Munden-Dixon, K.; Simpson, N.; Surls, R.; Van Soelen Kim, J. What difference does income make for Community Supported Agriculture (CSA) members in California? Comparing lower-income and higher-income households. *Agric. Hum. Values* **2017**, *34*, 435–452. [CrossRef]
40. Galt, R.E.; Bradley, K.; Christiensen, L.; Munden-Dixon, K. Exploring member data for Community Supported Agriculture (CSA) in California: Comparisons of former and current CSA members. *Data Brief* **2018**, *21*, 2082–2088. [CrossRef]
41. Rossi, J.; Woods, T. Understanding shareholder satisfaction and retention in CSA incentive programs. *J. Food Distrib. Res.* **2020**, *51*, 16–40.
42. Darnhofer, I. Farming from a Process-Relational Perspective: Making Openings for Change Visible. *Sociol. Rural.* **2020**, *60*, 505–528. [CrossRef]
43. Vroom, V.H. Organizational choice: A study of pre- and post-decision processes. *Organ. Behav. Hum. Perform.* **1966**, *1*, 212–225. [CrossRef]
44. Singh, R. Information integration theory applied to expected job attractiveness and satisfaction. *J. Appl. Psychol.* **1973**, *60*, 621–623. [CrossRef]
45. Belllet, M.; Colletis, G.; Lung, Y. Économie des proximités. *Rev. D'économie Régionale Urbaine* **1993**, *3*, 357–606.
46. Rallet, A.; Torre, A. Is geographical proximity necessary in the innovation networks in the era of global economy? *GeoJournal* **1999**, *49*, 373–380. [CrossRef]
47. Moodysson, J.; Jonsson, O. Knowledge collaboration and proximity: The spatial organization of Biotech innovation projects. *Eur. Urban Reg. Stud.* **2007**, *14*, 115–131. [CrossRef]
48. Coenen, L.; Raven, T.; Verbong, G. Local niche experimentation in energy transitions: A theoretical and empirical exploration of proximity advantages and disadvantages. *Technol. Soc.* **2010**, *32*, 295–302. [CrossRef]
49. Holloway, L.; Cox, R.; Venn, L.; Kneafsey, M.; Dowler, E.; Tuomainen, H. Managing sustainable farmed landscape through 'alternative' food networks: A case study from Italy. *Geogr. J.* **2006**, *172*, 219–229. [CrossRef]
50. Hayden, J.; Buck, D. Doing community supported agriculture: Tactile space, affect and effects of membership. *Geoforum* **2012**, *43*, 332–341. [CrossRef]

51. Michel-Villarreal, R.; Hingley, M.; Canavari, M.; Bregoli, I. Sustainability in Alternative Food Networks: A Systematic Literature Review. *Sustainability* **2019**, *11*, 859. [CrossRef]
52. Rossi, A. Beyond Food Provisioning: The Transformative Potential of Grassroots Innovation around Food. *Agriculture* **2017**, *7*, 6. [CrossRef]
53. Parker, G. Social innovation in local food in Japan: Choku-bai-jo markets and Teikei cooperative practices. In *Real Estate & Planning Working Papers*; University of Reading: Reading, UK, 2016.
54. Edquist, C.; Johnson, B. Institutions and organizations in systems of innovation. In *Systems of Innovation: Technologies, Institutions, and Organizations*; Edquist, C., Ed.; Pinter: London, UK, 1997; pp. 41–63.
55. Breschi, S.; Lissoni, F. *Mobility and Social Networks: Localised Knowledge Spillovers Revisited*; CESPRI Working Paper No. 142; University Bocconi: Milano, Italy, 2003. Available online: https://ideas.repec.org/p/cri/cespri/wp142.html (accessed on 18 January 2021).
56. Broekel, T.; Boschma, R. Knowledge networks in the Dutch aviation industry: The proximity paradox. *J. Econ. Geogr.* **2012**, *12*, 409–433. [CrossRef]
57. Samoggia, A.; Perazzolo, C.; Kocsis, P.; Del Prete, M. Community supported agriculture farmers' perceptions of management benefits and drawbacks. *Sustainability* **2019**, *11*, 3262. [CrossRef]
58. Hatano, J. The organic agriculture movement (teikei) and factors leading to its decline in Japan. *Rural Food Econ.* **2008**, *54*, 21–34.
59. Kondoh, K. The alternative food movement in Japan: Challenges, limits, and resilience of the teikei system. *Agric. Hum. Values* **2015**, *32*, 143–153. [CrossRef]
60. Hvitsand, C. Community supported agriculture (CSA) as a transformational act—Distinct values and multiple motivations among farmers and consumers. *Agroecol. Sustain. Food Syst.* **2016**, *40*, 333–351. [CrossRef]
61. Rømo Grande, E. Eating is an Agricultural Act: Community Supported Agriculture (CSA) in Norway. Master's Thesis, Norwegian University of Life Sciences, Ås, Norway, 2009.
62. Rømo Grande, E. Norway. In *Overview of Community Supported Agriculture in Europe*; European CSA Research Group: Aubagne, France, 2016; pp. 74–77.
63. Coff, C.; Korthals, M.; Barling, D. Ethical traceability and informed food choice. In *Ethical Traceability and Communicating Food*; The International Library of Environmental, Agricultural and Food Ethics; Coff, C., Barling, D., Korthals, M., Nielsen, T., Eds.; Springer: Dordrecht, The Netherlands, 2008; pp. 1–18.
64. Hofstede, G. *Culture's Consequences: Comparing Values, Behaviors, Institutions, and Organizations Across Nations*, 2nd ed.; Sage: Thousand Oaks, CA, USA, 2001.
65. Ferrin, D.L.; Gillespie, N. Trust differences across national-societal cultures: Much to do, or much ado about nothing? In *Organizational Trust: A Cultural Perspective*; Saunders, M.N., Skinner, D., Dietz, G., Gillespie, N., Lewicki, R.J., Eds.; Cambridge University Press: Cambridge, UK, 2010; pp. 42–86.
66. Storstad, O. The impact of consumer trust in the Norwegian food market. In *Food, Nature and Society*; Rural Life in Late, Modernity; Blanc, M., Tovey, H., Eds.; Routledge: Abingdon, UK, 2001; pp. 113–134.
67. Wohlmacher, E. Comparing Community Supported Agriculture in Vienna and Vancouver. Master's Thesis, University of Natural Resources and Life Sciences, Vienna, Austria, 2018.
68. Plank, C.; Hafner, R.; Stotten, R. Analyzing values-based modes of production and consumption: Community-supported agriculture in the Austrian Third Food Regime. *Österreichische Zeitschrift für Soziologie* **2020**, *45*, 49–68. [CrossRef]
69. Freyer, B.; Bingen, J. *Re-Thinking Organic Food and Farming in a Changing World*; The International Library of Environmental Agricultural and Food Ethics 22; Springer: Dordrecht, The Netherlands, 2015.
70. Guthmann, J. Regulating meaning, appropriating nature: The codification of California organic agriculture. *Antipode* **1998**, *30*, 135–154. [CrossRef]
71. Devik, A. *Håndbok for å Starte Andelslandbruk*; Oikos–Økologisk Norge: Oslo, Norway, 2015.
72. Organic Norway. Community Supported Agriculture in Norway (Translated). 2020. Available online: https://www.andelslandbruk.no/english-1/english (accessed on 18 January 2021).
73. Poças Ribeiro, A.; Harmsen, R.; Feola, G.; Rosales Carréon, J.; Worrell, E. Organising Alternative Food Networks (AFNs): Challenges and Facilitating Conditions of different AFN types in three EU countries. *Sociol. Rural.* **2021**, *61*, 491–517. [CrossRef]
74. Braukmann, I. Potenzial und Grenzen von Community Supported Agriculture als Gegenhegemoniales Projekt. Master's Thesis, University of Vienna, Vienna, Austria, 2015. [CrossRef]
75. McGreevy, S.R.; Akitsu, M. Steering sustainable food consumption in Japan: Trust, relationships, and the ties that bind. In *Sustainable Consumption: Design, Innovation, and Practice*; Genus, A., Ed.; Springer: Cham, Switzerland, 2016; Volume 3, pp. 101–117.
76. Akitsu, M.; Aminaka, N. The development of farmer-consumer direct relationships in Japan: Focusing on the trade of organic produce. In Proceedings of the 4th Asian Rural Sociology Association (ARSA) International Conference, Legazpi City, Philippines, 7–10 September 2010.
77. Hatano, T. *Economy of Organic Agriculture: TEIKEI Networks*; Nihon Keizai Hyoronsha: Tokyo, Japan, 1998.
78. Hatano, T.; Karasaki, T. *CSA, Agriculture for Sharing: Case Studies in US, Europe, and Japan*; Soshinsya: Tokyo, Japan, 2019.

79. Engel, A.; Pabst, S.; Steigberger, E.; Wellmann, L. Austria. In *Overview of Community Supported Agriculture in Europe*; European CSA Research Group: Aubagne, France, 2016; pp. 12–15. Available online: http://www.fao.org/family-farming/detail/en/c/416085/ (accessed on 18 January 2021).
80. Carifio, J.; Perla, R.J. Ten common misunderstandings, misconceptions, persistent myths and urban legends about Likert scales and Likert response formats and their antidotes. *J. Soc. Sci.* **2007**, *3*, 106–116. [CrossRef]
81. Field, A.P. *Discovering Statistics Using IBM SPSS Statistics: And Sex and Drugs and Rock 'n' Roll*, 15th ed.; Sage: Chicago, IL, USA, 2013.
82. Kaiser, H.F. The application of electronic computers to factor analysis. *Educ. Psychol. Manag.* **1960**, *20*, 141–151. [CrossRef]
83. Robinson, J.P.; Shaver, P.R.; Wrightsman, L.S. Criteria for scale selection and evaluation. In *Measures of Personality and Social Psychological Attitudes*; Robinson, J.P., Shaver, P.R., Wrightsman, L.S., Eds.; Academic Press: San Diego, CA, USA, 1991; pp. 1–16. [CrossRef]
84. Kline, P. *The Handbook of Psychological Testing*, 2nd ed.; Routledge Taylor & Francis Group: London, UK, 1999.

agriculture

MDPI

Article

Building an Agroecological Process towards Agricultural Sustainability: A Case Study from Southern Spain

Manuel González-Rosado, Luis Parras-Alcántara *, Jesús Aguilera-Huertas and Beatriz Lozano-García *

SUMAS Research Group, Department of Agricultural Chemistry, Soil Science and Microbiology, Faculty of Science, Agri-Food Campus of International Excellence—ceiA3, University of Cordoba, 14071 Cordoba, Spain; mgrosado@uco.es (M.G.-R.); a52aghuj@uco.es (J.A.-H.)

* Correspondence: qe1paall@uco.es (L.P.-A.); beatriz.lozano@uco.es (B.L.-G.)

Abstract: The urgent need to implement agricultural systems that provide greater sustainability and resilience to the challenges of the climate change process has meant that alternative paradigms for agri-food systems and agriculture have become more relevant in recent times. In this study, we present the building process and consolidation of an agro-ecological project (Extiercol) in a rural area of southern Spain, with a prolonged depopulation process and close connections to nearby urban areas. Through participatory action research, the specific objectives of this study are (1) to describe the agroecological collective process from its creation by a youth association to its establishment as a viable agricultural project; (2) to identify the drivers for the development of this type of transition process towards agricultural sustainability and (3) to analyse urban-rural alliances in the establishment of agroecological projects. Finally, the replicability of this project was assessed, with a special focus on the main barriers to be addressed in order to implement this agricultural system such as difficult to land access or a negative perception of sustainable management by farmers. Through this study we have shown how the connection between the food production area and nearby urban areas can be achieved through an agroecological project.

Keywords: sustainable agriculture; rural-urban interaction; agroecology; youth; human-nature connectedness; sustainability transitions; depopulation

Citation: González-Rosado, M.; Parras-Alcántara, L.; Aguilera-Huertas, J.; Lozano-García, B. Building an Agroecological Process towards Agricultural Sustainability: A Case Study from Southern Spain. *Agriculture* **2021**, *11*, 1024. https://doi.org/10.3390/agriculture11101024

Academic Editors: José Luis Vicente-Vicente, Cristina Quintas-Soriano and María D. López-Rodríguez

Received: 17 September 2021
Accepted: 17 October 2021
Published: 19 October 2021

1. Introduction

Agricultural land comprises 9.1% of the territory in Spain [1]; in these areas, the transition from organic to conventional agricultural systems took place within a short period of time in the 20th century [2]. In Spain, agricultural workers decreased from 48.5% of the working population in 1950 to 4% in 2019 [3], and the rural labour surplus in this process of agricultural system transformation was crucial in the different processes of industrial and tourist development that took place in Spanish and European urban areas. Thus, there was a link between urban growth and expansion; capital accumulation; concentration of the means of production, technologies, innovations and power and the configuration of decapitalised spaces characterised by unemployment, a lack of productive structure, sluggish economic activity and population decline in rural areas [4]. In this context, a territorial relationship of domination and inequality is re-created in which the rural areas are isolated and reduced to a dependent and submissive space for the extraction and production of resources and labour. On the other hand, with this continuous urbanization process, a separation between primary production and consumers is fostered over time [5], with a disconnection from local biological cycles of production and the traditional knowledge related to these production processes.

In rural areas, the agricultural sector and primary agri-food processing activities are key elements in territorial functionality, both in rural development and environmental terms. The fight against depopulation in economically disadvantaged rural areas and the promotion of agri-food activities that are linked to endogenous productive resources have long been among the objectives of EU policies [6]. More recently, the introduction of 'greening' measures in the 2013 EU's Common Agricultural Policy (CAP) reform was intended to reduce the environmental effects of agricultural intensification; however, these measures have had limited environmental impacts [7,8]. In European agricultural policies, the productive functionality of agricultural activities seemed to lose weight among these objectives [9] because there is a growing emphasis on the ecosystem services provision. In addition, there has recently been an increase in social awareness and policies towards more sustainable production models that are local or regional in scale and increase food security [10,11]. In this paradigm, the link between the supply of local agricultural products and the provision of ecosystem services may be one of the keys to a future shift towards more sustainable regional agroecosystems. In this sense, according to [12] "The agroecosystem is regarded as an epistemological tool for creating an ontology or representation of agriculture based on a system view". For this purpose, the ecosystem services valuation and an equitable value chain with farmers seem to be fundamental aspects.

Limiting global warming to 1.5 °C above preindustrial levels would require major reductions in greenhouse gas emissions in all economic sectors [13]. At the same time, climate change adaptation and mitigation is one of the greatest challenges facing food production and consumption in the world [14]. Increases in agricultural intensification on a regional and global scale generates serious environmental impacts, such as the reduction of biodiversity [15,16], an increase in greenhouse gas emissions [17], soil degradation [18], as well as a decrease in food security and an increase in the dependence on external resources [19]. Agriculture is one of the main sectors in the emission of greenhouse gases worldwide; in Spain, this sector is responsible for 12% of total emissions, and they have increased by 8% since 1990 [20]. However, the agricultural sector has a great potential to contribute to the reduction of emissions and become a key sector for mitigating the effects of climate change [21].Sustainable management practices implementation are not only important to mitigate CO_2 emissions and increase the C sequestration rate in agricultural soils [22] but also to enhance soil organic matter sustainable practices to improve soil quality by increasing soil fertility [23–25], water holding capacity [26,27] resistance to prolonged drought periods [28] and erosion processes [29]. The implementation of sustainable management practices will be of crucial importance to achieve the objectives of the European Green Deal [30] and Horizon Europe Mission on Soil health and Food [31]. However, a real assessment of the sustainability of these kinds of production models must be accompanied by a regionalization of production involving the reduction of the distance travelled by agricultural products from the production area to the consumption area, particularly when designing climate change mitigation strategies [30]. In addition, promoting regional consumption of local agricultural products would have a positive effect on the reduction of the regional carbon footprint since transporting local production produces lower greenhouse gas emissions [32–35].

More agricultural sustainability models are necessary; furthermore, agri-food systems and farming have gained greater importance in recent times in the fight against climate change [36–38]. In this sense, the implementation of agroecological production systems can be an interesting alternative in the search for sustainable production models that can generate employment in rural areas and serve to connect society with the sustainable use of agroecosystems as well as contribute to the Sustainable Development Goals [39–44]. Agroecology aims a holistic view of agricultural systems sustainability, based on the participatory interaction between traditional knowledge and modern science, in order not only to reduce environmental impacts but also to serve the socio-economic needs of farmers and the society [45]. Agroecology promotes the recovery of the logics and practices of local knowledge as a strategy for sustainability [46]. The potential for rural development and population fixation are among the virtues that this approach claims for itself [47]. Agroecological management systems would enhance the territorial functions of rural areas as a space for the ecosystem services provision as well as food quality.

In the Andalusian region, high urban-rural connectivity, a population spatial distribution with a large number of medium-sized cities that act as the backbone or articulators of the territory [48] and the high agronomic potential mean that sustainable agroecological strategies of local consumption and increased food security are a priori winning strategies. Although there is great potential and a large number of small-scale local agroecological experiences throughout the region, there is no regional strategy to strengthen these elements in order to develop local and sustainable consumption. In this sense, the expansion of monoculture, such as olive cultivation, and the decline of small and medium-sized livestock farms that provide local resources for the organic fertilisation of agricultural soils may be one of the major handicaps in the implementation of this regional strategy.

The purpose of this study was (i) to describe the agroecological collective process from its creation by a youth association to its establishment as a viable agricultural project; (ii) to identify the drivers for the development of this type of transition process towards agricultural sustainability and (iii) to analyse urban-rural alliances in the establishment of agroecological projects. Furthermore, this research aims to identify an agricultural system to promote more efficient types of farming within the regional agri-food system detailing practical aspects, addressing the problems and discussing the impact of the project, in order to improve territorial food security and connect society to the sustainable use of agro-ecosystems.

2. Materials and Methods

2.1. Study Area

The agroecological study *EXperiencias en TIERras COLectivas* (Extiercol) is located in the municipality of Cuevas del Becerro (Figure 1). The municipality has an extension of 16 km^2 and is in the northwest of the Málaga province (Andalusia, Spain), 80 km from the Málaga capital. Cuevas del Becerro has traditionally been considered the northern gateway to the Serranía de Ronda. It is an area at the intersection of two regional units, Serranía de Ronda and the Guadalteba-Antequera region, and has physical elements and characteristics common to both areas since it is an extension of the Antequera depression but includes limestone reliefs at medium altitudes characteristic of the eastern area of the Serranía de Ronda These common elements and characteristics extend beyond the physical level and are mixed with the administrative level; on the one hand, the municipality is part of the Comarca del Guadalteba in terms of joint services and strategic European development plans, and on the other hand, it is clearly under the sphere of influence of the municipality of Ronda (18 km away), although it is highly dependent on the economic dynamism of the two most important conurbations in the south of Spain, Costa del Sol (43 km away) and Málaga and its metropolitan area.

The study area is characterised by limestone relief with scarce vegetation and heights above 700 m above sea level, while in the lower areas there is a predominance of loamy and clayey soils that are crossed by De las Cuevas river and numerous small streams as a consequence of the water sources from the aquifers located in the limestone formations that form the relief of the municipality. The main agricultural land uses are dedicated to extensive rainfed crops such as olive groves, almond and cereal crops, although there are a great number of small recreational gardens around the village. The predominant livestock farming in the region is extensive sheep farms and to a lesser degree goats and horses. On the other hand, agrotourism is not a developed activity in the region although rural tourism is growing with a large number of rural accommodations. Annual rainfall is around 600 mm, the monthly distribution of rainfall is concentrated in the late autumn and winter months, decreasing during the spring and becoming very scarce during the summer season. The annual temperatures are characterised by a large thermal contrast between the summer and winter months with frequent frosts in the cold months. The average annual temperatures are around 14.3 °C, with a minimum of 7.5 °C in January and maximums of 23 °C in July on average.

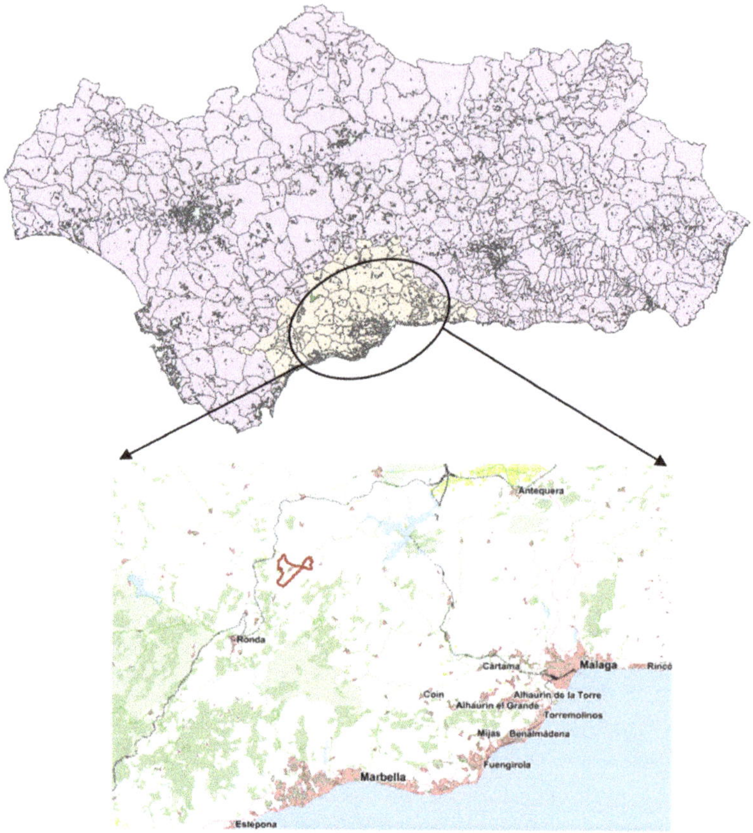

Figure 1. Map of Andalusian region (purple colour) Málaga providence (yellow colour) and Cuevas del Becerro municipality (red outline).

2.2. Study Area History

The demographic dynamics of the territory can be a useful indicator of the socio-economic dynamics and, therefore, of the context and situation in which this project was framed. The municipality had a total population of 1597 in 2020, a consequence of a continuously declining population in the last 80 years.

In order to understand the local population evolution (Figure 2), it is necessary to consider the economic dynamics of Cuevas del Becerro. This economic dynamic is characterised by its dependence, during most of the 20th century, exclusively on agricultural activity. In the municipality, this agricultural activity has not been characterised by its dynamism and diversification; rather, it has remained dedicated to a little diversified and constant production, dedicated especially to extensive cereal crops (mainly wheat) in rotation with anise or chickpea, which, in recent decades, has been displaced as the main crop by the olive tree.

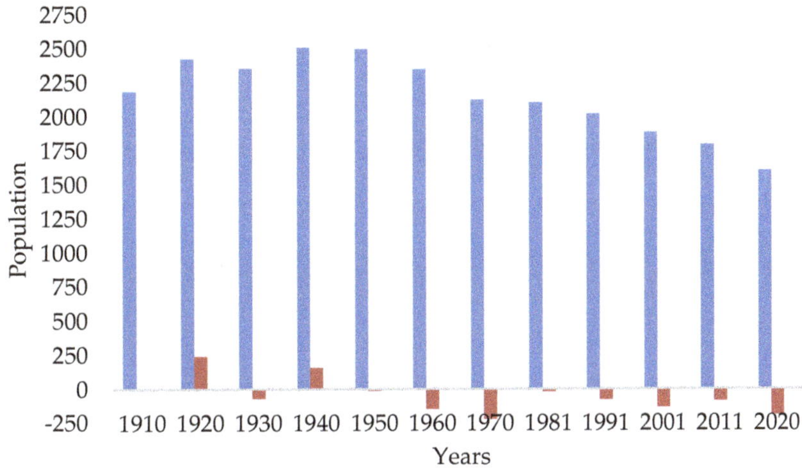

Figure 2. Total population (blue bars) and variation in total population in the given period (red bars) in Cuevas del Becerro (1910–2020).

Agricultural activity employed, and still employs, the vast majority of the municipality's population, most of them day labourers who only work during the harvesting season and therefore experience a high level of seasonal unemployment, a situation that continues to persist to the present day. Given the impossibility of employing the entire working population in agriculture and the poor working conditions in this sector, a continuous rural exodus began in the 1960s [49]. This situation was aggravated by the crisis of traditional agriculture in the 1970s, which deepened the continuous emigration that took place in the municipality. This migratory dynamic has continued to the present day, albeit with lower total values that have not broken the continuity of the population drain, due to a rapidly ageing population and low birth rates, which place the municipality, at present, at the lowest total population value in recorded history with a 10% population loss in the last decade (2011–2020).

2.3. The Creation of the Agroecological Project Experience on Colective Lands (Extiercol)

In the context analysed, the intention to initiate a project within agroecology was born within the youth association 'El Peñoncillo' of Cuevas del Becerro. El Peñoncillo is an association with a long tradition in the municipality since it was founded in the 1970s, in which young people from the village participate and carry out different activities throughout the year. In September 2012, the decision to carry out this project was taken in their assemblies where through this participation method weeklythe members of the association met to decide what type actions to develop. Using this method of participation a total of 14 members participated. The starting point was an initial diagnosis carried out at the meetings by the members of the association and in consensus of all members. Then in the assembly each member took the floor to express their evaluation, point of view and opinion on a subject or problem that is likely to change [50] in this case conventional agriculture and the possibility of starting an agro-ecological project. Finally, a working group was created where several members developed the idea of the project and made up the different sections to be formed. This group was formed by the voluntary subscription of interested members of the association, a total of 5 members formed the working group.

From the beginning of the Extiercol project, an associative culture was present and transmitted from 'El Peñoncillo' to Extiercol, involving the members of the association in the start-up of the project and transferring the collective responsibility and management of the resources, promoting participative, horizontal and collaborative processes as opposed to conventional models where individualistic, hierarchical and competitive dynamics are predominant. For this purpose, in Extiercol project weekly meetings were held where the work of the previous week was evaluated for the project members, and the work of the following week was planned. In these meetings the planned work was distributed among the members of the project so that each member was responsible for a task and then explained the evolution of the work in the weekly meeting.

With regard to the methods for the information gathering in this study a participatory action research methodology was applied [51] where the authors were an active part of the process together with project members from the beginning to the current day participating in the different processes, in decision-making and in the concrete actions to be developed during research. Using the principles of a participatory action research the data collection and analysis was carried out using a situational analysis tool, conducted face-to-face with project members, which involved the systematic collection of detailed information on everything related to the project from the workplace, procedures and strategies to production, planning and sales.

3. Results and Discussion

3.1. *Phase I: Project Initiation*

As a first stage in 2012 (Figure 3), the members of this association decided to collectively carry out a project within the Youth in Action programme established by the European Parliament and of the Council for the period 2007–2013 [52].

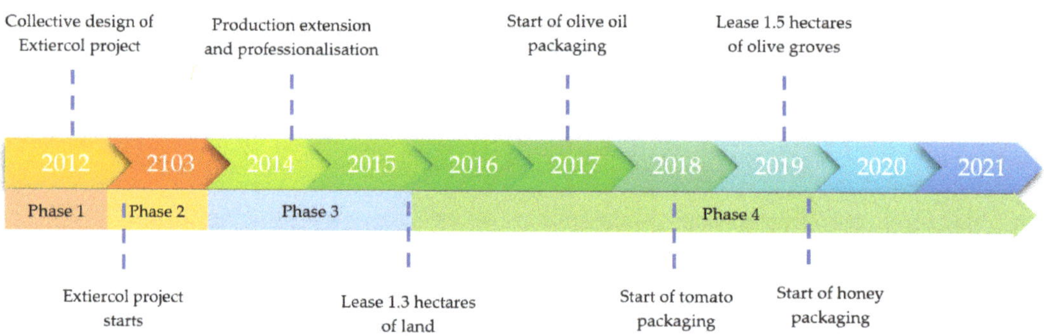

Figure 3. Historical steps in Extiercol project.

Youth in Action was a European programme for all young people aged 15–28 that included a wide variety of actions aimed at youth organisations and collectives. It funded different types of projects, cultural events, sports and non-formal educational activities, in addition to promoting the active participation of young people in society. In this programme, different projects were presented across the European Union, and after a selection process, the projects with the highest scores were funded.

Under the conditions of the Youth in Action programme, the members of the 'El Peñoncillo' youth association elaborated the Extiercol project. The design of Extiercol project consisted of one year of training in theoretical and mainly practical aspects of agro-ecological activity. The practical part covered the whole cycle of multiple vegetable production (tomato, potatoes, peppers, aubergines ...), covering all possible aspects from the most essential, such as collecting seeds, to the more complex, such as composting and processing vegetables.

The practical work was combined with theoretical and practical training mainly related to the production of compost. The staff of specialists who conducted the training of the Extiercol project was composed by professionals with long experience in implementing, developing and advising agroecological projects. Field visits to farms in the province of Málaga under sustainable management were also scheduled. An interesting activity for the Extiercol project members because it responds to their needs in the implementation of sustainable practices and helps them to learn.

In this initial phase, the objectives of the Extiercol project were of a training character with the subsequent possibility of generating employment. The aim was to train young people in agro-ecological techniques, which would allow the reformulation of a sector with historical importance in the economic and social evolution of the municipality. In January 2013, after going through the different administrative phases, the project was approved by the European Commission and funded with 5000 Euros.

3.2. *Phase II: Agroecological Process Begins: Training and Tool Access*

After the Extiercol project was approved, a period was opened for the articulation of a group of young people who would take the project forward; through open assemblies, the project, its strategies and its objectives were explained to all the young people of the village. The aim was to open up the project to the young people of the village, to integrate them and not to close it exclusively to the members of the association. In these spaces for participation, the group of young people who would develop the project was increased to 13 participants all of them without experience in agroecological activity, and the project was started. The economic crisis, which further increased the unemployment levels in the municipality (27.1% in 2011), favoured the interest of young people in Extiercol project.

With regard to the physical areas for the farm work (Figure 4), the project initially had 500 m^2 of self-consumption gardens in the village, as well as a disused and deteriorated municipally owned greenhouse which was added later and which the project members adapted and transformed for agro-ecological activity. In Andalusia, access to resources such as agricultural land and capital has been identified as the main barrier to initiation in the farming sector, especially by young people who do not come from a family tradition linked to agricultural activities [53]. This represents a great difficulty for generational renewal in a highly ageing sector in Europe where, in 2016, for every farm manager under 40, there were three farm managers over 65 [54]. In parallel to this, paradoxically, a long process of land abandonment is taking place in Europe, which has important environmental, socio-economic and landscape consequences [55–57]. The absence of generational renewal in the agricultural sector could aggravate this trend in the following years, especially in arid, semi-arid and mountain rural areas [58].

The project implementation process began with an intensive training process within agroecology, with the organisation by the members of numerous courses, for which experts in these alternative methods were used, dealing with all the aspects involved in agroecological activity, highlighting the learning of the production of bio-inputs, using local resources, which were intended to be applied as an input in agricultural production.

Through these methods and techniques, which were totally respectful with local resources, the aim was to put the farmer in the middle of the decisions he makes on his farm, breaking the dependence on inputs from industry. Therefore, the project initially focused on learning how to produce their own bio-inputs (compost, mineral broths) with local resources, which would allow them to achieve profitable and stable vegetable production. To produce compost, ingredients from local farms were used, mainly sheep and horse manure with abundant straw and mixed with wood ash and tree leaves collected in the village.

Figure 4. Aerial view of Extiercol project sites at Cuevas del Becerro (Málaga, Spain). (A) Greenhouse (B) Self Cosumption gardens (C) Current farmland.

The results of this phase showed the importance of training in agroecological management techniques and access to resources, especially land, as key factors for the initiation of this agroecological experience, as these initial elements can be decisive for the implementation of an agroecological project. In this sense, the development of sufficiently relevant strategies to promote these two key factors is currently lacking.

3.3. *Phase III: Production and Comercializatoin*

From 2014 onword, a crucial point in this experience took place, outside the funding programme, the members decided to progress from agroecological learning to professionalising the agricutural activity. In this process, some of the initial members of the project decided not to continue, leaving the total number of participants at six. An aspect that should be highlighted is the management of the collective project, which is a conflictive aspect in continuous transformation due to the different positions that can appear within a heterogeneous group of people. Working collectively and taking decisions in assembleis, aspects that were incorporated into this initiative seem fundamental, such as (i) a progressive professionalisation of the work, (ii) the construction of common rules that meet the needs of the members and the initiative and (iii) the creation of work spaces within the initiative, where each person identifies in a specific way with the task to be carried out.

At this point, good working dynamics were created and the cultivation space was kept under control, establishing weekly work tasks among the members of the project and keeping track of these activities in the meetings that took place within the project, moving towards professionalisation. A key factor in the progress from the learning phase to the professionalisation phase in Extiercol was the leasing of 1.3 hectares of farmland. This land was managed under the conventional system for rain-fed cereal cultivation, and the Extiercol project members transformed it into an agro-ecological space for multiple vegetable cultivation. The adaptation of the farmland to agro-ecological management required a significant effort not only in terms of economic resources but also in terms of personnel work, due to the spontaneous vegetation removal, the installation of the irrigation system and the application of compost to the farm. In this phase, the inflow of capital from the sale of horticultural products was largely used for reinvestment, determining the progress of the project since the cultivated land increase involved the sprinkler irrigation system installation from the system of irrigation ditches that runs through the vegetable gardens around the village and the purchase of agricultural machinery, a small tractor and equipment for working in the fields. Management change from conventional to agroecological farming systems involves a transition process [59] as a result of which farms are transformed into resilient and sustainable agroecosystems under sustainable agricultural practices implying crop diversification and low external inputs, and a reconnection between the farming system and its ecological and social environment. The efforts of the Extiercol project members were dedicated to a farmland that is not their property (five- year lease) with the insecurity that this implies, because at the end of the contract they could lose the farm, especially in a process of transition towards an agro-ecological model that requires the restoration and improvement of the quality of degraded soils.

These typologies of small farms are generally more productive and contain a greater diversity of crops and biodiversity than large farms, making them more resilient [60]. However, small farms have not been favoured by agricultural policies; in these sense, the proposal for the new CAP, which will be in force between 2023 and 2027, still favours larger farms because the bulk of payments are distributed on the basis of land area [61].

This farm lease increased the project's scale, with more than 40 varieties of horticultural crops grown throughout the year, and many of these varieties are local seeds. These seeds have been recovered from the surrounding areas, many of them from the seed bank of the Serranía de Ronda, this seed bank belongs to the Silvema Association, a non-profit organisation that has been working since 1988 for the protection of the environment, and others directly from farmers who maintain local varieties. Behind the use of these seeds was not only independence from the seed industry and the idea of the farmer as a central agent in decisions regarding his land, but there was also a differentiation from the conventional market through these native seeds which, on many occasions, are not interesting for the agri-food system for reasons that are far from the interests of consumers and are centred on logistical and economic issues of large supermarkets and distributors. On the other hand, local varieties of seeds have been demonstrated to be more resistant and adaptable to the impacts of climate change [62].

In agroecological processes, direct, clear and flexible marketing chains are essential, where trust and recognition of those involved play a decisive role [39]. In this case, the generation process of these alliances was gradual, first in circles of friends and acquaintances and then extending to other people who were reached through word of mouth or the many meetings, workshops, seminars etc. in which the Extiercol members participated. As a result of participating in these events, groups of people committed to the project were created who consume the food produced by the project on a regular basis, this being the main marketing chain for the garden's products. Currently, a network of 50 families supports the project, mainly in Ronda and Málaga and is working on a partnership between the project and consumers to create a formal organisation. The disconnection between production processes and consumers has been identified as one of the factors leading to unsustainable food consumption, especially in highly urbanised areas [63]. Under this consumption model, the connection to the production processes and farm sustainability is promoted, thus establishing interactions between rural and urban areas. Participation in these consumer groups can increase social identity and awareness of the impact of collective actions. These aspects are pointed out as determinants in increasing environmental awareness and thus the approach to environmental and sustainable processes [64].

Small shops, generally specialised, are an important ally, both as a point of contact with consumers and as a space for selling garden produce. Weaving these alliances is an essential aspect, as we find ourselves agents who, within the agri-food system, suffer the consequences of an unequal power correlation. In addition the process of direct farmer to consumer marketing outside of conventional sales channels is fundamental because with direct sales, the profitability of production is higher due to the increase in the profit margin of the product as a result of the elimination of intermediaries in the food chain.

The crop diversification on the farmland determines the marketing system, which involves two modalities, the box model with 6–7 vegetable varieties and the grouping of consumers with the possibility of choosing variety and quantity. This second modality is subject to the first, which the members understand to be more loyal and committed. With three weekly delivery routes, one local, one to Ronda and one to Malaga, for groups of consumers and shops, rural-urban alliances are essential to sustain an agro-ecosystem that is consistent with its philosophy and healthy, conscious and responsible food. In this sense, several studies have demonstrated the environmental and climatic benefits of diversified farming, as well as seasonally and proximally produced food products [65–67].

In this phase, the relationship between Extiercol project and the consumers, the approximation between the production and consumption phases have been shown to be essential in the development of the agro-ecological project, being important the construction of spaces for the meeting of both spheres. In these agro-ecological systems of production, distribution and consumption, new alliances between farmers and consumers seem fundamental as an alternative proposal to organise the current agro-food system and towards a new rural-urban relationship reconnected with nature. Consumer adaptation to this production and marketing model is part of the agroecological strategy where the consumer is aware of the process and maintains close contact with the production processes and the advantages that derive from them. In this sense, the trust relationship between Extiercol and the consumers has allowed the project to be supported from the beginning, without the need to obtain official accreditation as agroecological land. For this reason, this model is often restricted to consumers with a high level of environmental awareness; nevertheless, it achieves a strong connection between producers and consumers [68].

3.4. Phase IV: Value Chain Increase

Since 2017, Extiercol has been committed to the processing of primary products as a strategy to add new products that open up new marketing chains, complementing the current ones and increasing the profitability of its productions. The objective was to have a diverse offering of products, initially with small productions that allow not only diversification of the agricultural activity but also expansion of the commercial capacity by offering consumers a greater number of products.

The first step in this process of processing primary products was taken with olive oil production. Since the 2016/2017 campaign, where the first tests were carried out, there has been a continuous increase in the production and marketing of this product, which has led to an increase from 538 litres of oil in the 2016/2017 campaign to 1556 litres in the 2020/21 campaign (Figure 5). The incorporation of new farmers from the area who, coming from industrial agriculture, changed their olive grove management model to market their production together with that of Extiercol, as well as the increase in the project's olive grove land with the leasing of 1.5 ha in 2019, was fundamental in this process of production increase. In this sense, the incorporation of new farmers into the experience and the replication of agroecological management on their farms was, without a doubt, a really positive aspect towards the sustainability of local agricultural management.

Figure 5. Annual olive oil production in Extiercol project. Green bars represent total production (Litres) and yellow line annual increase (%).

An added element in the processing process was the accompanying of the oil bottles with a label indicating the individual costs of the production steps for each bottle: processing process, workers' salaries, reinvestment in the Extiercol project or land leasing. This value label brought the consumer closer to the project, showed its influence and capacity to reverse the current ways with its consumption. In Spain, the reasons for which the consumer buys organic olive oil are usually related to health, food safety and taste [69]. However, environmental preservation does not appear among the current consumer priorities. Due to the economic and territorial importance of olive groves in Spain and especially in Andalusia, it seems necessary to place a greater emphasis on the environmental qualities that olive groves under organic management can offer. In this sense several studies have shown how organic olive groves management in Andalusia improves the capacity to provide ecosystem services [70–73]. Since the 2018/2019 campaign, the Extiercol project has incorporated among its products the tomato jars packaged under the brand 'Tomate Rosado' (Figure 6), with which the project team have had positive experiences. This is a local variety of tomato (Pink Tomato); with this process, as well as extending the range of products on offer and generating added value, it also provides an outlet for possible summer surpluses as well as incorporating the product in the winter months when the tomato plant in open air conditions in the climatic conditions where the study area is located does not produce, so consumers cannot obtain this product. In 2020, 550 jars of tomatoes were packaged. Both the processing of olive oil and tomato jars are carried out in small processing plants close to the municipality where the products are processed under organic and sanitary standards.

Figure 6. Products packaged in Extiercol project: (**A**) Olive oil, (**B**) Tomato jars, (**C**) Honey.

In order to expand the range of products in the consumer network, honey has been added to the consumer offering since 2019. As in olive oil production, other beekeepers were involved in the production of honey, as the number of own hives in the project is not very large. At present, 520 kg of honey is marketed annually, and the intention is to continue increasing this product, as well as oil and tomato packaging.

In this phase, the multifunctionality and diversification of the agro-ecological project has been highlighted. The relationship with other farmers and agents in the food chain, in this case in the processing sector, has been fundamental. These relationships not only favour an increase in the scale of the project but also improve efficiency, as in the case of tomatoes, where there was a reduction in harvest losses. In these sense, future challenges for the project include the involvement of a greater number of farmers to bring about a significant change in local production methods and the extension of processing to a wider range of horticultural products.

3.5. The Extiercol Project as a Model Case

The process and methodology developed can be extrapolated to any other place, although fundamental aspects such as 1. an analysis that identifies local opportunities, problems and needs, as well as the assessment of traditional knowledge 2. participatory and collaborative work that aims to include the concerns and aspirations of the participants so that they feel that the initiative is their own and 3. a training process that allow them to learn sustainable production methods are important.

In the replicability of this agroecological model and its extension to a larger scale a possible risk is its possible conventionalisation, i.e., the exclusive adoption of certain aspects that comply with the organic regulations but not substantially modify the structures and methods of conventional agriculture [74,75] that have resulted in the current critical situation of the agricultural sector. For example, the monoculture crops are allowed under organic regulations, which barely contribute to the agroecosystem's sustainability [76]. It conventionalisation also implies the integration of the organic sector in conventional processing and marketing structures in which the production, distribution and consumption of organic food implies the processing of organic food into homogenous mass-produced commodities, controlled by audited certification and distributed through the mass food chain dominated by large companies far removed from the farmers and rural areas [77].

Similarly, in line with [78] by simply replacing the synthetic inputs with organic ones on the farm, the methods and actions of conventional agriculture are not questioned, with the intention that a given input or technique should replace another that was practised in conventional agriculture, and, furthermore, that the same results should be obtained as before. Thus, in the reconversion of production areas, there is no reconversion in the forms of action, nor is there a questioning of previous methods, this process has been detected in Andalusia region [79]. In short, the aim is to replace inputs individually, when there should be a change in the relationships and ways of acting as a whole, which implies a holistic vision of the agro-ecosystem, since, if this is not the case, there is no complete and diverse ecological action, and we will be exposed to the risks posed by the imbalances in these types of production.

Throughout the process of this experience, there has been an evolution, a transformation in the perception of the activity or at least of the components that have developed this project with these agro-ecological methods, which has taken the members of the project from 'ignorant to experts'. This means that they are transferring knowledge and other farmers are taking on board the management and operating methods that the project has set out since its beginnings, with more farmers appearing who are discordant with conventional methods. In this sense, the implementation of sustainable management including vegetation cover has been negatively perceived by farmers in some areas [80], and this perception could be an obstacle to the inclusion of these management systems. In this case study, the inclusion of farmers' crops could be the beginning of a transition towards sustainable models on a larger scale in the analysed area.

In the success of this project, several aspects can be highlighted; among them is the quality of the products. This is due to their proximity, which allows us to maintain their freshness and harvest them in an optimal state of maturity, enhancing the flavour with which they reach the consumer. Consumers are attracted to organic food because it is produced without synthetic chemicals, and there is a growing interest in healthy consumption patterns. However, in a significant transformation of production systems towards more sustainable systems, a high degree of consumer awareness is essential to demand a real change in the methods applied in conventional agriculture. On the other hand, the implementation of public policies that encourage the regionalisation of production and local consumption could be an interesting way to reduce our carbon footprint and emissions and to maintain populations in rural areas while promoting food security.

4. Conclusions

The challenge for agriculture posed by the changes arising from the climate change process or the loss of biodiversity has led to a growing interest in undertaking actions to promote sustainable agroecosystem models with a new emphasis on agroecology. Small diversified farms under sustainable management practices such as those analysed by this study may be typologies that can play an important role in the mitigation of the effects derived from the climate change process, and thus further attention seems necessary.

Throughout this work, we have shown an agroecological experience in the rural area that, taking into account its multifunctional character, has developed an agroecological project to enhance its functions, such as the provision of healthy and quality products, the maintenance of the population and sustainable local production systems that aim to restructure agri-food production. On the other hand, this process has shown how the closer relations between food production area and nearby urban areas have been fundamental in connecting both territories and the people who inhabit them and thus connect society with the sustainable use of agroecosystems. In this study elaborates on important criteria for the agroecological project development, detailed practical aspects, addresses problems and discusses the impact of the project in a wider socio-economic setting.

The content of this article could be interest regarding the UN SDGs, the Green Deal, the Horizon Europe Mission on Soil Health and Food, and the recent activities of the SCAR strategic working group on Agroecology. In this regard, we think that our practical aspects and experience as described in the article could be high value to the research community.

Author Contributions: Conceptualization, B.L.-G. and M.G.-R.; methodology, B.L.-G., L.P.-A., J.A.-H. and M.G.-R.; validation, B.L.-G., L.P.-A. and M.G.-R.; formal analysis, B.L.-G. and L.P.-A.; investigation, L.P.-A., B.L.-G., J.A.-H. and M.G.-R.; resources, B.L.-G. and M.G.-R.; data curation, M.G.-R.; writing—original draft preparation, M.G.-R.; writing—review and editing, M.G.-R., B.L.-G. and L.P.-A.; visualization, B.L.-G., M.G.-R. and L.P.-A.; supervision, B.L.-G., M.G.-R. and L.P.-A. All authors have read and agreed to the published version of the manuscript.

Funding: This research received no external funding.

Institutional Review Board Statement: Not applicable.

Data Availability Statement: Data sharing not applicable.

Acknowledgments: We are grateful to the members of Extiercol project (Cristobal, José, Pedro, Heredia and Juan Ramón) for their collaboration in the realization of this research.

Conflicts of Interest: The authors declare no conflict of interest.

References

1. Informe Anual de Indicadores. Agricultura, Pesca y Alimentación. 2019. Available online: https://www.mapa.gob.es/es/ministerio/servicios/analisis-y-prospectiva/indicadores_semestre.aspx (accessed on 25 May 2021).
2. Aguilera, E.; Guzmán, G.I.; de Molina, M.G.; Soto, D.; Infante-Amate, J. From animals to machines. The impact of mechanization on the carbon footprint of traction in Spanish agriculture: 1900–2014. *J. Clean. Prod.* **2019**, *221*, 295–305. [CrossRef]
3. Medina, L.J.G.; Rodrigo, E.C. La ocupación en el sector agrario: Trayectoria y actualidad. *Panor. Soc.* **2020**, *31*, 113–124.
4. Taylor, P.J.; Flint, C. *Geografía Política: Economía-Mundo, Estado-Nación y Localidad*, 2nd ed.; Trama: Madrid, Spain, 2002; p. 447.
5. Camilleri, A.R.; Larrick, R.P.; Hossain, S.; Patino-Echeverri, D. Consumers underestimate the emissions associated with food but are aided by labels. *Nat. Clim. Chang.* **2019**, *9*, 53–58. [CrossRef]
6. European Commission. *Employment and Social Developments in Europe: 2019 Review*; European Commission: Luxembourg, 2019; Available online: https://ec.europa.eu (accessed on 10 May 2021).
7. Alons, G. Environmental Policy Integration in the EU's Common Agricultural Policy: Greening or Greenwashing? *J. Eur. Public Policy* **2017**, *24*, 1604–1622. [CrossRef]
8. Gocht, A.; Ciaian, P.; Bielza, M.; Terres, J.M.; Röder, N.; Himics, M.; Salputra, G. EU-Wide Economic and Environmental Impacts of CAP Greening with High Spatial and Farm-Type Detail. *J. Agric. Econ.* **2017**, *68*, 651–681. [CrossRef]
9. Hristov, J.; Clough, Y.; Sahlin, U.; Smith, H.G.; Stjernman, M.; Olsson, O.; Brady, M.V. Impacts of the EU's Common Agricultural Policy "Greening" Reform on Agricultural Development, Biodiversity, and Ecosystem Services. *Appl. Econ. Perspect. Policy* **2020**, *42*, 716–738. [CrossRef]
10. European Union. *From Farm to Fork Strategy: For a Fair, Healthy and Environmentally-Friendly Food System*; European Commission: Brussels, Belgium, 2020.
11. Piorr, A.; Zasada, I.; Doernberg, A.; Zoll, F.; Ramme, W. *Research for AGRI Committee—Urban and Peri-Urban Agriculture in the EU*; European Parliament: Brussels, Belgium, 2018.
12. Caporali, F. History and development of agroecology and theory of agroecosystems. In *Law and Agroecology*; Springer: Berlin/Heidelberg, Germany, 2015; pp. 3–29.
13. IPCC. Special Report on Global Warming of 1.5. Available online: https://www.ipcc.ch/sr15/ (accessed on 20 May 2021).
14. Dai, T.; Yang, Y.; Lee, R.; Fleischer, A.S.; Wemhoff, A.P. Life cycle environmental impacts of food away from home and mitigation strategies—A review. *J. Environ. Manag.* **2020**, *265*, 110471. [CrossRef]
15. Sullivan, G.T.; Ozman-Sullivan, S.K. Alarming evidence of widespread mite extinctions in the shadows of plant, insect and vertebrate extinctions. *Austr. Ecol.* **2021**, *46*, 163–176. [CrossRef]
16. Singh, J.; Schädler, M.; Demetrio, W.; Brown, G.G.; Eisenhauer, N. Climate change effects on earthworms—A review. *SOIL Org.* **2019**, *91*, 114–138. [CrossRef]
17. Ghosh, A.; Misra, S.; Bhattacharyya, R.; Sarkar, A.; Singh, A.K.; Tyagi, V.C.; Kumar, R.V.; Meena, V.S. Agriculture, dairy and fishery farming practices and greenhouse gas emission footprint: A strategic appraisal for mitigation. *Environ. Sci. Pollut. Res.* **2020**, *27*, 10160–10184. [CrossRef]
18. González-Rosado, M.; Parras-Alcántara, L.; Aguilera-Huertas, J.; Lozano-García, B. Long-term evaluation of the initiative 4‰ under different soil managements in Mediterranean olive groves. *Sci. Total. Environ.* **2021**, *758*, 143591. [CrossRef]
19. Hendrickson, M.K.; Heffernan, W.D. Opening spaces through relocalization: Locating potential resistance in the weaknesses of the global food system. *Sociol. Rural.* **2002**, *42*, 347–369. [CrossRef]
20. Inventario Nacional de Gases de Efecto Invernadero (GEI). 2019. Available online: https://www.miteco.gob.es/es/calidad-y-evaluacion-ambiental/temas/sistema-espanol-de-inventario-sei-/es-2021-nir_tcm30-523942.pdf (accessed on 2 May 2021).
21. Sanz-Cobena, A.; Lassaletta, L.; Aguilera, E.; del Prado, A.; Garnier, J.; Billen, G.; Smith, P. Strategies for greenhouse gas emissions mitigation in Mediterranean agriculture: A review. *Agric. Ecosyst. Environ.* **2017**, *238*, 5–24. [CrossRef]
22. Paustian, K.; Lehmann, J.; Ogle, S.; Reay, D.; Robertson, G.P.; Smith, P. Climate-smart soils. *Nature* **2016**, *532*, 49–57. [CrossRef]
23. Oldfield, E.E.; Wood, S.A.; Bradford, M.A. Direct effects of soil organic matter on productivity mirror those observed with organic amendments. *Plant Soil* **2018**, *423*, 363–373. [CrossRef]
24. González-Rosado, M.; Parras-Alcántara, L.; Aguilera-Huertas, J.; Lozano-García, B. Soil Productivity Degradation in a Long-Term Eroded Olive Orchard under Semiarid Mediterranean Conditions. *Agronomy* **2021**, *11*, 812. [CrossRef]
25. González-Rosado, M.; Lozano-García, B.; Aguilera-Huertas, J.; Parras-Alcántara, L. Short-term effects of land management change linked to cover crop on soil organic carbon in Mediterranean olive grove hillsides. *Sci. Total Environ.* **2020**, *744*, 140683. [CrossRef]

26. Bogunovic, I.; Telak, L.J.; Pereira, P. Agriculture management impacts on soil properties and hydrological response in Istria (Croatia). *Agronomy* **2020**, *10*, 282. [CrossRef]
27. Morugán-Coronado, A.; Linares, C.; Gómez-López, M.D.; Faz, A.; Zornoza, R. The impact of intercropping, tillage and fertilizer type on soil and crop yield in fruit orchards under Mediterranean conditions: A meta-analysis of field studies. *Agric. Syst.* **2020**, *178*, 102736. [CrossRef]
28. Zomer, R.J.; Bossio, D.A.; Sommer, R.; Verchot, L.V. Global sequestration potential of increased organic carbon in cropland soils. *Sci. Rep.* **2017**, *7*, 1–8. [CrossRef]
29. Bombino, G.; Denisi, P.; Gómez, J.A.; Zema, D.A. Mulching as best management practice to reduce surface runoff and erosion in steep clayey olive groves. *Int. Soil Water Conserv. Res.* **2020**, *9*, 26–36. [CrossRef]
30. Wiesmeier, M.; Mayer, S.; Paul, C.; Helming, K.; Don, A.; Franko, U.; Steffens, M.; Kögel-Knabner, I. CO_2 certificates for carbon sequestration in soils: Methods, management practices and limitations. *BonaRes Ser.* **2020**, *4*, 1–23. [CrossRef]
31. Veerman, C.; Pinto Correia, T.; Bastioli, C.; Biro, B.; Bouma, J.; Cienciela, E. Caring for soil is caring for life. In *EU Soil Health and Food Mission Board*; European Commission: Brussels, Belgium, 2020.
32. Puigdueta, I.; Aguilera, E.; Cruz, J.L.; Iglesias, A.; Sanz-Cobena, A. Urban agriculture may change food consumption towards low carbon diets. *Glob. Food Secur.* **2021**, *28*, 100507. [CrossRef]
33. Pichler, P.P.; Zwickel, T.; Chavez, A.; Kretschmer, T.; Seddon, J.; Weisz, H. Reducing urban greenhouse gas footprints. *Sci. Rep.* **2017**, *7*, 1–11. [CrossRef]
34. Kriewald, S.; Pradhan, P.; Costa, L.; Ros, A.G.C.; Kropp, J.P. Hungry cities: How local food self-sufficiency relates to climate change, diets, and urbanisation. *Environ. Res. Lett.* **2019**, *14*, 094007. [CrossRef]
35. Vicente-Vicente, J.L.; Piorr, A. Can a shift to regional and organic diets reduce greenhouse gas emissions from the food system? A case study from Qatar. *Carbon Balance Manag.* **2021**, *16*, 2. [CrossRef]
36. C40 Food Systems Network. 2019. Available online: www.c40.org/networks/food_systems (accessed on 5 June 2021).
37. Fabbri, K. Food 2030: Future-proofing our food systems through research and innovation. *Brussels* **2017**, *1*, 4–44. [CrossRef]
38. Van der Ploeg, J.D. The political economy of agroecology. *J. Peasant. Stud.* **2021**, *48*, 274–297. [CrossRef]
39. Anderson, C.R.; Bruil, J.; Chappell, M.J.; Kiss, C.; Pimbert, M.P. From Transition to Domains of Transformation: Getting to Sustainable and Just Food Systems through Agroecology. *Sustainability* **2019**, *11*, 5272. [CrossRef]
40. Altieri, M.A.; Nicholls, C.I. Agroecology and the reconstruction of a post-COVID-19 agriculture. *J. Peasant. Stud.* **2020**, *47*, 881–898. [CrossRef]
41. Vaarst, M.; Escudero, A.G.; Chappell, M.J.; Brinkley, C.; Nijbroek, R.; Arraes, N.A.M.; Andreasen, L.; Gattinger, A.; De Almeida, G.F.; Bossio, D. Exploring the concept of agroecological food systems in a city-region context. *Agroecol. Sustain. Food Syst.* **2018**, *42*, 686–711. [CrossRef]
42. De Bernardi, P.; Azucar, D. The Food System Grand Challenge: A Climate Smart and Sustainable Food System for a Healthy Europe. In *Innovation in Food Ecosystems*; Springer: Cham, Switzerland, 2020; pp. 1–25. [CrossRef]
43. Bouma, J. Soil security as a roadmap focusing soil contributions on sustainable development agendas. *Soil Secur.* **2020**, *1*, 100001. [CrossRef]
44. Isgren, E.; Ness, B. Agroecology to Promote Just Sustainability Transitions: Analysis of a Civil Society Network in the Rwenzori Region, Western Uganda. *Sustainability* **2017**, *9*, 1357. [CrossRef]
45. Altieri, M.A. Agroecology: A new research and development paradigm for world agriculture. *Agric. Ecosyst. Environ.* **1989**, *27*, 37–46. [CrossRef]
46. Gliessman, S.R. *Agroecology: The Ecology of Sustainable Food Systems*; CRC Press: Boca Raton, FL, USA, 2014.
47. Altieri, M.A. *Agroecology: The Science of Sustainable Agriculture*; CRC Press: Boca Raton, FL, USA, 2018.
48. Del Valle Ramos, C.; Almoguera, P. Envejecimiento demográfico y (des) población en las ciudades medias interiores de Andalucía (2008–2018). *Cuad. Geográficos De La Univ. De Granada* **2020**, *59*, 263–286. [CrossRef]
49. González-Rosado, M. Desarrollo Local en Zonas Rurales Atrasadas: El Caso de Cuevas del Becerro. Master's Thesis, University of Málaga, Málaga, Spain, 2013.
50. Martínez, M. *Ciencia y Arte en la Metodología Cualitativa*; Trillas: México, Mexico, 2009.
51. McIntyre, A. *Participatory Action Research*; Sage Publications: Thousand Oaks, CA, USA, 2007.
52. Decision No 1719/2006/EC of the European Parliament and of the Council of 15 November 2006 Establishing the Youth in Action Programme for the Period 2007 to 2013. Available online: https://eur-lex.europa.eu/eli/dec/2006/1719/oj (accessed on 18 July 2021).
53. Estudio Sobre el Acceso a la Tierra: Documento Final del Grupo Focal de Acceso a la Tierra. Ministerio de Agricultura, Pesca y Alimentación. 2020. Available online: https://cpage.mpr.gob.es/producto/estudio-sobre-el-acceso-a-la-tierra/ (accessed on 12 May 2021).
54. Commission Staff Working Document. Executive Summary of the Evaluation of the Impact of CAP on Generational Renewal. *Local Development and Jobs in Rural Areas.* Available online: https://ec.europa.eu/info/sites/default/files/food-farming-fisheries/key_policies/documents/ext-eval-cap-gene-renewal-study-exe-summary_2021_en.pdf (accessed on 18 May 2021).
55. Lasanta, T.; Nadal-Romero, E.; Arnáez, J. Managing abandoned farmland to control the impact of re-vegetation on the environment. The state of the art in Europe. *Environ. Sci. Policy* **2015**, *52*, 99–109. [CrossRef]

56. Levers, C.; Schneider, M.; Prishchepov, A.V.; Estel, S.; Kuemmerle, T. Spatial variation in determinants of agricultural land abandonment in Europe. *Sci. Total. Environ.* **2018**, *644*, 95–111. [CrossRef] [PubMed]
57. Dolton-Thornton, N. How should policy respond to land abandonment in Europe? *Land Use Policy* **2021**, *102*, 105269. [CrossRef]
58. Lasanta, T.; Arnáez, J.; Pascual, N.; Ruiz-Flaño, P.; Errea, M.P.; Lana-Renault, N. Space–time process and drivers of land abandonment in Europe. *Catena* **2017**, *149*, 810–823. [CrossRef]
59. Gliessman, S.R.; Rosemeyer, M. *The Conversion to Sustainable Agriculture: Principles, Processes and Practices*; CRC Press: Boca Raton, FL, USA, 2009.
60. Ricciardi, V.; Mehrabi, Z.; Wittman, H.; James, D.; Ramankutty, N. Higher yields and more biodiversity on smaller farms. *Nat. Sustain.* **2021**, *4*, 651–657. [CrossRef]
61. Pe'er, G.; Bonn, A.; Bruelheide, H.; Dieker, P.; Eisenhauer, N.; Feindt, P.H.; Lakner, S. Action needed for the EU Common Agricultural Policy to address sustainability challenges. *People Nat.* **2020**, *2*, 305–316. [CrossRef]
62. Westengen, O.T.; Brysting, A.K. Crop adaptation to climate change in the semi-arid zone in Tanzania: The role of genetic resources and seed systems. *Agric. Food Secur.* **2014**, *3*, 3. [CrossRef]
63. Goldstein, B.; Birkved, M.; Fernández, J.; Hauschild, M. Surveying the environmental footprint of urban food consumption. *J. Ind. Ecol.* **2016**, *21*, 151–165. [CrossRef]
64. Bamberg, S.; Rees, J.; Seebauer, S. Collective climate action: Determinants of participation intention in community-based pro-environmental initiatives. *J. Environ. Psychol.* **2015**, *43*, 155–165. [CrossRef]
65. Tobarra, M.A.; Lopez, L.A.; Cadarso, M.A.; Gomez, N.; Cazcarro, I. Is seasonal households' consumption good for the nexus carbon/water footprint? The Spanish fruits and vegetables case. *Environ. Sci. Technol.* **2018**, *52*, 12066–12077. [CrossRef]
66. Ivanova, D.; Barrett, J.; Wiedenhofer, D.; Macura, B.; Callaghan, M.; Creutzig, F. Quantifying the potential for climate change mitigation of consumption options. *Environ. Res. Lett.* **2020**, *15*, 093001. [CrossRef]
67. Pradhan, P.; Kriewald, S.; Costa, L.; Rybski, D.; Benton, T.G.; Fischer, G.; Kropp, J.P. Urban food systems: How regionalization can contribute to climate change mitigation. *Environ. Sci. Technol.* **2020**, *54*, 10551–10560. [CrossRef]
68. Wezel, A.; Bellon, S.; Doré, T.; Francis, C.; Vallod, D.; David, C. Agroecology as a science, a movement and a practice. *A Review. Agron. Sustain. Dev.* **2009**, *29*, 503–515. [CrossRef]
69. Vega-Zamora, M.; Parras, R.M.; Murgado, A.E.M.; Torres, R.F.J. The influence of the term organic on food purchasing behaviour. *Procedia Soc. Behav. Sci.* **2013**, *81*, 660–671. [CrossRef]
70. Parras-Alcántara, L.; Díaz, J.L.; Lozano-García, B. Organic farming affects C and N in soils under olive groves in Mediterranean areas. *Land Degrad Dev.* **2013**, *26*, 800–806. [CrossRef]
71. Parras-Alcántara, L.; Lozano-García, B. Conventional tillage vs. organic farming in relation to soil organic carbon stock in olive groves inMediterranean rangelands (southern Spain). *Solid Earth Discuss* **2014**, *6*, 35–70. [CrossRef]
72. Soriano, M.A.; Álvarez, S.; Landa, B.B.; Gómez, J.A. Soil properties in organic olive orchards following different weed management in a rolling landscape of Andalusia, Spain. *Renew. Agric. Food Syst.* **2013**, *29*, 83–91. [CrossRef]
73. Pleguezuelo, C.R.R.; Zuazo, V.H.D.; Martínez, J.R.F.; Peinado, F.J.M.; Martín, F.M.; Tejero, I.F.G. Organic olive farming in Andalusia, Spain. A review. *Agron. Sustain. Dev.* **2018**, *38*, 1–16. [CrossRef]
74. Best, H. Organic agriculture and the conventionalization hypothesis: A case study from West Germany. *Agric. Hum. Values* **2008**, *25*, 95–106. [CrossRef]
75. Darnhofer, I.; Lindenthal, T.; Bartel-Kratochvil, R.; Zollitsch, W. Conventionalisation of organic farming practices: From structural criteria towards an assessment based on organic principles. A review. *Agron. Sustain. Dev.* **2010**, *30*, 67–81. [CrossRef]
76. Aguilera, E.; Díaz-Gaona, C.; García-Laureano, R.; Reyes-Palomo, C.; Guzmán, G.I.; Ortolani, L.; Rodríguez-Estévez, V. Agroecology for adaptation to climate change and resource depletion in the Mediterranean region. A review. *Agric. Syst.* **2020**, *181*, 102809. [CrossRef]
77. Navarrete, M. How do farming systems cope with marketing channel requirements in organic horticulture? The case of market-gardening in south eastern France. *J. Sustain. Agric.* **2009**, *33*, 552–565. [CrossRef]
78. Gliessman, S. Transforming food systems with agroecology. *Agroecol. Sustain. Food Syst.* **2016**, *40*, 187–189. [CrossRef]
79. Ramos García, M.; Guzmán, G.I.; González De Molina, M. Dynamics of organic agriculture in Andalusia: Moving toward conventionalization? *Agroecol. Sustain. Food Syst.* **2018**, *42*, 328–359. [CrossRef]
80. Cerdà, A.; Rodrigo-Comino, J.; Giménez-Morera, A.; Keesstra, S.D. Hydrological and erosional impact and farmer's perception on catch crops and weeds in citrus organic farming in Canyoles river watershed, Eastern Spain. *Agric. Ecosyst. Environ.* **2018**, *258*, 49–58. [CrossRef]

agriculture

MDPI

Article

Evaluation of Farmers' Ecological Cognition in Responses to Specialty Orchard Fruit Planting Behavior: Evidence in Shaanxi and Ningxia, China

Zhe Chen [1,2,†], Apurbo Sarkar [1,2,†], Ahmed Khairul Hasan [3], Xiaojing Li [1,2] and Xianli Xia [1,*]

1 College of Economics and Management, Northwest Agriculture and Forestry University, Yangling, Xianyang 712100, China; xncz@nwafu.edu.cn (Z.C.); apurbo@nwafu.edu.cn (A.S.); xnlxj@nwafu.edu.cn (X.L.)

2 Sixth Industry Research Institute, Northwest Agriculture and Forestry University, Yangling, Xianyang 712100, China

3 Department of Agronomy, Bangladesh Agricultural University, Mymensingh 2202, Bangladesh; akhasan@bau.edu.bd

* Correspondence: xnxxli@nwafu.edu.cn

† These authors contributed equally to work.

Abstract: Developing specialties in orchard fruits productions with ecological and economic benefits is a practical and effective way to guarantee eco-friendliness and increase farmers' income in the Loess Plateau area. Therefore, to understand these factors, the study constructs an agriculture ecological cognition index from three dimensions of eco-agriculture cognitions (increase income cognition, water conservation cognition and eco-product price cognition). Our analysis was based on micro survey data from 416 farmers in Shaanxi and Ningxia, China. The study used two main econometric models, double-hurdle and Interpretative Structural Modeling (ISM), to examine the relationship and influence pathways between cognition of ecological agriculture and farmers' specialty orchard fruit planting behavior. The results show that: (i) the cognition of eco-agriculture affects whether farmers plant specialty fruits (participation decision). The cognition of eco-agriculture increases income and the cognition of eco-product price significantly affect the scale of specialty orchard fruits planting (quantity decision). (ii) Household resource endowments influence specialty orchard fruit planting responses through ecological farming cognitions. (iii) The factors influencing the participation and quantity decisions of orchard fruit planting are significantly different. Therefore, when the government actively encourages farmers to participate in specialty orchard planting, it should fully consider the cognitive factors of ecological agriculture of the growers and develop targeted training strategies.

Keywords: ecological agriculture; water conservation; double-hurdle model; interpretative structural modeling; adoptions

Citation: Chen, Z.; Sarkar, A.; Hasan, A.K.; Li, X.; Xia, X. Evaluation of Farmers' Ecological Cognition in Responses to Specialty Orchard Fruit Planting Behavior: Evidence in Shaanxi and Ningxia, China. *Agriculture* **2021**, *11*, 1056. https://doi.org/10.3390/agriculture11111056

Academic Editors: José Luis Vicente-Vicente, Cristina Quintas-Soriano and María D. López-Rodríguez

Received: 2 September 2021
Accepted: 4 October 2021
Published: 28 October 2021

1. Introduction

In the new era of modernization and globalization, agribusiness, especially orchards management, becomes a challenging venture as there is a pressing demand regarding the quality of products [1]. The overexploitation of natural resources and agriculture intensification are two major drivers which significantly threaten natural landscapes and global sustainability [2]. All the fundamental components of agricultural production, from the seed or plant planting to culture and nourishing them, until harvesting and marketing, need to be managed carefully with a higher intensity for coping with the challenges of current food demands without hampering the ecological balances and diversity. Nowadays, the careful management of farms has become a focal point that supports the current trends of production intensification in a specialized way while facilitating ecological friendliness [3]. However, facilitating specialized fruits production tactics has become a prominent way to promote ecological construction while enabling farmers' poverty alleviation and economic development [4].

Interestingly, specialty fruit crops represent an innovative production method that enhances the substantial portion of agricultural production value [5]. The United States Department of Agriculture (USDA) defines specialty crops by covering fruits and vegetables, tree nuts, dried fruits, horticulture and nursery crops. Specifically, the study focuses on fruit production because it represents many specialty crops [6]. However, China's orchard fruit industry mainly covers cultivating, managing and processing grapes, citrus, apples, pears, peaches and other related fruit production and processing industries [7]. Seemingly, the orchard fruit industry is an essential component of China's agricultural industry structure [8] which has higher competitive advantages, fosters benefits than conventional agriculture and helps farmers achieve rapid growth in agricultural income [9,10]. The government is also highlighting the importance of specialty crops in various ways. For example, in November 2016, the State Council of China issued the notice regarding the 13th five-year plan for poverty alleviation to combat poverty, which proposed combining the national ecological construction project and highlighted the importance of several orchard industries with ecological and economic benefits [11]. Moreover, in 2018, the "No. 1 Central Document of China" emphasized to "further promote the greening, quality supervising, specializing and branding the specialty agricultural products [12]".

However, as the main agribusiness agent, the behavioral responses of the farmers should be captured effectively for understanding the development of the special orchard-based fruit industry [13,14]. According to Corris [15], farmers' ecological cognition mostly relies on their interpersonal understanding, perception and plan of action, which is mostly altered by several externalities. Yang et al. [16] defined farmer ecological responses behavior as "the set of knowledge, skills and thought that can alters or minimize the negative externalities" which lead them to face external environmental changes spontaneously for taking the planting decisions and behavior accordingly. Some scholars have roughly divided the key factors affecting farmers' behavioral decisions into external and individual factors [17,18]. While some scholars highlighted that individual characteristics such as household characteristics, household heads perceptions, social impacts, educational status, training facilities and interpersonal innovativeness could be decisive factors in understanding farmers' behavior [19–21]. However, some academics have different opinions on whether farmers' cognition influences their decision-making behavior [22,23]. Some scholars believe that there is a positive correlation between behavioral cognition and behavioral actions, which leads behavioral cognition directly to the actor's behavioral intention and decision [24,25]. Seemingly, some scholars also point out the inconsistency between farmers' cognition and behavioral decision-making process and they also pointed there is no significant causal relationship between farmers' cognition and decision-making [26,27]. The divergence between cognition and behavior of economic agents is reflected as cognitive conflict [28,29].

The existing studies on farmers' responses and decision-making behavior towards new technology and its influence have been relatively wealthy [30–32]. In contrast, very few publications have been traced to quantify the farmers' ecological cognition in response to special orchard fruit planting behavior. There is a lack of research on whether a specific technology or measure will affect farmers' decision-making behavior [33,34]. However, maximizing the orchards fruit farmer's economic return and the ecological benefits of specialty orchard fruit planting still need to be explored compressively [35]. Fewer studies have focused on the ecological factors on farmers' decision-making and response behavior within the context of orchard farmers [36]. Several external and internal factors frequently influence farmers' decision-making behavior and these variables should be explored cohesively [37]. Seemingly, the key factors that affect farmers' ecological behavior regarding specialty orchard fruit planting have not been explored adequately yet. The inner relationship between these critical factors has not been explored critically also by existing pieces of literature.

Therefore, the study intends to analyze the following research questions: (i) Does farmers' cognition of ecological agriculture influence their response to specialty fruit productions? (ii) Does farmers' adoption of water conservation measures influence their response to specialty orchard fruit planting? (iii) What other factors influence farmers' response to specialty orchard fruit planting? (iv) Finally, which factors are the deep-rooted root causes of constraints on farmers' response to specialty orchard fruits planting? The answers to the above questions are convenient in screening the potential driving forces affecting farmers' planting of specialty orchard farming and opening up the channel to increase farmers' income and protect the ecology simultaneously. The study selects Shaanxi and Ningxia provinces as the research area covering the Loess Plateau region of China. The research focuses on how the adoption behavior of planting specialty fruits and its degree impacts the farmers' income, water conservation and eco-product price cognition, which quantifies as the prime strength and novelty of the study. Interestingly, to the best of our knowledge, the inner relationships between specialty fruit productions behavior and farmers' ecological cognition have not been studied previously.

2. Conceptual Framework

The specialty forestry and fruit industry and its planting decision have a significant relationship between economic benefits and ecological protection maximization [38]. The primary purpose of planting any sort of crops or orchards is to sell products to gain income, so the study takes the theory of farmers' behavior as the primary theoretical basis [39]. According to the theory, the rational farmer can be further subdivided into complete rational and limited rational farmers. The complete rational farmers believe that the rational person's goal depends on optimization or utility maximization, but the hypothesis of complete rationale is relatively complicated [40]. Therefore, Russell and Simon [41] proposed the "limited rationality hypothesis," which argues that farmers' decision-making behavior is "subjectively perfectly rational, but objectively limited to do so." Therefore, from the most basic gist of the limited rationality hypothesis, the maximization of benefits in farmers' decision-making process is only for the subjective knowledge of decision-makers [42].

In contrast, cognition plays a vital role in farmers' decision-making process and, specifically, the level of ecological agriculture cognition is an essential factor influencing farmers' special forestry and fruit planting [43]. Different scholars have different definitions of ecological agriculture cognition. For example, Tang et al. [44] defined farmers' cognition as the interpersonal concern and perception regarding any specific situation that impacts their interests. Zhu and Wang [45] defined ecological agriculture cognition as farmers' subjective knowledge and thought about the ecological agriculture production models. By evaluating the above definition, the study defines ecological agriculture cognition as "how farmers obtain information through various channels, analyze and understand it in order to capture the maximum value within limited resources". We evaluate farmers' cognition of ecological agriculture as three distinct criterion (cognition of eco-agriculture in increasing income, water conservation and eco-product price).

The cognition of eco-agriculture in increasing income reflects the objective reality of farmers' cognition by capturing the household's economic solvency from the ecological development [46,47]. Farmers who understand this issue deeply will be optimistic about the future income increase brought by planting unique orchard fruits and then paying more attention to ecological agriculture and specialty orchards fruits industry [48]. Mouron et al. [49] studied Swiss Apple orchards and found that environmental cognition substantially helps choose the best pesticides and organic farming tactics, which eventually helps farmers' increase household income. As a result, it could be estimated that farmers will be more enthusiastic about planting specialty orchard fruits and expanding the planting rate. Based on this, the study proposes Hypothesis 1:

Hypothesis 1 (H1). *The cognition of eco-agriculture increase income positively influences farmers' response to specialty forestry and fruits planting.*

The cognition of eco-agriculture water conservation reflects the result of farmers' awareness of the objective reality that the development of eco-agriculture can maintain maximum use of soil and water resources [50]. Therefore, the development of ecological agriculture, especially in the unique forestry and fruit industry, farmers' ecological cognition can positively affect soil and water conservation [51]. The more farmers know about the importance of ecological soundness, the more they can understand the criticality of developing specialty forestry and fruits for soil and water conservation and ecological protection [52]. Therefore, it can be assumed that the more the farmer possesses a positive attitude regarding ecological safety, the more they will be willing to develop unique forestry and fruits and expand the planting rate. Based on this, the article proposes Hypothesis 2:

Hypothesis 2 (H2). *The cognition of eco-agriculture water conservation positively influences farmers' response to specialty forestry and fruits planting.*

The cognition of eco-product price reflects the result of farmers' objective reality that the price of ecological agricultural products is different from the other conventional products [53]. Product price is an important driving force for farmers to improve the mode of the agricultural operation and adjust the structure of agricultural operation [54,55]. Specialty orchard fruit products are an essential type of ecological product that is found to gain more price than the other fruit as it is widely recognized as organic and relatively safer food [4]. In several studies, it has been found that ecological friendly oriented fruit successfully refers to high-value fruit than the other conventional fruits (such as Weibel et al. [56] and Canavari et al. [57]). The higher the price recognition of unique orchard fruit products farmers can get, the more they will develop their particular orchard fruit industry and expand the planting scale [58,59]. Based on this, the study proposes Hypothesis 3:

Hypothesis 3 (H3). *The cognition of eco-product price positively influences farmers' response to specialty forestry and fruits planting.*

The above hypotheses are graphically illustrated in Figure 1, which we used as the study's conceptual framework.

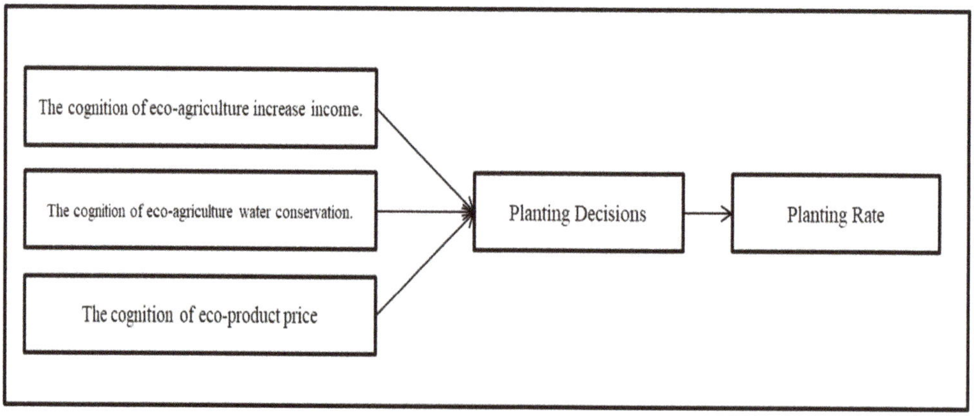

Figure 1. Conceptual framework of ecological cognition and response to specialty orchard fruit planting.

3. Materials and Methods

3.1. Data Collection

The study developed a cross-sectional survey in Shaanxi Province and Ningxia Hui Autonomous Region, China (Figure 2), to capture the empirical data. Geographically the two regions are sound for orchards farming. The largest river in China, the Yellow River, flows through Shaanxi Province and Ningxia Hui Autonomous Region. In addition, Shaanxi and Ningxia are located in the Loess Plateau region of China, where the climate is arid and soil erosion is more severe than in other regions. However, the Loess Plateau region is not fertile enough for conventional farming with severe soil erosion, serious sanding, salinization, stone desertification and arable land with low and unstable grain yield. According to local conditions, the Chines government encourages the farmers of these regions to exercise planned and systematic cultivation and relace the vegetation land by afforestation and grass planting. Moreover, Shaanxi and Ningxia are important pilot areas of China's "Returning Farmland to Forestry Project".

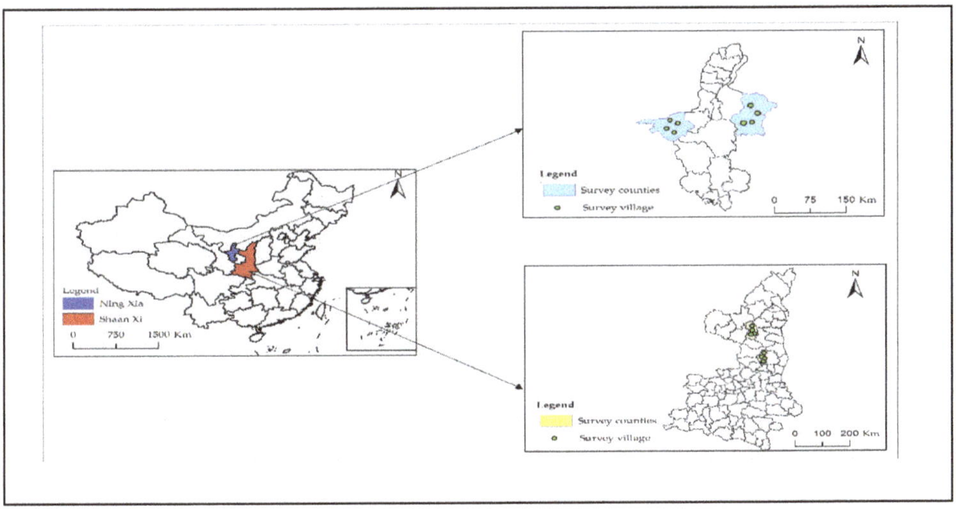

Figure 2. The study area.

The study utilized multi-stage stratified random sampling methods to select the sample. First, two counties were selected from Ningxia and Shaanxi provinces according to the size of the specialty orchard planting (out of the two largest scale specialty orchard fruit planting counties). Second, four towns were selected from each county (out of the four largest scale specialty orchard fruit planting towns). Finally, we selected four villages with sound planting characteristics for orchard farming. The final investigation includes 10 to 15 farmers from each village, which leads us to 476 respondents. After eliminating invalid samples and samples with significant problems, the final sample consisted of 309 farmers engaged in specialty orchard fruit cultivation and 107 farmers not engaged in specialty orchard fruit cultivation. We conducted face-to-face interviews with farmers who planted specialty orchard fruit. However, the sample distribution of the farmers in this study follows the basic principles of random sampling and stratified sampling. In the questionnaire, the study uses the five-level Likert scale to measure the responses. A high score means better farmers' cognition of ecological agriculture.

According to the respondents' essential characteristics (Table A1), the respondents were mainly male, with a proportion of 94.47%. Fewer growers were under 50 years old and most of the growers were above 50 years old. The educational background of the interviewed farmers was mostly below junior high school education and the overall education level was relatively low. There were not many farmers with village cadres and party members among the interviewees, of which only 46 were members of village cadres and 63 were party members. In addition, most of the respondents had a total household size of fewer than six people and fewer (1.93%) had a total household size of more than ten people.

3.2. Methods

The study first uses the double hurdle model to analyze the influencing factors of farmers' specialty orchard fruit planting response focuses on whether the cognition of ecological agriculture increases income, water conservation and product price influences farmers' specialty orchard fruit planting response. Then, according to the influencing factors extracted by the double-hurdle model, the Interpretative Structural Model (ISM) was used to evaluate the hierarchical structure and the relationship among the influencing factors as suggested by Cheung et al. [60]. The study uses STATA 12.0 software (StataCorp LLC, College Station, TX, USA) to analyze the sample data empirically. The explanatory variables' variance inflation factor (VIF) was calculated to test the collinearity among explanatory variables and avoid biased results due to multicollinearity issues, as suggested by Wang et al. [61].

3.2.1. Double-Hurdle Model

The double-hurdle model is derived from the Probit and truncreg models [62], which correspond to the two decision-making stages of farmers' response to specialty orchard fruit planting. The selected model is participation decision (whether to plant specialty orchard fruits) and quantity decision (planting rate of specialty orchard fruits). The participation decision is described in Equations (1) and (2).

$$Z_i^* = \alpha_0 + \sum_i \alpha_i Z_i + \sum_i \alpha_i' control_i + D_i + \varepsilon_i \quad \varepsilon_i \sim N(0,1) \tag{1}$$

$$P_i = \begin{cases} 1 & Z_i^* > 0 \\ 0 & Z_i^* \leq 0 \end{cases} \quad i = 1,2\ldots n \tag{2}$$

Among them, Z_i^* in Equation (1) is the potential variable to participate in decision-making, which cannot be directly observed. While P_i in Equation (2), the decision-making participation and represents whether farmers plant specialty orchard fruits, which is a binary choice variable. When $Z_i^* > 0$, $P_i = 1$, it means the ith farmer planting specialty orchard fruits and when $Z_i^* \leq 0$, $P_i = 0$, it means that the ith farmer does not plant specialty orchard fruits. Seemingly, Z_i is the core explanatory variable or potential variable, $control_i$ is the control variable of potential variable, D_i is the regional dummy variable of potential variable, ε_i is the error term and obeys the standard normal distribution $\varepsilon_i \sim N(0,1)$. Here n represents the number of variables, α_0, α_i, α_j' are the parameters to be estimated and the decision is described in Equations (3) and (4).

$$Y_i^* = \beta_0 + \sum_i \beta_i X_i + \sum_i \beta_i' control_i + D_i + \mu_i; \quad \mu_i \sim N(0,\sigma^2) \tag{3}$$

$$Y_i = \begin{cases} Y_i^* & P_i = 1 \\ 0 & P_i = 0 \end{cases}; \quad i = 1,2\ldots n \tag{4}$$

If $Z_i^* > 0$ and $P_i = 1$, then $Y_i = Y_i^* = \beta_0 + \sum_i \beta_i X_i + \sum_i \beta_i' control_i + D_i + \mu_i$. In Equation (3), Y_i^* is the planting rate of the specialty orchard fruits of the ith farmer is the continuous variable. Seemingly, X_i represents the core explanatory variable and μ_i is the error term and obeys the normal distribution. If $Z_i^* \leq 0$ and $P_i = 0$, then $Y_i = 0$; β_0, β_i, β_j' and σ are the parameters to be evaluated.

3.2.2. ISM Analysis Method

In recent years, the ISM method has been widely used to analyze and identify influencing factors of farmers' behavior [63]. The study's basic principle comprises a combination of incidence matrix and computer technology principle to clarify the correlation and hierarchy among factors [64]. The methodology is also helpful for determining the main influencing factors and exploring their internal relationships [65]. The specific steps are as follows:

The first step is to establish the adjacency matrix between the factors. We assume that there are k significant influencing factors, denoted by S_i $(i = 0, 1, \ldots, k)$, then S_0 denotes the farmer's characteristic orchard fruit planting response. The Delphi method is used to determine the logical relationship between the significant factors, represented by the adjacent order matrix R. The element $r_{ij} = 1$ in the matrix indicates that the factor S_i has a direct impact on S_j and $r_{ij} = 0$ means that factor S_i has no effect on S_j, where $i = 0, 1, \ldots, k$; $j = 0, 1, \ldots, k$.

The second step is to establish the reachability matrix among the factors. The calculation of the reachability matrix has portrayed in Equation (5), where I denotes the identity matrix $2 \leq \lambda \leq k$ and the matrix is obtained by Boolean operations using Matlab (R2019, MathWorks, Inc., Natick, MA, USA) software for power operations (for more details, please check Yang et al. [66]).

$$M = (S+I)^{\lambda+1} = (S+I)^{\lambda} \neq (S+I)^{\lambda-1} \neq \ldots (S+I)^{2} \neq (S+I)^{1} \quad (5)$$

The third step is to determine the level-by-level division. First, the reachability matrix is divided into the reachable set $M(S_i)$ and antecedent set $A(S_i)$. Among them, the following two equations have been used: (i) $M(S_i) = \{S_i | n_{ij} = 1\}$ and (ii) $A(S_i) = \{S_j | n_{ji} = 1\}$, where n_{ij} and n_{ji} are factors in the reachability matrix. Seemingly, the set expression derived by the following equation has been used to find each layer's feature set: $M(S_i) = \{S_i | M(S_i) = M(S_i) \cap A(S_i); i = 1, 2, \ldots, k\}$. More specifically, the following steps have been taken as per the suggestion of Sarkar et al. [67]: First, find the highest element set, then cross out the corresponding rows and columns from the reachable matrix and then find the new highest element (i.e., the second layer element) from the remaining reachable matrix to find the set of elements of each layer. The fourth step is to determine the hierarchical structure of factors according to the level. The hierarchical structure of the influencing factors of the response of the specialty orchard fruits planting of farmers is obtained by connecting the factors between the adjacent layers and the same level with directional arrows.

4. Results

4.1. Variables and Description Statistics

The farmers' response to specialty orchard fruits planting was the behavioral interaction of farmers, including whether to plant the fruits and the planting rate. Among the sample farmers, 309 households (74.28%) planted specialty orchard fruits, with an average planting scale of 4.29 mu and the average planting rate of specialty orchard fruits was 49.86%. However, another vital issue that reflects the behavior of farmers is endowment impact. Farmer endowment refers to the family members' natural and acquired resources and abilities, representing the whole family [68]. As the endowment of farmers played an essential role in the response of farmers to the planting of specialty orchard fruits [69], the study endorsed the variables from three dimensions: (i) individual characteristics of the head of household, (ii) family characteristics and (iii) production and operation characteristics. Table 1 shows all the variables used in the study and the corresponding descriptive statistics.

4.2. Correlations among Farmers' Responses to Specialty Orchard Fruit Planting and Influencing Factors

Figure 3 shows the heat map of the correlation between the specialty orchard fruit planting behavior and its influencing factors. The darker color denotes a more excellent absolute value of the correlation coefficient between the variables. According to Figure 3, cognition of eco-agriculture increase income, cognition of eco-agriculture water conservation and cognition of eco-product price were positively correlated with whether to plant special orchard fruits. The findings suggest that the cognition of eco-agriculture has a positive influence on farmers' response to planting specialty fruits. In addition, annual household income, agricultural planting scale, degree of agricultural specialization and effective irrigation rate were positively correlated with whether to plant unique orchard fruits. However, weaker correlations were found between age, gender and whether to plant unique orchard fruits. These findings suggest that age and gender may not have a substantial effect on whether to plant unique orchard fruits.

Table 1. Variable meaning and description statistics.

Serial Number	Variables	Definition	Mean	SD	Min	Max
S01	Whether to plant special orchard fruits	No = 0, Yes = 1	0.74	0.44	0.00	1.00
S02	Planting rate of specialty orchard fruits	The proportion of planting area of family specialty orchard Fruits in its actual cultivated land area (%)	49.86	30.38	1.79	100.00
S1	Cognition of eco-agriculture increase income	Can the development of ecological agriculture increase income? No effect = 1, small effect = 2, general = 3, large effect = 4, very large effect = 5	3.37	1.63	1.00	5.00
S2	Cognition of eco-agriculture water conservation	Can the development of ecological agriculture maintain soil and water? No effect = 1, small effect = 2, general = 3, large effect = 4, very large effect = 5	3.21	1.75	1.00	5.00
S3	Cognition of eco-product price	Is the price of ecological agricultural products higher than that of general products? No action = 1, less action = 2, general = 3, more action = 4, very big action = 5	4.23	1.23	1.00	5.00
S4	Age	The actual age of the head of household	55.10	10.24	27.00	83.00
S5	Ecological agriculture training	Have you participated in ecological agriculture training? No = 0, yes = 1	0.55	0.50	0.00	1.00
S6	Annual household income	Net income of the family in 2016 (RMB 10,000)	6.42	5.17	0.25	32.95
S7	Agricultural planting scale	The actual cultivated land area of households in 2016 (mu)	13.35	12.91	0.00	100.00
S8	Degree of agricultural specialization	The proportion of annual household planting income to annual household income (%)	33.77	27.11	0.00	99.50
S9	Province	Ningxia = 0, Shaanxi = 1	0.51	0.501	0.00	1.00
S10	Gender	Female = 0, Male = 1	0.95	0.23	0.00	1.00
S11	Education	Actual educational years of the head of household (years)	6.62	3.92	0.00	15.00
S12	Effective irrigation rate	The proportion of effective irrigation area in total cultivated land	18.42	35.70	0.00	100.00
S13	Agricultural technicians	Are they agricultural technicians? No = 0, yes = 1	0.05	0.21	0.00	1.00
S14	Number of family workers	Number of the labor force engaged in agricultural production in the family (person)	2.95	1.46	0.00	8.00

4.3. Analysis of Factors Influencing Farmers' Response to Specialty Orchard Fruit Planting

4.3.1. The Effect of the Cognition of Eco-Agriculture Increases Income on Farmers' Response to Planting Characteristic Orchard Fruits

The specific regression results obtained by fitting the double-hurdle model are shown in Table 2. The cognition of eco-agriculture increase income positively affected whether farmers planted characteristic orchard fruits at the 1% significance level. The cognition of eco-agriculture increased income positively affected whether farmers planted specialty orchard fruits and positively affected the rate of planting specialty orchard fruits at a 5% significance level. This indicates that farmers were more willing to develop eco-agriculture and plant specialty orchard fruits to gain increased income from eco-agricultural products. Second, farmers' awareness of ecological agriculture income increase was a decisive factor in the perceived usefulness of ecological agriculture and determining farmers' acceptance of planting specialty orchard fruits. Based on the above discussion, Hypothesis 1 is supported.

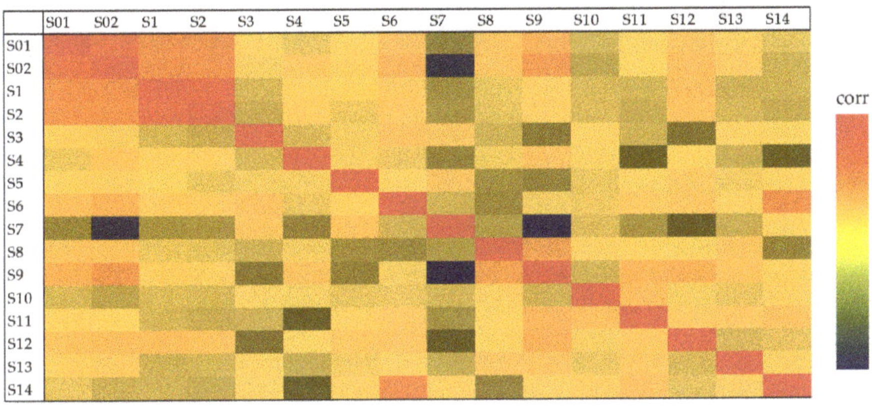

Figure 3. The heatmap of correlation coefficient (corr) matrix among variables. Note: S01–S14 is in the same order as the variable names in Table 1. Darker colors imply larger absolute values of the correlation coefficients among variables.

Table 2. Results of ecological agriculture cognition on farmers' response to specialty orchard fruit planting.

Variables	Participation Decision Model (Probit)		Quantitative Decision Models (Truncreg)	
	Marginal Effects	Standard Error	Coefficient	Standard Error
Cognition of eco-agriculture increase income	0.057 ***	0.015	4.976 **	2.188
Cognition of eco-agriculture water conservation	0.043 ***	0.015	1.607	1.925
Cognition of eco-product price	0.030 *	0.015	2.753 *	1.651
Age	−0.003 *	0.002	0.135	0.194
Gender	−0.052	0.088	−8.105	7.569
Education	−0.001	0.005	−0.213	0.509
Ecological agriculture training	0.093 **	0.038	11.832 ***	3.827
Agricultural technicians	0.084	0.102	7.450	7.523
Number of family workers	−0.018	0.014	−1.609	1.509
Annual household income	0.014 ***	0.005	1.348 ***	0.373
Agricultural planting scale	0.001	0.001	−2.353 ***	0.310
Degree of agricultural specialization	0.002 **	0.001	0.151 **	0.072
Effective irrigation rate	0.001	0.001	−0.068	0.049
Province	Control	Control	Control	Control
Constant			19.744	17.114
Observations	416		309	
Sigma	--		27.472 ***	
Log-Likelihood	−169.059		−1386.628	
Wald-chi2 (14)	--		115.96	
Prob > chi2	0.000		0.000	

Note: *** $p < 0.01$, ** $p < 0.05$, * $p < 0.1$.

4.3.2. The Effect of the Cognition of Eco-Agriculture Water Conservation on Farmers' Response to Planting Characteristic Orchard Fruits

The cognition of eco-agriculture water conservation positively affects farmers who planted specialty orchard fruits and passed the test at a 1% significance level. However, the effect on the planting rate of characteristic orchard fruits was not significant, indicating that the higher the farmers' cognition of eco-agriculture water conservation, the more they could realize the importance of eco-agriculture for soil and water conservation. Therefore, ecological agriculture water conservation cognition promotes farmers' specialty orchard fruit planting response. Hypothesis 2 is supported based upon the above discussion.

4.3.3. The Effect of the Cognition of Eco-Product Price on Farmers' Response to Specialty Orchard Fruit Planting

The cognition of eco-product price positively affected whether farmers planted specialty orchard fruit and the rate of specialty orchard fruit planting at the 10% significance level. It indicates that farmers' perception of eco-friendliness and the cognition of eco-product price was a crucial factor influencing farmers' production and planting decisions. Therefore, farmers are more sensitive to their prices and their ecological agricultural price cognition was positively related to the planting degree of characteristic orchard fruits. Based on the discussion mentioned above, Hypothesis 3 is verified.

4.4. Mechanism Analysis of Influencing Factors of Farmers' Specialty Orchard Fruits Planting Response

The farmers' decision-making process is a complex system, where each element is independent of the other and connected layer by layer and it constitutes a complete system of influencing factors [67]. Therefore, according to the logical relationship among elements, the logical relationship diagram is constructed using the Delphi method, as shown in Figure 4. It represents that the column factors impact the row factors, V represents that the row factors impact the column factors and 0 represents no relationship between them.

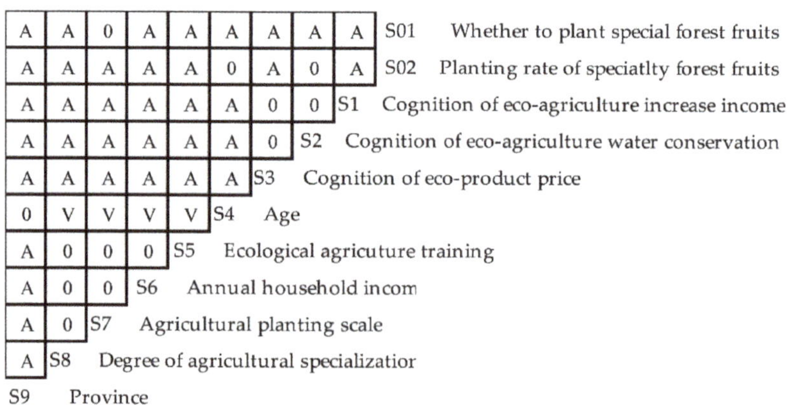

Figure 4. Relationship between factors affecting response to the planting of specialty orchard fruits.

According to the logical relationship of the factors affecting farmers' response to the planting of specialty orchard fruits, as shown in Figure 1. From Figure 1, we can obtain whether to plant specialty orchard fruits and the adjacency matrix of the planting rate within specialty orchard fruits. Combined with Equation (5), the study calculates the reachability matrix and then determine the method of level according to the level division and can obtain whether the farmers have planted specialty orchard fruits in each level as follows: $L_1 = \{S_{01}\}$, $L_2 = \{S_1, S_2, S_3\}$, $L_3 = \{S_5, S_6, S_8\}$, $L_4 = \{S_4, S_9\}$. The critical elements of planting rate of specialty orchard fruits of farmers are as follows: $H_1 = \{S_{02}\}$,

$H_2 = \{S_1, S_3\}$,$H_3 = \{S_5, S_6, S_7, S_8\}$. The reachability matrix after reordering is shown in Figures 5 and 6.

According to the reachability matrix sorted in Figures 5 and 6, the factors at the same level are represented by a box at the same level. According to the logical relationship among the influencing factors, the explanatory structure model that affects farmers' response to planting specialty orchard fruits can be obtained, as shown in Figures 7 and 8. The surface factors that directly affect whether farmers plant specialty orchard fruits are the cognition of eco-product price, eco-agriculture increased income, eco-agriculture water conservation (Figure 7). Among them, the deeper root factors of influence are age, ecological agriculture training, the annual income of families and the degree of agricultural specialization. It can be seen that whether farmers plant unique orchard fruits or not are as follows: "age and province" → "training in ecological agriculture, annual household income, degree of agricultural specialization" → "cognition of eco-agriculture increase income, cognition of eco-agriculture water conservation, cognition of eco-product price" → "farmers planting special orchard fruits." Therefore, it is an effective measure to promote the motivation of farmers to plant orchard fruits by providing relevant training and formulating corresponding incentive measures according to their individual and family endowment differences.

	S01	S1	S2	S3	S9	S4	S5	S6	S8
S01	1	0	0	0	0	0	0	0	0
S1	1	1	0	0	0	0	0	0	0
S2	1	0	1	0	0	0	0	0	0
S3	1	0	0	1	0	0	0	0	0
S9	1	0	0	0	1	0	0	0	0
S4	1	1	1	1	0	1	0	0	0
S5	1	1	1	1	0	0	1	0	0
S6	1	1	1	1	0	0	0	1	0
S8	1	1	1	1	0	0	0	0	1

Figure 5. Reachability matrix after participating in decision ranking.

	S02	S1	S3	S5	S6	S7	S8
S02	1	0	0	0	0	0	0
S1	1	1	0	0	0	0	0
S3	1	0	1	0	0	0	0
S5	1	1	1	1	0	0	0
S6	1	1	1	0	1	0	0
S7	1	1	1	0	0	1	0
S8	1	1	1	0	0	0	1

Figure 6. Reachability matrixes after ranking of quantitative decision making.

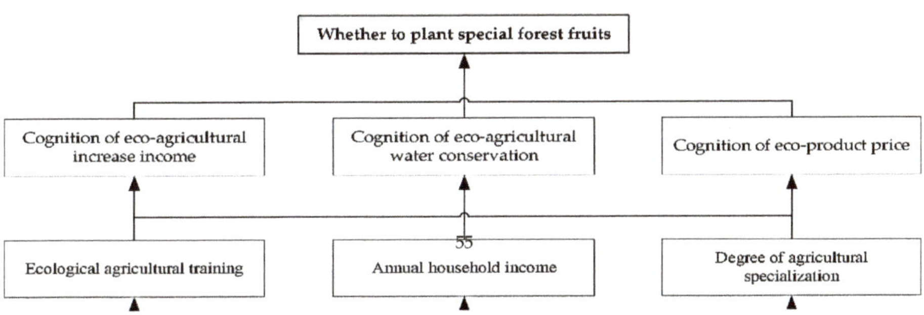

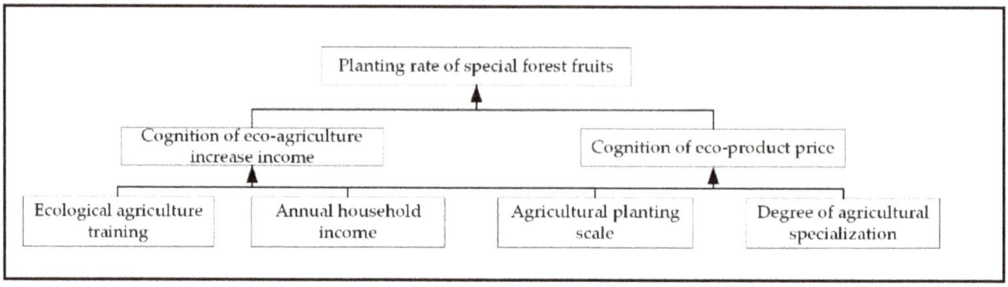

Figure 8. Explanatory structural model of planting rate of specialty orchard fruits.

As shown in Figure 8, it can be seen that the direct factors influencing the cultivation rate of specialty orchard fruits are the cognition of eco-agriculture increased income and the cognition of eco-product price. In contrast, ecological agriculture training, annual household income, degree of agricultural specialization and agricultural cultivation scale are significant influencing factors. As can be seen above, the critical paths influencing the cultivation rate of specialty orchard fruits by farmers are mainly along with the following relationship: "ecological agriculture training, annual household income, degree of agricultural specialization, agricultural cultivation scale" → "cognition of eco-agriculture increase income, cognition of eco-product price" → "Planting rate of specialty orchard fruits".

5. Discussion

This study crafted its findings based on research data from 416 farmers in specialty forest fruit growing areas in China's Shaanxi and Sichuan provinces. Regression analysis was conducted using an econometric model to explore the influence of ecological agriculture cognition on the response behavior of specialty forest fruit growing. The study first found that ecological agriculture cognition significantly influenced farmers' specialty forest fruit planting and quantity decisions. The finding also highlights that farmers' ecological agriculture cognition could dramatically improve farmers' specialty forest fruit planting behavior. The findings of this study are consistent with Xue et al. [70], Wang et al. [71], Li et al. [72], Azadi et al. [73] and Das V. et al. [74], who also found that farmers' cognition is an essential factor in farmers' behavioral decisions. The above findings are also consistent with the theory of planned behavior [75], which suggests that attitudes, subjective norms influence individuals' actual behavior and perceived behavioral control, which influences individuals' cognition and rectifies their actual decision-making behavior [25,76]. In particular, the study by Zhang et al. [77] indicated that farmers' perceptions of pesticide residues would positively impact farmers' adoption of eco-friendly agricultural production, which is consistent with the study's findings.

The effect of the cognition of eco-agriculture increases income on farmers' response to planting specialty orchard fruits is positive. It shows that the higher the expectation of ecological agriculture income increase, the more farmers are willing to develop ecological agriculture. The possible explanations are as follows: first, ecological agriculture improves the economic benefits of farmers by improving agricultural land-use efficiency and labor productivity. The economic benefit is the primary factor to stimulate farmers to engage in ecological agriculture, which determines farmers' planting behavior [78]. Seemingly, the effect of the cognition of eco-agriculture water conservation on farmers' response to planting specialty orchard fruits is positive. The possible explanation is that ecological agriculture is resource-saving agriculture, which can improve the land-use rate, output rate and have a water-saving effect [79]. In developing ecological agriculture, the "green" vegetation cover reduces water evaporation and conserves water sources, essential for soil and water conservation. However, soil and water conservation can protect scarce cultivated land resources, reduce crop yield risk, bring long-term benefits to farmers [80] and improve the level of ecological agriculture specialization [50]. Therefore, soil and water conservation and ecological agriculture promote each other. Specialty orchard fruits are typical representatives of commercialized ecological agriculture [81]. The effect of the cognition of eco-product price on farmers' response to specialty orchard fruit planting is also positive. The possible explanations are as follows: first, the market demand for ecological products is increasing with the improvement of social and economic movement, green transition and healthier food supply options. On the other hand, the market price is also relatively higher. Thus, price cognition of ecological products is steadily improving, promoting ecological agriculture and gradually transforming the ecological advantages of ecological agriculture into economic advantages [82].

The production mechanism and style of smallholder farmers have their particularity. In pursuing utility maximization, it should meet the consumption needs of family members and obtain market profits by participating in market transactions [83]. Typically, farmers seek a balance between consumer needs and market profits. With the implementation of ecological agriculture, the family planting structure has been adjusted and farmers increase their total income by planting crops with relatively high market prices. Compared with other agricultural products, the commercialization rate of specialty orchard fruits is higher [84], which means that the proportion of the specialty forest and fruits used in the market transaction is relatively large and the marketization degree is also high [85].

However, the study differs from some of the existing studies. For example, our study showed that gender did not affect farmers' specialty forest fruit growing behavior. This is not consistent with the investigations of He et al. [86] and Abdulai et al. [87]. The main reason for this difference is that with the increasing labor exodus in China, the labor force for agricultural production in rural areas has shifted mainly from male to female producers, thus leading to a gradual dilution of the gender factor [88,89]. In addition, our study found significant differences in the factors influencing farmers' decision-making behavior and quantity decisions for specialty forest fruit planting, where the scale of agricultural planting was not the main factor influencing whether farmers planted specialty forest fruit. In contrast, ecological agricultural training was an essential factor influencing farmers' specialty forest fruit planting rate. Zakaria et al. [90] found that farmers can learn about new technologies through training and application courses and by learning to promote new technologies, they can enhance their agricultural operations. It is similar to our study. Therefore, the government should consider strengthening the empowerment of decision-makers, raising their awareness of environmental protection by planting special forest fruits and encouraging their active participation to improve the decision-making behavior of farmers in the planting of unique forest fruits.

6. Conclusions

Based on micro survey data of 416 orchard farmer's households in Shaanxi and Ningxia provinces, the study uses the bounded rationality theory as a theoretical framework. A double-hurdle model was used to analyze farmers' responses for quantitative decision-making behavior. Moreover, the study uses the ISM model to analyze how the cognition of eco-agriculture increases income, the cognition of eco-agriculture water conservation and the cognition of eco-product price affecting farmers' behavior regarding specialty orchard fruit planting. Seemingly, the study also constructs a hierarchical structure relationship among the influencing factors and profoundly explores the root factors affecting orchards farmers' characteristics by using ISM. The main conclusions of this paper are as follows: first, the farmers who planted specialty orchard fruits accounted for 74.28% of the total sample farmers. The average planting scale was 4.29 mu and the average planting rate of characteristic orchard fruits was 49.86%. Second, farmers' ecological agriculture cognition has directly affected farmers' behavior and it has acted as a root factor to influence the farmer's behavior. The higher the degree of farmers' cognition of eco-agriculture increase income, eco-agriculture water conservation and eco-product price, the more inclined they are to plant specialty orchard fruit, which also verifies the correctness of hypotheses 1 and 2. The higher cognition level regarding eco-agriculture increases income and eco-product price lead the farmer to expand the specialty orchard fruit planting and it verifies the correctness of Hypothesis 3. Third, farmers' endowment differences and regional factors are found as root factors affecting farmers' responses to specialty orchard fruit planting. Fourth, regional variables, farmers with younger household heads, more training in ecological agriculture, higher annual household income and a higher degree of agricultural specialization have a higher probability of planting specialty orchard fruits. At the same time, farmers with more training in ecological agriculture, higher annual household income, smaller agricultural planting scale and a higher degree of agricultural specialization develop specialty orchard fruits on a larger scale.

The development of specialty orchard fruits has both ecological and economic benefits, which is a practical and effective way to ensure ecological security and increase farmers' income in the Loess Plateau area. However, how to promote farmers' response to the planting of specialty orchard fruits has become a vital issue. Therefore, the government departments should introduce policies to strengthen government guidance and improve farmers' awareness of ecological agriculture based on farmers' diversity characteristics. The specific recommendations are as follows:

The government should highlight the benefit of ecological products and the betterment of ecological agriculture. The government should also uphold the special characteristic of the ecological orchard to produce a brand effect, economic benefit and social benefit. For this thrives, concerned authorities should promptly arrange cultural festivals and experience exchange meetings to capture the added value of ecological products. The government should extend the supports of agricultural demonstration zone to practically displays the innovative tactics, methods and another technological advancement should also be properly circulated. The concerned authorities should also arrange specialized training facilities to enhance farmers' expectations of the rising price of characteristic orchard fruits, improve the ability to capture market equilibrium power and promote the peaceful development of characteristic orchard fruits. The government should strengthen the information-sharing platform to minimize the knowledge gap. Modern planting techniques and management concepts should also be highlighted via agricultural skills training programs. The farmers and agricultural service providers should be integrated for solving technical problems in agricultural production to improve farmers' specialization in specialty fruit production. There is a rising concern to refine the existing agro-environmental policies based on differences in individual farm household characteristics. The farmers' diversity and micro incentive measures should be introduced from the regional capital structure, technology, land and water use. The policies should focus on promoting large-scale operations and give small farmers space for being developed.

Author Contributions: Conceptualization, Z.C. and A.S.; methodology, Z.C. and A.S.; software, Z.C., X.L. and X.X.; validation, Z.C., X.L. and X.X.; formal analysis, A.S., Z.C. and X.L.; investigation, Z.C. and A.S.; resources, Z.C., A.K.H. and A.S.; data curation, A.S. and X.L.; writing—original draft preparation, A.S. and Z.C.; writing—review and editing, Z.C., A.S., A.K.H. and X.X.; visualization, A.S. and A.K.H.; supervision, X.X.; project administration, X.X. and X.L.; funding acquisition, X.X. All authors have read and agreed to the published version of the manuscript.

Funding: The authors are grateful to the financial support by the National Key Research and Development Program of China (NOS. 2016YFC0501707) and the Basic research business expenses of Northwest A & F University (NOS. 2452020055).

Institutional Review Board Statement: Not applicable.

Informed Consent Statement: Not applicable.

Data Availability Statement: The associated dataset of the study is available upon request to the corresponding author.

Conflicts of Interest: The authors declare no conflict of interest.

Appendix A. Demographic Data

Table A1. Basic characteristics of interviewed farmers.

Variable	Category	Count	Frequency (%)
Gender	Male	393	94.47
	Female	23	5.53
Age	[1, 30]	3	0.72
	(30, 50]	147	35.33
	(50, 60]	128	30.77
	>60	138	33.18
Educational background	None	61	14.66
	Primary school	139	33.41
	Junior high school	154	37.03
	Senior high school	55	13.22
	College and higher	7	1.68
Village cadre member	Yes	46	11.06
	No	370	88.94
Party member	Yes	63	15.14
	No	353	84.86
Total household population	[1, 3]	136	32.69
	[4, 6]	218	52.40
	[7, 9]	54	12.98
	≥10	8	1.93

References

1. Barnard, F.L.; Foltz, J.; Yeager, E.A.; Brewer, B. *Agribusiness Management*, 6th ed.; Routledge: London, UK, 2020; ISBN 978-0-429-32442-0.
2. Demestihas, C.; Plénet, D.; Génard, M.; Raynal, C.; Lescourret, F. Ecosystem services in orchards. A review. *Agron. Sustain. Dev.* **2017**, *37*, 12. [CrossRef]
3. Haggblade, S. Modernizing African agribusiness: Reflections for the future. *J. Agribus. Dev. Emerg. Econ.* **2011**, *1*, 10–30. [CrossRef]
4. Houston, L.; Capalbo, S.; Seavert, C.; Dalton, M.; Bryla, D.; Sagili, R. Specialty fruit production in the Pacific Northwest: Adaptation strategies for a changing climate. *Clim. Change* **2018**, *146*, 159–171. [CrossRef]
5. Josiah, S.J.; St-Pierre, R.; Brott, H.; Brandle, J. Productive Conservation: Diversifying Farm Enterprises by Producing Specialty Woody Products in Agroforestry Systems. *J. Sustain. Agric.* **2004**, *23*, 93–108. [CrossRef]
6. Lei, C.; Hai, J.; JiaXin, S.; XiuYun, M. The path and policy choice of agricultural branding: A case study of the brand construction of the specialty fruit industry in Heilin Town, Jiangsu Province. *Res. Agric. Mod.* **2018**, *39*, 203–210.
7. Zhang, X.; Zhu, Y. Envaluation on the prosperity level of Xinjiang's characteristic forest and fruit industry. *Chin. J. Agric. Resour. Reg. Plan.* **2020**, *41*, 189–197. [CrossRef]
8. Cen, Y.; Li, L.; Guo, L.; Li, C.; Jiang, G. Organic management enhances both ecological and economic profitability of apple orchard: A case study in Shandong Peninsula. *Sci. Hortic.* **2020**, *265*, 109201. [CrossRef]
9. Liu, Y.; Gao, M.; Wu, W.; Tanveer, S.K.; Wen, X.; Liao, Y. The effects of conservation tillage practices on the soil water-holding capacity of a non-irrigated apple orchard in the Loess Plateau, China. *Soil Tillage Res.* **2013**, *130*, 7–12. [CrossRef]
10. Suo, G.-D.; Xie, Y.-S.; Zhang, Y.; Luo, H. Long-term effects of different surface mulching techniques on soil water and fruit yield in an apple orchard on the Loess Plateau of China. *Sci. Hortic.* **2019**, *246*, 643–651. [CrossRef]
11. State Council of the PRC. *Notice on the Issuance of the 13th Five-Year Plan. for Poverty Alleviation to Combat Poverty*; State Council of the People's Republic of China: Beijing, China, 2016.
12. State Council of the PRC. *Opinions on the Implementation of Rural Revitalization Strategy*; State Council of the People's Republic of China: Beijing, China, 2018.
13. Rezaei, R.; Safa, L.; Ganjkhanloo, M.M. Understanding farmers' ecological conservation behavior regarding the use of integrated pest management- an application of the technology acceptance model. *Glob. Ecol. Conserv.* **2020**, *22*, e00941. [CrossRef]
14. Sottile, F.; Massaglia, S.; Peano, C. Ecological and Economic Indicators for the Evaluation of Almond (*Prunus dulcis* L.) Orchard Renewal in Sicily. *Agriculture* **2020**, *10*, 301. [CrossRef]
15. Corris, A. Defining the Environment in Organism–Environment Systems. *Front. Psychol.* **2020**, *11*, 1285. [CrossRef] [PubMed]
16. Yang, Q.; Nan, Z.; Tang, Z. Influencing factors of the grassland ecological compensation policy to herdsmen's behavioral response: An empirical study in Hexi corridor. *Acta Ecol. Sin.* **2021**, in press. [CrossRef]

17. Bekele, W.; Drake, L. Soil and water conservation decision behavior of subsistence farmers in the Eastern Highlands of Ethiopia: A case study of the Hunde-Lafto area. *Ecol. Econ.* **2003**, *46*, 437–451. [CrossRef]
18. Illukpitiya, P.; Gopalakrishnan, C. Decision-making in soil conservation: Application of a behavioral model to potato farmers in Sri Lanka. *Land Use Policy* **2004**, *21*, 321–331. [CrossRef]
19. Clarkson, G.; Dorward, P.; Osbahr, H.; Torgbor, F.; Kankam-Boadu, I. An investigation of the effects of PICSA on smallholder farmers' decision-making and livelihoods when implemented at large scale—The case of Northern Ghana. *Clim. Serv.* **2019**, *14*, 1–14. [CrossRef]
20. Hayden, M.T.; Mattimoe, R.; Jack, L. Sensemaking and the influencing factors on farmer decision-making. *J. Rural Stud.* **2021**, *84*, 31–44. [CrossRef]
21. Li, L.; Paudel, K.P.; Guo, J. Understanding Chinese Farmers' Participation Behavior Regarding Vegetable Traceability Systems. *Food Control.* **2021**, *130*, 108325. [CrossRef]
22. Findlater, K.M.; Satterfield, T.; Kandlikar, M. Farmers' Risk-Based Decision Making Under Pervasive Uncertainty: Cognitive Thresholds and Hazy Hedging. *Risk Anal.* **2019**, *39*, 1755–1770. [CrossRef]
23. Lamarque, P.; Meyfroidt, P.; Nettier, B.; Lavorel, S. How Ecosystem Services Knowledge and Values Influence Farmers' Decision-Making. *PLoS ONE* **2014**, *9*, e107572. [CrossRef]
24. Rezaei, R.; Seidi, M.; Karbasioun, M. Pesticide exposure reduction: Extending the theory of planned behavior to understand Iranian farmers' intention to apply personal protective equipment. *Saf. Sci.* **2019**, *120*, 527–537. [CrossRef]
25. Tama, R.A.Z.; Ying, L.; Yu, M.; Hoque, M.M.; Adnan, K.M.; Sarker, S.A. Assessing farmers' intention towards conservation agriculture by using the Extended Theory of Planned Behavior. *J. Environ. Manage.* **2021**, *280*, 111654. [CrossRef]
26. Li, Q.; Wagan, S.A.; Wang, Y. An analysis on determinants of farmers' willingness for resource utilization of livestock manure. *Waste Manag.* **2021**, *120*, 708–715. [CrossRef] [PubMed]
27. Liu, H.; Zhou, Y. Farmers' Cognition and Behavioral Response towards Cultivated Land Quality Protection in Northeast China. *Sustainability* **2018**, *10*, 1905. [CrossRef]
28. Kuang, Y.; Lin, B. Public participation and city sustainability: Evidence from Urban Garbage Classification in China. *Sustain. Cities Soc.* **2021**, *67*, 102741. [CrossRef]
29. Gao, F.; Shi, L. Convergence of labor productivity growth in Chinese provinces:1978 to 2006. *Manage World.* **2009**, *1*, 49–60. [CrossRef]
30. Adnan, N.; Nordin, S.M.; Bahruddin, M.A.; Tareq, A.H. A state-of-the-art review on facilitating sustainable agriculture through green fertilizer technology adoption: Assessing farmers behavior. *Trends Food Sci. Technol.* **2019**, *86*, 439–452. [CrossRef]
31. Djenontin, I.N.S.; Zulu, L.C.; Ligmann-Zielinska, A. Improving Representation of Decision Rules in LUCC-ABM: An Example with an Elicitation of Farmers' Decision Making for Landscape Restoration in Central Malawi. *Sustainability* **2020**, *12*, 5380. [CrossRef]
32. Teng, Y.; Chen, X.; Yu, Z.; Wei, J. Research on the Evolutionary Decision-Making Behavior Among the Government, Farmers, and Consumers: Based on the Quality and Safety of Agricultural Products. *IEEE Access* **2021**, *9*, 73747–73756. [CrossRef]
33. Bukchin, S.; Kerret, D. The role of self-control, hope and information in technology adoption by smallholder farmers—A moderation model. *J. Rural Stud.* **2020**, *74*, 160–168. [CrossRef]
34. Nigussie, Z.; Tsunekawa, A.; Haregeweyn, N.; Adgo, E.; Nohmi, M.; Tsubo, M.; Aklog, D.; Meshesha, D.T.; Abele, S. Factors influencing small-scale farmers' adoption of sustainable land management technologies in north-western Ethiopia. *Land Use Policy* **2017**, *67*, 57–64. [CrossRef]
35. Burton, R.J.F. The influence of farmer demographic characteristics on environmental behaviour: A review. *J. Environ. Manage.* **2014**, *135*, 19–26. [CrossRef] [PubMed]
36. Zhang, M.; Jin, Y.; Qiao, H.; Zheng, F. Product quality asymmetry and food safety: Investigating the "one farm household, two production systems" of fruit and vegetable farmers in China. *China Econ. Rev.* **2017**, *45*, 232–243. [CrossRef]
37. Singh, C.; Dorward, P.; Osbahr, H. Developing a holistic approach to the analysis of farmer decision-making: Implications for adaptation policy and practice in developing countries. *Land Use Policy* **2016**, *59*, 329–343. [CrossRef]
38. Wang, H.; Wang, X.; Sarkar, A.; Zhang, F. How Capital Endowment and Ecological Cognition Affect Environment-Friendly Technology Adoption: A Case of Apple Farmers of Shandong Province, China. *Int. J. Environ. Res. Public. Health* **2021**, *18*, 7571. [CrossRef]
39. Annes, A.; Bessiere, J. Staging agriculture during on-farm markets: How does French farmers' rationality influence their representation of rurality? *J. Rural Stud.* **2018**, *63*, 34–45. [CrossRef]
40. Li, F.; Ren, J.; Wimmer, S.; Yin, C.; Li, Z.; Xu, C. Incentive mechanism for promoting farmers to plant green manure in China. *J. Clean. Prod.* **2020**, *267*, 122197. [CrossRef]
41. Russell, C.M.; Simon, H.A. Administrative Behavior. *Am. J. Nurs.* **1950**, *50*, 46. [CrossRef]
42. Bagheri, A.; Emami, N.; Damalas, C.A. Farmers' behavior towards safe pesticide handling: An analysis with the theory of planned behavior. *Sci. Total Environ.* **2021**, *751*, 141709. [CrossRef]
43. Savari, M.; Abdeshahi, A.; Gharechaee, H.; Nasrollahian, O. Explaining farmers' response to water crisis through theory of the norm activation model: Evidence from Iran. *Int. J. Disaster Risk Reduct.* **2021**, *60*, 102284. [CrossRef]
44. Tang, J.; Folmer, H.; Xue, J. Estimation of awareness and perception of water scarcity among farmers in the Guanzhong Plain, China, by means of a structural equation model. *J. Environ. Manage.* **2013**, *126*, 55–62. [CrossRef] [PubMed]

45. Zhu, C.; Wang, S. Farmers' cognition of eco agriculture in the area of returning farmland to forest in Western China. *Rural Econ.* **2019**, *9*, 53–57.
46. Li, F.J.; Dong, S.C.; Li, F. A system dynamics model for analyzing the eco-agriculture system with policy recommendations. *Ecol. Model.* **2012**, *227*, 34–45. [CrossRef]
47. Liu, M.; Liu, W.; Yang, L.; Jiao, W.; He, S.; Min, Q. A dynamic eco-compensation standard for Hani Rice Terraces System in southwest China. *Ecosyst. Serv.* **2019**, *36*, 100897. [CrossRef]
48. Mausolff, C.; Farber, S. An economic analysis of ecological agricultural technologies among peasant farmers in Honduras. *Ecol. Econ.* **1995**, *12*, 237–248. [CrossRef]
49. Mouron, P.; Scholz, R.W.; Nemecek, T.; Weber, O. Life cycle management on Swiss fruit farms: Relating environmental and income indicators for apple-growing. *Ecol. Econ.* **2006**, *58*, 561–578. [CrossRef]
50. Amfo, B.; Ali, E.B.; Atinga, D. Climate change, soil water conservation, and productivity: Evidence from cocoa farmers in Ghana. *Agric. Syst.* **2021**, *191*, 103172. [CrossRef]
51. Huang, Y.; Tao, B.; Xiaochen, Z.; Yang, Y.; Liang, L.; Wang, L.; Jacinthe, P.-A.; Tian, H.; Ren, W. Conservation tillage increases corn and soybean water productivity across the Ohio River Basin. *Agric. Water Manag.* **2021**, *254*, 106962. [CrossRef]
52. Valizadeh, N.; Bijani, M.; Karimi, H.; Naeimi, A.; Hayati, D.; Azadi, H. The effects of farmers' place attachment and identity on water conservation moral norms and intention. *Water Res.* **2020**, *185*, 116131. [CrossRef]
53. Fernández Sánchez, J.L.; Fernández Polanco, J.M.; Llorente García, I. Evidence of price premium for MSC-certified products at fishers' level: The case of the artisanal fleet of common octopus from Asturias (Spain). *Mar. Policy* **2020**, *119*, 104098. [CrossRef]
54. Dorce, L.C.; da Silva, M.C.; Mauad, J.R.C.; de Faria Domingues, C.H.; Borges, J.A.R. Extending the theory of planned behavior to understand consumer purchase behavior for organic vegetables in Brazil: The role of perceived health benefits, perceived sustainability benefits and perceived price. *Food Qual. Prefer.* **2021**, *91*, 104191. [CrossRef]
55. Brown, E.; Dury, S.; Holdsworth, M. Motivations of Consumers That Use Local, Organic Fruit and Vegetable Box Schemes in Central England and Southern France. *Appetite* **2009**, *53*, 183–188. [CrossRef]
56. Weibel, F.P.; Tamm, L.; Wyss, E.; Daniel, C.; Häseli, A.; Suter, F. Organic Fruit Production in Europe: Successes in Production and Marketing in the Last Decade, Perspectives and Challenges for the Future Development. In Proceedings of the I International Symposium on Organic Apple and Pear 737, Wolfville, NS, Canada, 28 February–2 March 2006; pp. 163–172.
57. Canavari, M.; Bazzani, G.M.; Spadoni, R.; Regazzi, D. Food Safety and Organic Fruit Demand in Italy: A Survey. *British Food Journal* **2002**, *104*, 220–232. [CrossRef]
58. Zhao, Y. *Specialty Foods: Processing Technology, Quality, and Safety*; CRC Press: Boca Raton, FL, USA, 2012; ISBN 1-4398-5423-8.
59. Lee, W.S.; Alchanatis, V.; Yang, C.; Hirafuji, M.; Moshou, D.; Li, C. Sensing technologies for precision specialty crop production. *Comput. Electron. Agric.* **2010**, *74*, 2–33. [CrossRef]
60. Cheung, L.T.O.; Ma, A.T.H.; Lam, T.W.L.; Chow, A.S.Y.; Fok, L.; Cheang, C.C. Predictors of the environmentally responsible behaviour of participants: An empirical investigation of interpretative dolphin-watching tours. *Glob. Ecol. Conserv.* **2020**, *23*, e01153. [CrossRef]
61. Wang, W.; Wang, L.; Gu, L.; Zhou, G. Understanding farmers' commitments to carbon projects. *Sci. Total Environ.* **2021**, *784*, 147112. [CrossRef]
62. Kaplan, J.; Travadon, R.; Cooper, M.; Hillis, V.; Lubell, M.; Baumgartner, K. Identifying economic hurdles to early adoption of preventative practices: The case of trunk diseases in California winegrape vineyards. *Wine Econ. Policy* **2016**, *5*, 127–141. [CrossRef]
63. Lin, X.; Cui, S.; Han, Y.; Geng, Z.; Zhong, Y. An improved ISM method based on GRA for hierarchical analyzing the influencing factors of food safety. *Food Control* **2019**, *99*, 48–56. [CrossRef]
64. Attri, R.; Ashishpal; Khan, N.Z.; Siddiquee, A.N.; Khan, Z.A. ISM-MICMAC Approach for Evaluating the Critical Success Factors of 5S Implementation in Manufacturing Organisations. *Int. J. Bus. Excell.* **2020**, *20*, 521–548. [CrossRef]
65. Cai, Y.; Xia, C. Interpretive Structural Analysis of Interrelationships among the Elements of Characteristic Agriculture Development in Chinese Rural Poverty Alleviation. *Sustainability* **2018**, *10*, 786. [CrossRef]
66. Yang, W.Y.; Cao, W.; Kim, J.; Park, K.W.; Park, H.-H.; Joung, J.; Ro, J.-S.; Lee, H.L.; Hong, C.-H.; Im, T. *Applied Numerical Methods Using MATLAB*; John Wiley & Sons: Hoboken, NJ, USA, 2020; ISBN 1-119-62680-3.
67. Sarkar, A.; Qian, L.; Peau, A.K.; Shahriar, S. Modeling drivers for successful adoption of green business: An interpretive structural modeling approach. *Environ. Sci. Pollut. Res.* **2021**, *28*, 1077–1096. [CrossRef]
68. Wachenheim, C.; Fan, L.; Zheng, S. Adoption of unmanned aerial vehicles for pesticide application: Role of social network, resource endowment, and perceptions. *Technol. Soc.* **2021**, *64*, 101470. [CrossRef]
69. Ndlovu, P.N.; Thamaga-Chitja, J.M.; Ojo, T.O. Factors influencing the level of vegetable value chain participation and implications on smallholder farmers in Swayimane KwaZulu-Natal. *Land Use Policy* **2021**, *109*, 105611. [CrossRef]
70. Xue, Y.; Guo, J.; Li, C.; Xu, X.; Sun, Z.; Xu, Z.; Feng, L.; Zhang, L. Influencing factors of farmers' cognition on agricultural mulch film pollution in rural China. *Sci. Total Environ.* **2021**, *787*, 147702. [CrossRef]
71. Wang, Y.; Wang, J.; Wang, X.; Li, Q. Does policy cognition affect livestock farmers' investment in manure recycling facilities? Evidence from China. *Sci. Total Environ.* **2021**, *795*, 148836. [CrossRef] [PubMed]
72. Li, B.; Ding, J.; Wang, J.; Zhang, B.; Zhang, L. Key factors affecting the adoption willingness, behavior, and willingness-behavior consistency of farmers regarding photovoltaic agriculture in China. *Energy Policy* **2021**, *149*, 112101. [CrossRef]

73. Azadi, Y.; Yazdanpanah, M.; Mahmoudi, H. Understanding smallholder farmers' adaptation behaviors through climate change beliefs, risk perception, trust, and psychological distance: Evidence from wheat growers in Iran. *J. Environ. Manage.* **2019**, *250*, 109456. [CrossRef] [PubMed]
74. Das V., J.; Sharma, S.; Kaushik, A. Views of Irish Farmers on Smart Farming Technologies: An Observational Study. *AgriEngineering* **2019**, *1*, 164–187. [CrossRef]
75. Ajzen, I. The Theory of Planned Behavior. *Organ. Behav. Human Dec. Proc.* **1991**, *50*, 179–211. [CrossRef]
76. Bagheri, A.; Bondori, A.; Allahyari, M.S.; Damalas, C.A. Modeling farmers' intention to use pesticides: An expanded version of the theory of planned behavior. *J. Environ. Manage.* **2019**, *248*, 109291. [CrossRef]
77. Zhang, L.; Li, X.; Yu, J.; Yao, X. Toward cleaner production: What drives farmers to adopt eco-friendly agricultural production? *J. Clean. Prod.* **2018**, *184*, 550–558. [CrossRef]
78. Liu, S.; Zhang, P.; Marley, B.; Liu, W. The Factors Affecting Farmers' Soybean Planting Behavior in Heilongjiang Province, China. *Agriculture* **2019**, *9*, 188. [CrossRef]
79. Belder, P.; Bouman, B.A.M.; Cabangon, R.; Guoan, L.; Quilang, E.J.P.; Yuanhua, L.; Spiertz, J.H.J.; Tuong, T.P. Effect of water-saving irrigation on rice yield and water use in typical lowland conditions in Asia. *Agric. Water Manag.* **2004**, *65*, 193–210. [CrossRef]
80. Blanco-Canqui, H.; Lal, R. Soil and Water Conservation. In *Principles of Soil Conservation and Management*; Blanco-Canqui, H., Lal, R., Eds.; Springer Netherlands: Dordrecht, The Netherlands, 2008; pp. 1–19, ISBN 978-1-4020-8709-7.
81. Aguilera, E.; Guzmán, G.; Alonso, A. Greenhouse gas emissions from conventional and organic cropping systems in Spain. II. Fruit tree orchards. *Agron. Sustain. Dev.* **2015**, *35*, 725–737. [CrossRef]
82. Mfitumukiza, D.; Nambasa, H.; Walakira, P. Life cycle assessment of products from agro-based companies in Uganda. *Int. J. Life Cycle Assess.* **2019**, *24*, 1925–1936. [CrossRef]
83. Zhang, B.; Fu, Z.; Wang, J.; Tang, X.; Zhao, Y.; Zhang, L. Effect of Householder Characteristics, Production, Sales and Safety Awareness on Farmers' Choice of Vegetable Marketing Channels in Beijing, China. *Br. Food J.* **2017**, *119*, 1216–1231. [CrossRef]
84. Xu, G.; Sarkar, A.; Qian, L. Does organizational participation affect farmers' behavior in adopting the joint mechanism of pest and disease control? A study of Meixian County, Shaanxi Province. *Pest Manag. Sci.* **2021**, *77*, 1428–1443. [CrossRef]
85. Jiang, Y. Key Technology and Industrial Applications of Storage and Transport Specialty Fruits and Vegetables in Southern China. *Sci. Technol. Dev.* **2016**, *5*, 234–238.
86. He, K.; Zhang, J.; Zeng, Y.; Zhang, L. Households' willingness to accept compensation for agricultural waste recycling: Taking biogas production from livestock manure waste in Hubei, P.R. China as an example. *J. Clean. Prod.* **2016**, *131*, 410–420. [CrossRef]
87. Ma, W.; Abdulai, A. IPM adoption, cooperative membership and farm economic performance: Insight from apple farmers in China. *China Agric. Econ. Rev.* **2019**, *11*, 218–236. [CrossRef]
88. Xu, D.; Deng, X.; Huang, K.; Liu, Y.; Yong, Z.; Liu, S. Relationships between labor migration and cropland abandonment in rural China from the perspective of village types. *Land Use Policy* **2019**, *88*, 104164. [CrossRef]
89. Zhou, H.; Yan, J.; Lei, K.; Wu, Y.; Sun, L. Labor migration and the decoupling of the crop-livestock system in a rural mountainous area: Evidence from Chongqing, China. *Land Use Policy* **2020**, *99*, 105088. [CrossRef]
90. Zakaria, A.; Azumah, S.B.; Appiah-Twumasi, M.; Dagunga, G. Adoption of climate-smart agricultural practices among farm households in Ghana: The role of farmer participation in training programmes. *Technol. Soc.* **2020**, *63*, 101338. [CrossRef]

agriculture

Article

Effects of Agricultural Use on Endangered Plant Taxa in Spain

José Luis Molina-Pardo [1,*], Emilio Rodríguez-Caballero [2], Miguel Cueto [3], Pablo Barranco [3], Manuel Sánchez-Robles [1], Azucena Laguía-Allué [1] and Esther Giménez-Luque [3]

[1] CECOUAL, Universidad de Almería, Carretera de Sacramento s/n, 04120 Almería, Spain; msr725@ual.es (M.S.-R.); ala147@ual.es (A.L.-A.)
[2] Departamento de Agronomía, CECOUAL, Universidad de Almería, Carretera de Sacramento s/n, 04120 Almería, Spain; rce959@ual.es
[3] Departamento de Biología y Geología, CECOUAL, Universidad de Almería, Carretera de Sacramento s/n, 04120 Almería, Spain; mcueto@ual.es (M.C.); pbvega@ual.es (P.B.); egimenez@ual.es (E.G.-L.)
* Correspondence: jmp648@ual.es

Abstract: Agriculture is one of the most widespread human activities and has the greatest impact on terrestrial ecosystems, as it transforms natural ecosystems into artificial landscapes using, in many cases, large amounts of pesticides as well as overexploiting natural resources. Therefore, for effective biodiversity conservation, it is necessary to include agricultural systems in conservation programs. In this work, the 50 plant taxa described for Spain as threatened by agricultural use were selected. These were divided according to the type of threat into those affected by crop extension, intensification, or abandonment. In addition, information was obtained concerning their conservation status, level of protection and functional traits (life form, pollination, and dispersal). Finally, the evolution of land use, in the areas near the populations of the selected species, was identified. The selected taxa belong to 21 families and present different life forms and modes of dispersal or pollination. Forty-six percent are endangered (EN) and most are included in legal protection lists. Nearly three-quarters are threatened by crop expansion and land use dynamics, reflecting an expansion of cultivated areas, which adds further pressure to these species. In addition to agricultural expansion, taxa are also at risk, due to important rates of agricultural land abandonment, and mention agricultural intensification. Nevertheless, conservation measures do exist to promote biodiversity in agricultural landscapes that may help to reverse the negative effect of land use dynamics on selected species, but few are specific to threatened flora. Therefore, if threatened plants are to be conserved in agricultural areas, it is necessary to promote a profound transformation of our socioecological systems. One of these transformative changes could come from the human-nature reconnection.

Keywords: threatened plant; agriculture; Spain; land use; conservation; human-nature reconnection

Citation: Molina-Pardo, J.L.; Rodríguez-Caballero, E.; Cueto, M.; Barranco, P.; Sánchez-Robles, M.; Laguía-Allué, A.; Giménez-Luque, E. Effects of Agricultural Use on Endangered Plant Taxa in Spain. *Agriculture* **2021**, *11*, 1097. https://doi.org/10.3390/agriculture11111097

Academic Editors: José Luis Vicente-Vicente, Cristina Quintas-Soriano and María D. López-Rodríguez

Received: 9 October 2021
Accepted: 2 November 2021
Published: 4 November 2021

1. Introduction

Anthropogenic activities have been altering the natural environment for thousands of years, affecting the structure and functioning of ecosystems [1,2]. Anthropogenic biomes occupy more than 75% of the terrestrial land surface [3], and humans currently appropriate more than one third of global net primary productivity [4]. This has contributed to overcoming several of the planetary boundaries proposed as a safe operating space for humanity [5]. In order to provide resources, food, and contribute to global food security, agriculture has extended during the last decades and actually occupies one-third of the ice-free land surface and almost half of potentially productive land area [2,6]. Thus, it is considered one of the most widespread human activities worldwide [7]. Agriculture transforms natural ecosystems into artificial ones created and managed by humans [8]. This has, in many cases, severe environmental impacts such as soil degradation [9], greenhouse gas emissions [10], depletion and degradation of water resources [11–13], pollution [14,15], or habitat loss [16]. Indeed, agriculture is a major contributor to the transgressing of four planetary boundaries:

biosphere integrity, biogeochemical flows, land-system change, and freshwater use [17]. For example, crop fertilization is the largest anthropogenic perturbation of global N and P cycles [5].

It is estimated that the world population could reach 9.1 billion by 2050 [18]. Increasing population growth, and the continuing development of global trade and the world economy, will increase food demand by 70% [18]. This would imply an increase of 100–110% of the global cultivated land area by 2050 [19]. Within this context of current population growth and increasing food demand, during the 1950s and 1960s, the "Green Revolution" began. This, led to change in the production system that extended for many countries all over the world [20,21] and lead to an increase in world agricultural production mainly by one third in 50 years, with reduced agricultural land expansion (only 12%) [22]. The scientific and technological improvement achieved during the "Green Revolution" was possible because of the intensification of agriculture [23], the use of agrochemicals, the breeding of high-yielding varieties, and innovations in irrigation systems [23–25]. These advances provide us with the possibility to increase productivity by limiting the conversion of natural ecosystems to crops and to prevent the release of huge amounts of greenhouse gases [18].

Traditional agricultural systems or agroecosystems, although less productive than intensive systems developed after the green revolution, had the capacity to preserve natural values [26]. In general, modern intensive agricultural practices cause a simplification and homogenization of the landscape at different scales. For example, at a local scale, the use of agrochemicals and increased mechanization leads to the elimination of trees and shrubs presented in crop fields and a loss or simplification of herbaceous diversity. At a landscape level, the planting of large extensions of monocultures and the elimination of unproductive areas (boundaries, patches of natural vegetation, water points, etc.), leads to the loss of natural habitats and their disconnection. This, together with long-lasting damage to soil and water availability and the large amount of waste generated, is causing an unprecedented loss of global biodiversity [15,25,27–31]. Biodiversity-aggressive practices also lead to a decrease in agroecosystem resilience [32] and the modification of its capacity to provide key ecosystem services [33–38]. One example is the loss of pollinators that affects more than one third of the crops used for food production. Pollinator losses caused by agricultural intensification is not only an emerging risk for ecosystems but also for the economy, as this ecosystem service improves productivity and represents a profit of USD 235–577 billion per year worldwide [39,40]. In Europe, where that local plant diversity co-existed with traditional agriculture over centuries, agricultural intensification is also one of the main causes of biodiversity losses [41–43]. Therefore, one of the current challenges is to find a balance between long-term sustainable agricultural production for the increasing population growth and the effective conservation of biodiversity and its associated ecological processes [44].

Concern about the negative impacts of intensive agriculture on the environment has stimulated interest in alternative agricultural systems, such as those proposed by agroecology and organic farming [23,45–48]. New policy initiatives have also emerged, such as the Agri-Environment Schemes (AES) of the European Union (EU) Common Agricultural Policy (CAP), which provide economic incentives for farmers to undertake agrobiodiversity-friendly practices [49]. The number of scientific studies on biodiversity conservation in agroecosystems has also increased in recent years [27,50–55]. These studies reinforce the idea that with proper management, agricultural areas can be rich in native taxa and key sites for their conservation [44,56,57]. Moreover, according to Storkey et al. [58], the intrinsic ecological value of endangered taxa and their delicate conservation status justify their priority conservation target.

Scientific literature shows agriculture affects some threatened taxa in cultivated areas [41,59,60], either by crop expansion, management change, or agricultural abandonment [58,61,62]. However, not all taxa respond equally to these changes; some are simply not able to adapt to living in cultivation, while for others, agroecosystems are important and sometimes essential for their survival [44,62], being strongly affected by agricultural intensification [58] or by land abandonment [63]. Among all of the different biological groups, plants are a key component of agroecosystems as they provide resources to a wide variety of organisms [64], and also to humans. Plant functional traits, in addition to environmental characteristics, may be responsible for vulnerability to local extinction in agricultural landscapes [65,66] and are frequently used in studies on land-use change or management and their effects [67].

In Spain, as in other Mediterranean countries, major agricultural land transformations have taken place during the last decades. Agriculture has expanded in some areas and the most profitable agricultural areas have intensified while marginal areas have been abandoned [68]. These changes have led to an unfavorable conservation status for part of its biodiversity and a loss of associated ecosystem services [69]. This has happened despite the existence of legal tools for their conservation, and the fact that Spain has an extensive network of protected areas, whose boundaries were established taking into account the presence of endangered species [70]. In order to better understand the conservation status of threatened vascular plants in Spain, since 2000, their conservation status has been evaluated in the Atlas y Libro Rojo de Flora Vascular Amenazada de España (AFA), and its addenda [62,71–74]. However, the effects of different agricultural changes on their populations have never been deeply analyzed. Different types of threats derived from agricultural use and their effects may vary according to the functional traits of the threatened taxa. In addition, the category of threat or the level of legal protection may condition the survival of threatened flora in agricultural environments now or in the future. Thus, we proposed a study aimed at evaluating the state of plant taxa threatened by changes in agricultural practices in Spain. To achieve this, we proposed the following specific objectives: (1) to identify taxa threatened by agricultural activities, and to determine their type of threat and their degree of protection; (2) to analyze the relationship between different threats and key functional traits of plants; (3) to evaluate land use changes in areas close to populations of the endangered taxa in Spain. Finally, we performed an assessment of the current state and expected trend of the endangered taxa threatened by agriculture in Spain and we have drawn up a list of potential actions for conservation.

2. Materials and Methods

2.1. Study Area

In this study, we focused on continental Spain, (the Canary Islands, Balearic Islands, Ceuta and Melilla were not included), which has an extension of 493–486 km^2 [75]. Spain is a European country located in the Mediterranean basin, which is one of the world's main biodiversity hotspots [76] and, therefore, a priority area for conservation [77]. The great diversity of biomes, types of vegetation, relief, climates and microclimates, soil types, and human activity, give it an environmental heterogeneity that confers enormous biodiversity [78,79], with high conservation value. Its flora is remarkable in the European context as it hosts more than 7000 taxa [76] and approximately 80% of the flowering plants living in the European Union [80,81]. Threatened flora represents 17% of the total plant taxa [82].

2.2. Studied Taxon

For this work, we selected all taxa described for Spain as currently or potentially threatened by agricultural use in the AFA and its addenda [62,71–74]. The selected taxa were then divided into three categories, according to the type of threat from agriculture: (i) crop extension (CE), which represents a threat to taxa that is not typical in agricultural areas, but whose populations inhabit bordering areas or other areas that may change to agricultural use due to the extension of crops; (ii) crop intensification (CI), as a threat to taxa living in agricultural areas where land management practices change, mainly to an intensified production system; and (iii) crop abandonment (CA), which represents a threat to taxa whose survival depends on agricultural activities (i.e., taxa well specialized to coexist with crops in agricultural areas). All scientific names listed in AFA have been revised and some have been updated according to bibliography [83–87]. For each taxon, we explored the conservation status and level of legal protection and we obtained information with reference to several plant functional traits related to the tolerance of threatened taxa to agricultural changes. Finally, for each of the identified species, we collected occurrence records from the Global Biodiversity Information Facility (GBIF). Before using the spatial data, we cleaned the dataset to minimize common errors in GBIF occurrence data [88]. From the preliminary list, wrong records (e.g., records whose coordinates were outside the possible range values or those in which latitude or longitude were equal to 0), records whose presence was outside the study area, and those outside their known distribution were removed.

2.2.1. Conservation Status

We retrieved the threat level of each specie according to IUCN classes: (i) CR, critically endangered; (ii) EN, endangered; and (iii) VU, vulnerable. In addition, we identified the level of legal protection of each taxon. For this purpose, we used the information related to the legal protection and threat level collected in the AFA and its addenda [62,71–74], and in the Dríada database (https://www.conservacionvegetal.org/drtest/, accessed on 1 July 2021).

2.2.2. Trait Data

For each of the taxon studied, a search was carried out in the AFA and its addenda [62,71–74], on plant functional traits related to the tolerance of threatened taxa to agricultural changes [89]. The life form was selected as a taxon's response to disturbances [67], whereas the type of pollination and dispersal mode are indicators of the dispersal capacity and recruitment success of the plants [67]. According to Raunkiaer [90], we classify selected species into six life forms (chamaephytes, geophytes, hemicryptophytes, hydrophytes, phanerophytes, and therophytes). This classification has been used to determine the response of some taxa to different intensities of agricultural management [65,66]. Given the diversity of pollination type and mode of dispersal of plant taxa threatened by agriculture, they have been classified into three categories: abiotic, biotic, and unknown. Pollination was classified as abiotic when autogamous or anemophilous taxa were involved, and as biotic if the mode of pollination was by zoogamy (entomophilous). The dispersal and pollination mechanism was not determined for the identified threatened taxa. In these cases, as well as in the cases not presenting obvious adaptations, the pollination mechanism was classified as unknown. In the case of the mode of dispersal, it was included in the abiotic category when the mode of dispersal of the taxon was autochory, baricory, anemochory, or hydrochory, and as biotic, if the mode of dispersal was by zoochory (myrmecochory and zoochory without specifying the vector). Again, taxa with unknown mechanisms or with no obvious adaptations were classified as "unknown".

2.3. *Agricultural Use Evolution*

Using species records obtained from GBIF (Section 2.2), we identified the main land use in a buff area of 500 m radius around each location using Coordination of Information of the Environment (CORINE). To reduce land use complexity, the original legend was reclassified into Urban land, Natural ecosystems and seven agricultural classes: (i) rainfed agriculture; (ii) irrigated lands; (iii) rice plantation; (iv) tree plantation; (v) other crops (including areas with a mix of different crops); (vi) pasture; and (vii) mixed crop-natural (including agroforestry areas and areas occupied by agriculture but with a significant extension of natural lands); see supplementary Table S1 for further details. This process was repeated for the land use classification of 1990 and 2018 and the total change of the different uses in each of the influence areas of each record was calculated as the difference between both dates. Finally, we analyzed, as a reference, the total change of each of the identified classes for the complete study area.

3. Results

3.1. Threatened Plant Taxa and Level of Protection

Of the 1233 plant taxa included in AFA for continental Spain, 591 are in the threatened categories (CR, EN, and VU). Of these taxa, 50 have been classified as threatened by some type of agriculture-related activity (Table 1), Seventy four percent (n = 274) of their populations are threatened for this reason. The total number of taxa belongs to 21 families, although more than 25% belong to two families, Plumbaginaceae (14%) and Compositae (12%). These families, together with Cruciferae (8%), Caryophyllaceae (8%), Leguminosae (6%), Marsileaceae (6%), and Scrophulariaceae (6%), comprise more than 50% of the selected taxa (Table 1). Of these, 24% are classified as vulnerable (VU) (n = 12), 40% as endangered (EN) (n = 20), and 36% as critically endangered (CR) (n = 18). Most of the taxa (90%) are included on legal protection lists (n = 45; 32 at regional level, 5 at regional-national level, 7 at regional-national-supranational level, and 1 at supranational level only). The predominant life form is hemicryptophytes, corresponding to this category 42% (n = 21) of identified taxa; 24 % are therophytes (n = 12); 14 % geophytes (n = 7); 8 % chamaephytes (n = 4); 6 % phanerophytes (n = 3), and 6 % hydrophytes (n = 3). For most, taxa pollination is biotic (82%, n = 41), while dispersal is mainly abiotic (78%, n= 37) (Figure 1).

Table 1. Taxa included in this study. The table shows the taxa studied. Family, specie and subspecies are indicated. The reference is indicated when the taxonomic status has been updated according to the AFA and not implying a change in the number of populations or individuals. In addition, the type of threat that mainly affects the taxa is indicated (CE, crop extension; CI, crop intensification; CA, crop abandonment). The following are also indicated are: P, number of populations; % TP, percentage of threatened populations; threat category in IUCN Red List (CR, critically endangered; EN, endangered; VU, vulnerable); PR, degree of legal protection (-, absent; R, regional; N, national; RN, regional-national; S, supranational; RNS, national, regional, and supranational).

Family	Taxon	Threat	P	% TP	IUCN	PR
Alliaceae	*Allium scaberrimum* M.Serres [84]	CA	16	100	VU	R
Amaryllidaceae	*Narcissus nevadensis* Pugsley subsp. *nevadensis* [83]	CE	2	50	EN	RN
Amaryllidaceae	*Narcissus bujei* (Fern. Casas) Fern. Casas	CI	14	100	VU	R
Caryophyllaceae	*Dianthus inoxianus* Gallego	CE	16	56	EN	R
Caryophyllaceae	*Silene sennenii* Pau	CE	3	67	EN	RN
Caryophyllaceae	*Silene stockenii* Chater	CE	4	100	CR	R
Caryophyllaceae	*Silene diclinis* (Lag.) M. Laínz	CI	5	80	EN	R
Colchicaceae	*Androcymbium europaeum* (Lange) K. Richt.	CE	5	80	VU	RNS
Compositae	*Anthemis bourgaei* Boiss. & Reut.	CE	2	50	EN	R
Compositae	*Centaurea kunkelii* N. García	CE	2	50	CR	R
Compositae	*Centaurea ultreiae* Silva Pando	CE	1	100	CR	R
Compositae	*Jacobaea auricula* (Coss.) Pelser [86]	CE	7	100	VU	R
Compositae	*Leucanthemum gallaecicum* Rodr. Oubiña & S. Ortiz	CE	4	75	EN	R
Compositae	*Pentanema bifrons* (L.) D. Gut. Larr. Santos-Vicente, Anderb., E. Rico & M.M. Mart. Ort. [85]	CE	1	100	CR	R
Cruciferae	*Clypeola eriocarpa* Cav.	CE	2	50	CR	R
Cruciferae	*Coincya longirostra* (Boiss.) Greuter & Burdet	CE	10	100	EN	R
Cruciferae	*Vella pseudocytisus* L. subsp. *pseudocytisus*	CE	2	100	EN	R
Cruciferae	*Isatis aptera* (Boiss. & Heldr.) Al-Shehbaz, Moazzeni & Mumm. [87]	CA	6	100	EN	-
Dipsacaceae	*Succisella carvalhoana* (Mariz) Baksay	CE	4	25	VU	R
Geraniaceae	*Erodium paularense* Fern. Gonz. & Izco	CE	11	9	EN	RNS
Geraniaceae	*Erodium recoderi* Auriault & Guitt.	CE	6	17	VU	-
Gramineae	*Puccinellia pungens* (Pau) Paunero	CE	9	11	EN	RNS
Gramineae	*Enneapogon persicus Boiss.*	CI	2	100	CR	R
Labiatae	*Nepeta hispanica* Boiss. & Reut.	CE	8	62.5	VU	R
Labiatae	*Teucrium edetanum* M.B. Crespo, Mateo & T. Navarro	CE	2	50	VU	R
Leguminosae	*Astragalus oxyglottis* M. Bieb.	CE	10	30	EN	R

Table 1. *Cont.*

Family	Taxon	Threat	P	% TP	IUCN	PR
Leguminosae	*Ononis azcaratei* Devesa	CE	4	50	CR	R
Leguminosae	*Astragalus nitidiflorus* Jiménez Mun. & Pau	CI	1	100	CR	RN
Lythraceae	*Lythrum baeticum* Gonz. Albo	CE	24	83	EN	R
Lythraceae	*Lythrum flexuosum* Lag.	CE	57	100	EN	RNS
Malvaceae	*Malvella sherardiana* (L.) Jaub. & Spach	CA	4	100	VU	-
Marsileaceae	*Marsilea batardae* Launert	CE	17	53	EN	RNS
Marsileaceae	*Marsilea strigosa* Willd.	CE	33	97	VU	RNS
Marsileaceae	*Pilularia minuta* Durieu	CE	4	100	CR	RNS
Plantaginaceae	*Plantago notata Lag.*	CI	1	100	CR	R
Plumbaginaceae	*Armeria merinoi* (Bernis) Nieto Fel. & Silva Pando	CE	6	50	CR	R
Plumbaginaceae	*Limonium aragonense* (Debeaux) Font Quer	CE	1	100	CR	R
Plumbaginaceae	*Limonium quesadense* Erben	CE	2	100	EN	R
Plumbaginaceae	*Limonium soboliferum* Erben	CE	1	100	CR	R
Plumbaginaceae	*Limonium squarrosum* Erben	CE	1	100	CR	R
Plumbaginaceae	*Limonium ugijarense* Erben	CE	2	50	EN	-
Plumbaginaceae	*Limonium mansanetianum* M.B. Crespo & Lledó	CI	4	100	CR	R
Polygalaceae	*Polygaloides balansae* (Coss.) O. Schwarz	CE	1	100	CR	-
Ranunculaceae	*Delphinium bolosii* C. Blanché & Molero	CE	2	50	EN	RN
Ranunculaceae	*Ranunculus lingua* L.	CE	1	100	CR	R
Scrophulariaceae	*Scrophularia herminii* Hoffmanns. & Link	CE	31	32	EN	S
Scrophulariaceae	*Linaria nigricans* Lange	CI	6	50	EN	R
Scrophulariaceae	*Verbascum fontqueri* Benedí & J.M. Monts.	CA	8	100	VU	R
Thymelaeaceae	*Thymelaea lythroides* Barratte & Murb.	CE	1	100	CR	RN
Umbelliferae	*Hohenackeria polyodon* Coss. & Durieu	CE	4	100	VU	R

3.2. Current State of Taxa Endangered by Agricultural Threat Categories and Trends

3.2.1. Taxa Endangered by Crop Extension

Almost three-quarters of the total plant taxa classified as threatened by agriculture-related changes in land use ($n = 39$) are threatened by crop extension (Table 1). Of these, 41.03% ($n = 16$) have all their populations threatened by crop extension and 43.59% ($n = 17$) have at least half of their populations affected due to this reason (Table 1). Moreover, 43.6% of the taxa threatened by agricultural extension ($n = 17$) are endangered (EN), 35.9% ($n = 14$) are critically endangered (CR), and 20.5% ($n = 8$) are vulnerable (VU). Most of the selected taxa (92.3%; $n = 47$) are protected, except *Erodium recoderi* (VU), *Limonium ugijarense* (EN), and *Polygaloides balansae* (CR). However, 61.54% of them ($n = 24$) are protected only at the regional level, 10.26% ($n = 4$) are protected at the national-regional level, and 17.95% ($n = 7$) are also protected at the supranational level (Table 1). One taxon (*Scrophularia herminii)* is protected only at the supranational level by the Habitats Directive (Table 1).

A detailed analysis of the different life forms of the plant taxa threatened by the expansion of agricultural use in Iberian Spain revealed that 43.6 % of them are hemicryptophytes ($n = 17$), 23% therophytes 23% ($n = 9$), while the other types (geophytes, chamaephytes, phanerophytes, hydrophytes) account for only about 10% each ($n = 3$–4). Pollination of plants in this group is mainly biotic (84.62%, $n = 33$) and the predominant mode of dispersal is abiotic (76.92%, $n = 30$) (Table 2).

Figure 2 shows the agricultural uses in the areas of influence of the plant taxa classified as threatened by crop extension, as well as the trend of expansion or reduction of agricultural use between 1990 and 2018 according to CORINE land cover. As observed, there is large variability among taxa. Some of them, are located in areas occupied by large extensions of agricultural use (more than 50% of the surface), under both increasing (e.g., *Ononis azcaratei, Anthemis bourgaei, Pilularia minuta* and *Jacobaea auricula*) and decreasing (e.g., *Limonium aragonense, Lythrum flexuosum* and *Vella pseudocytisus*) trends. There are also taxa located in areas with reduced agricultural extension but with a large proportion of intensive practices (irrigated crops) and with a positive trend to increase agricultural extension (e.g., *Delphinium bolosii*). Others, such as *Centaurea kunkelii*, showed the opposite pattern. Finally, regarding several taxa located in heavily cultivated areas (e.g., *Silene sennenii*) or lightly cultivated areas (e.g., *Dianthus inoxianus*), we did not find a significant change in the cultivation extension. However, in most of these cases, there are important changes in the agricultural practices, with a dominant trend toward agricultural intensification or irrigation.

Table 2. Summary of the trial for each of the taxon included in the three threat types (crop extension, crop intensification, and crop abandonment). The trial data included are: life form (C, chamaephytes; G, geophytes; H, hemicryptophytes; Hy, hydrophytes; P, phanerophytes; T, therophytes), pollination (-, unknown; abiotic; biotic) and dispersal mode (-, unknown; abiotic; biotic).

Threat	Taxon	Life Form	Pollination Mode	Dispersal Mode
Crop extension	*Androcymbium europaeum*	G	Biotic	Abiotic
	Anthemis bourgaei	T	Biotic	Abiotic
	Armeria merinoi	H	Biotic	Abiotic
	Astragalus oxyglottis	T	Biotic	Abiotic
	Centaurea kunkelii	H	Biotic	Abiotic
	Centaurea ultreiae	H	Biotic	Biotic
	Clypeola eriocarpa	T	Biotic	Abiotic
	Coincya longirostra	H	Biotic	Abiotic
	Delphinium bolosii	G	Biotic	Abiotic
	Dianthus inoxianus	C	Biotic	Abiotic
	Erodium paularense	C	Biotic	Abiotic
	Erodium recoderi	T	Biotic	Abiotic
	Hohenackeria polyodon	T	Abiotic	Abiotic
	Jacobaea auricula	H	Biotic	Abiotic
	Leucanthemum gallaecicum	H	Biotic	Biotic
	Limonium aragonense	H	Biotic	Abiotic
	Limonium quesadense	H	Biotic	Abiotic
	Limonium soboliferum	H	Abiotic	Biotic
	Limonium squarrosum	H	Biotic	Abiotic
	Limonium ugijarense	H	Biotic	Abiotic
	Lythrum baeticum	T	Biotic	-
	Lythrum flexuosum	T	Biotic	-
	Marsilea batardae	Hy	Abiotic	Abiotic
	Marsilea strigosa	Hy	Abiotic	Biotic
	Narcissus nevadensis nevadensis	G	Biotic	Abiotic
	Nepeta hispanica	G	Biotic	Abiotic
	Ononis azcaratei	T	Biotic	Abiotic
	Pentanema bifrons	H	Biotic	Abiotic
	Pilularia minuta	H	Abiotic	Abiotic
	Polygaloides balansae	P	Biotic	Abiotic
	Puccinellia pungens	H	Abiotic	Abiotic
	Ranunculus lingua	Hy	Biotic	-
	Scrophularia herminii	H	Biotic	-
	Silene sennenii	C	Biotic	Abiotic
	Silene stockenii	T	Biotic	Abiotic
	Succisella carvalhoana	H	Biotic	Abiotic
	Teucrium edetanum	H	Biotic	Abiotic
	Thymelaea lythroides	P	Biotic	Biotic
	Vella pseudocytisus pseudocytisus	P	Biotic	Abiotic
Agricultural intensification	*Astragalus nitidiflorus*	H	Biotic	-
	Enneapogon persicus	G	Abiotic	Abiotic
	Limonium mansanetianum	H	-	-
	Linaria nigricans	T	Biotic	Abiotic
	Narcissus bujei	G	Biotic	Abiotic
	Plantago notata	T	Abiotic	Biotic
	Silene diclinis	C	Biotic	Abiotic
Crop abandonment	*Allium scaberrimum*	G	Biotic	Abiotic
	Isatis aptera	T	Biotic	Abiotic
	Malvella sherardiana	H	Biotic	Abiotic
	Verbascum fontqueri	H	Biotic	Abiotic

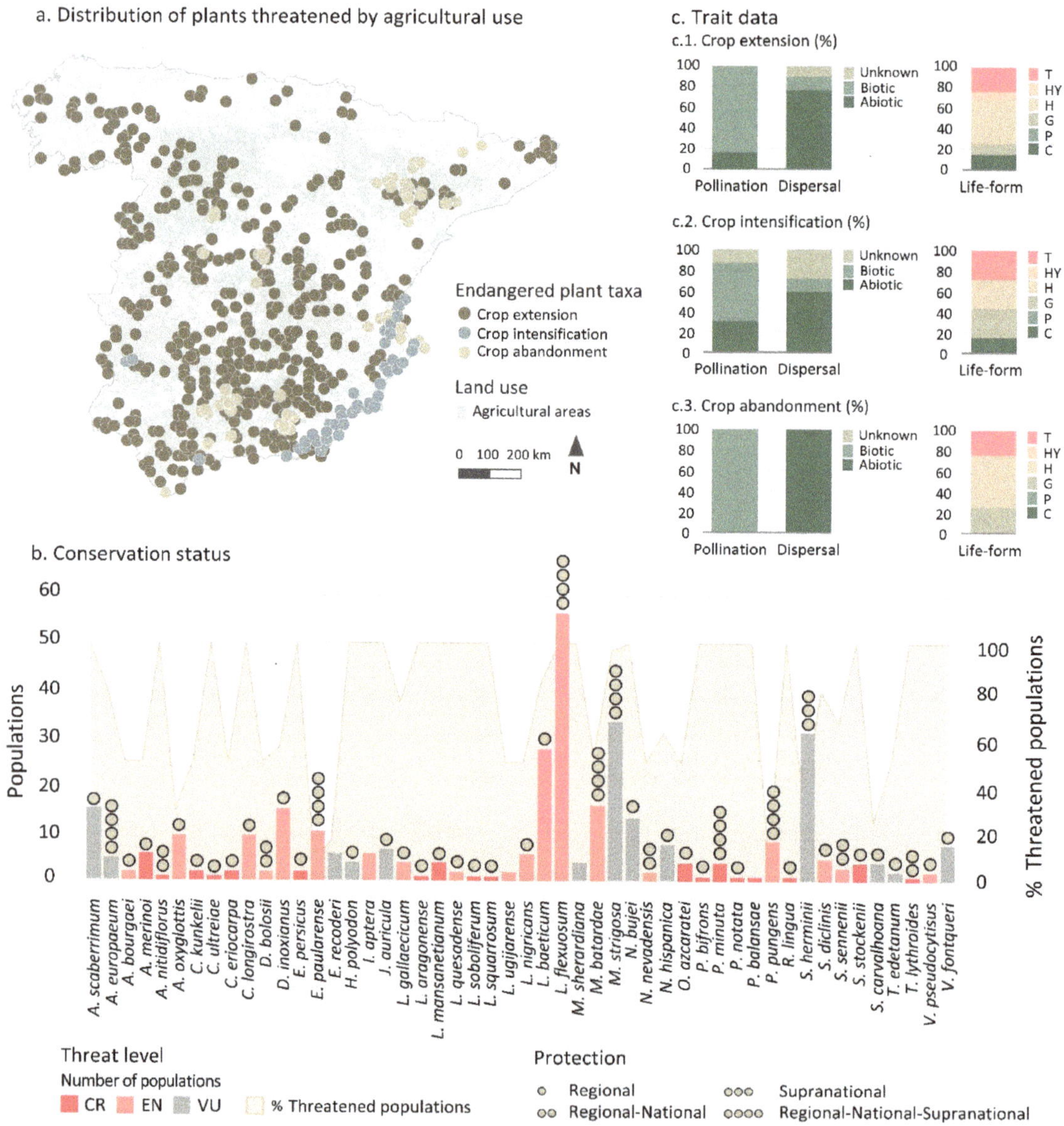

Figure 1. Summary of plants threatened by agricultural use in continental Spain, their conservation status, and functional traits. (**a**) Distribution of plants threatened by agricultural use. Map of the presence of taxa threatened by agriculture included in each category (represented in three different colors) and base map with the area of agricultural use present in the study area. (**b**) Conservation status. The graph shows the taxa identified as threatened by agricultural use; the height of the histogram bar shows the number of populations (values indicated on the left axis), the color of the bar shows the threat category (CR, EN, VU) and the brown area shows the percentage of threatened populations (values indicated on the right axis). In addition, the level of protection is indicated by circles on the histogram bar (1, regional; 2, regional-national; 3, supranational; 4, regional-national-supranational). (**c**) Trait data. Plant trait includes for each threat category (**c1–c3**): % pollination mode (unknown; abiotic; biotic); % dispersal mode (unknown; abiotic; biotic); and % life form (C, chamaephytes; G, geophytes; H, hemicryptophytes; Hy, hydrophytes; P, phanerophytes; T, therophytes).

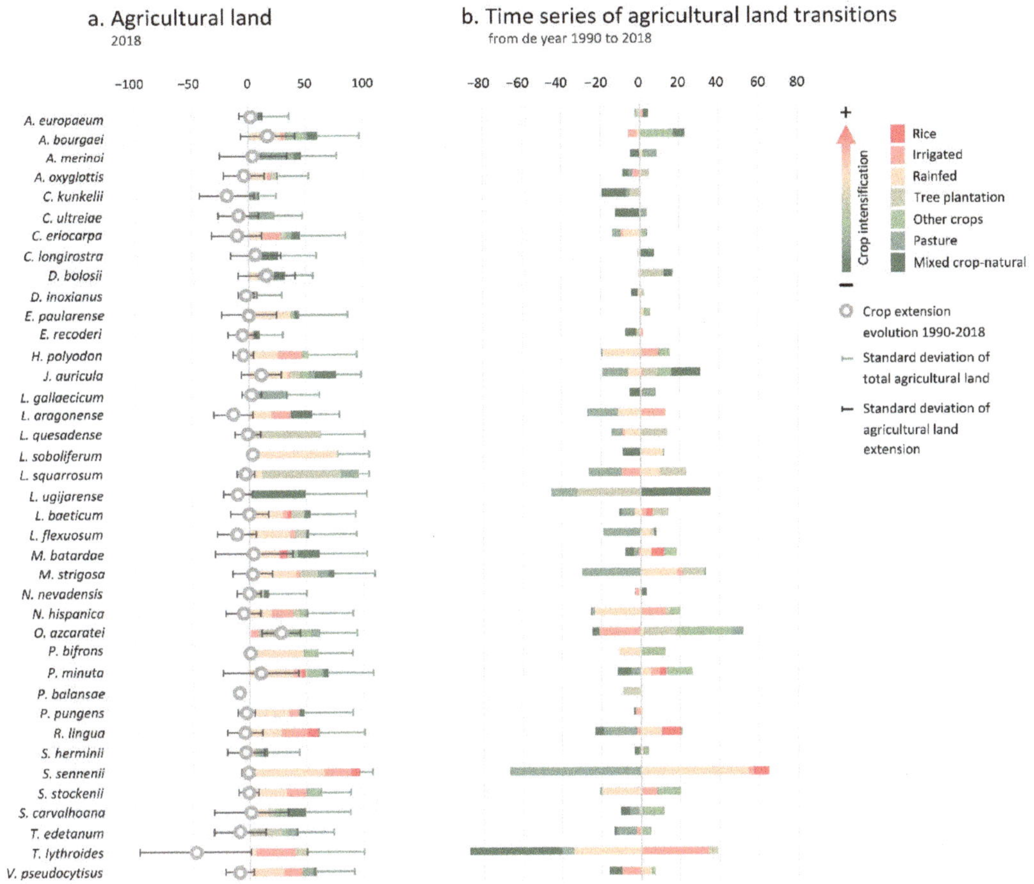

Figure 2. Taxa threatened by the extension of cultivation. (**a**) Agricultural land. The figure shows the total area of agricultural uses, according to CORINE, of the area of influence of taxa threatened by the extension of cultivation in 2008. The dot indicates the evolution of the agricultural land use area for each taxon between 1990 and 2018. (**b**) Time series of agricultural land uses. The figure shows, for each taxon, the trend of expansion or reduction of agricultural use between 1990 and 2018 according to CORINE.

3.2.2. Taxa Endangered by Crop Intensification

Seven plants are threatened by agricultural intensification. Of these, *Astragalus nitidiflorus, Enneapogon persicus, Limonium mansanetianum, Narcissus bujei,* and *Plantago notata,* have all their populations threatened by agricultural intensification, while *Silene diclinis* and *Linaria nigricans* have four (80%) and three (50%) of their populations threatened by agricultural intensification, respectively. Most are critically endangered (CR), except *Linaria nigricans* and *Silene diclinis,* which are listed as endangered (EN), and *Narcissus bujei,* which is, listed as vulnerable (VU). All are protected; *Enneapogon persicus, Limonium mansanetianum, Linaria nigricans, Narcissus bujei,* and *Plantago notata* only at the regional level, while *Astragalus nitidiflorus* is protected at the regional-national level.

This group includes plants with different life forms, such as hemicryptophytes (*Astragalus nitidiflorus* and *Limonium mansanetianum*), geophytes (*Enneapogon persicus* and *Narcissus bujei*), therophytes (*Linaria nigricans* and *Plantago notata*) and chamaephytes (*Silene diclinis*). More than half have biological pollination (*Astragalus nitidiflorus, Linaria nigricans*, *Narcissus bujei* and *Silene diclinis*) and the mode of dispersal is abiotic in almost all known cases ($n = 5$) (Table 2).

A detailed analysis of land use evolution in the area of influence of these taxa revealed that most of the taxa included in this category have been found in areas occupied by a large extension of crops, with *Enneapogon persicus* having more than 80% of the surface area dedicated to this use (Figure 3). However, there is no dominance of intensive practices. The taxa located in regions with a higher dominance of intensive agriculture are *Enneapogon persicus*, *Plantago notata* and *Linaria nigricans* with 24.07%, 11.13%, and 6.57% of their areas of influence covered by irrigated crops, respectively.

In most cases, the area of agricultural use has changed minimally between 1990 and 2018, and more traditional and less aggressive uses such as rainfed or mixed crops, have increased. Irrigated crops have only slightly increased around some populations of *Plantago notata* and *Linaria nigricans* (Figure 3).

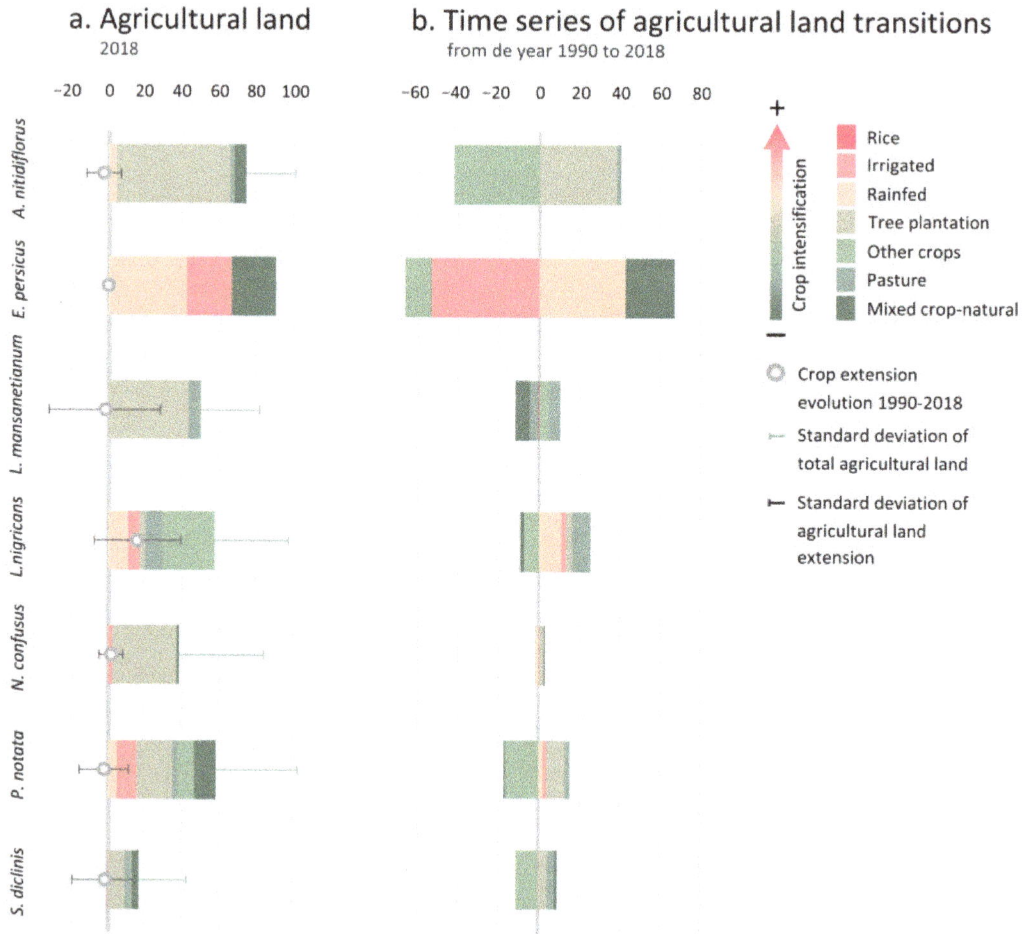

Figure 3. Taxa threatened by crop intensification. (**a**) Agricultural land. The figure shows the total area of agricultural uses according to CORINE of the area of influence of taxa threatened by crop intensification in 2008. The dot indicates the evolution of the area of agricultural use for each taxon between 1990 and 2018. (**b**) Time series of agricultural uses. The figure shows, for each taxon, the trend of expansion or reduction of agricultural use between 1990 and 2018, according to CORINE.

3.2.3. Taxa Endangered by Crop Abandonment

Only four of the identified plant taxa endangered by agricultural practices are threatened by crop abandonment (Table 1). All of them have 100% of their populations threatened for this reason. However, most of the taxa included in this group (*Allium scaberrimum*, *Malvella sherardiana*, and *Verbascum fontqueri*) are listed as vulnerable (VU) and only *Isatis aptera* is listed as endangered. This, as well as *Malvella sherardiana* (VU), have no direct legal protection, whereas *Allium scaberrimum* and *Verbascum fontqueri* are protected by national and supranational regulations. *Malvella sherardiana* and *Verbascum fontqueri* are hemicryptophytes, *Isatis aptera* is a therophyte, and *Allium scaberrimum* is a geophyte. All of them have biotic pollination and abiotic modes of dispersal (Table 2).

According to Figure 4, the four species of this group are located in areas with a large extension of crops, especially *Isatis aptera*, which occupies areas with an average cover of crops of around 80%. The taxon with the lowest representation of agricultural use is *Verbascum fontqueri*, (20%). Net Agriculture extension in the buffer area of the different population of these taxa has changed minimally in most cases with the exception of *Allium scaberrimum*. In this case, a net decrease of about 20% of the agriculture extension has been observed between 1990 and 2018. Though the net area covered by crops did not experience large modifications, there is an important rate of change between different agricultural practices with a clear trend to increase the area dedicated to the most intensive land uses (Figure 4b). This is the case for *Allium scaberrimum*, in the areas close to their populations, rainfed crops, other crops, and pastures have been abandoned and replaced by more intensive crops, such as irrigated crops. Something similar has occurred with *Malvella sherardiana*, although in this case, the pasture area has increased (change identified with the abandonment of agriculture according to CORINE) and there has been a greater fluctuation between the losses and gains of the different types of crops.

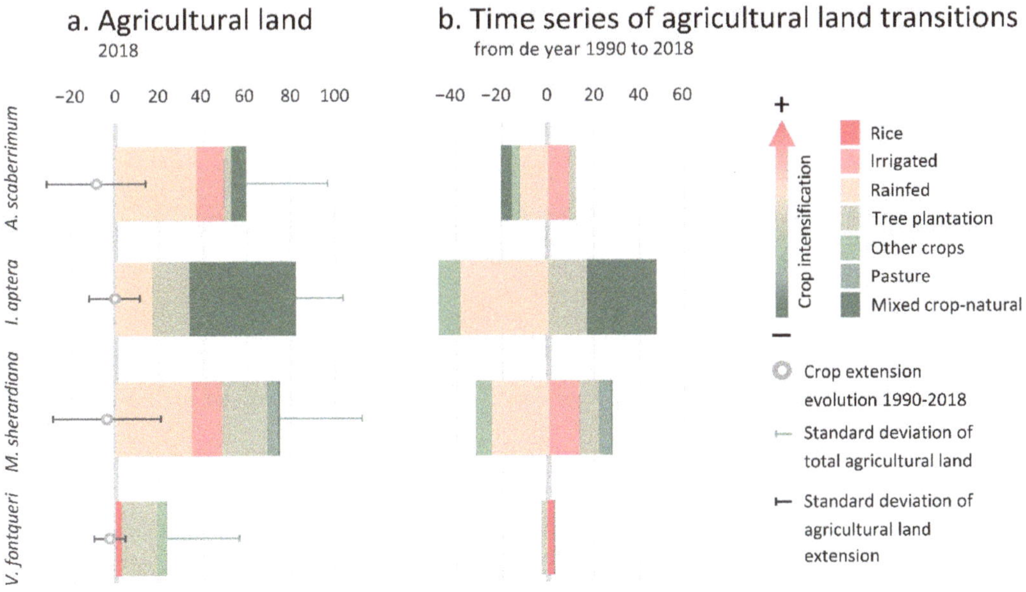

Figure 4. Taxa threatened by Crop abandonment. (**a**) Agricultural land. The figure shows the total area of agricultural use according to CORINE of the area of influence of taxa threatened by crop abandonment in 2008. The dot indicates the evolution of the area of agricultural uses for each taxon between 1990 and 2018. (**b**) Time series of agricultural uses. The figure shows, for each taxon, the trend of expansion or reduction of agricultural use between 1990 and 2018, according to CORINE.

4. Discussion

Agricultural land use changes, such as the conversion of natural areas to agricultural land, crop intensification or abandonment, are considered to be one of the main threats for endangered plant taxa on a global scale [58,61,62,91]. Our review reveals that [92] in continental Spain, there are 50 taxa threatened by any of these land use changes, representing 8.5% of the total number of threatened taxa in Spain. A list has been made based on current knowledge and the number appears to be lower than in other countries [92]. Nonetheless, these numbers may be underestimated as threat assessment efforts frequently focused on endemic and rare taxa and the actual number of plants threatened by agricultural practices may be higher than those provided in the official red list. For example, in Greece, the red data book includes few species threatened by agriculture, but according to [93] numerous widespread species are reducing their populations until levels that do merit a threat status, due to modernization of agricultural practices. Moreover, these numbers may increase in the near future and, as already described by other biological groups, such as steppe birds [91]. This is especially relevant as changes towards higher threat categories in Spain are mostly related to human activities [94]. For these reasons, it is very important to assess the threat status of the native flora of agricultural land, and not focus only on rare and endemic species, which is typically the case in red list assessments.

The probability of persistence of plant taxa in agricultural areas is related, among other plant traits, to those affecting their tolerance to disturbance [95]. As expected, one of the most common life forms among the taxa identified are therophytes, considered as indicators of disturbed ecosystems, regardless of whether they are active or abandoned crops [96]. In addition, there is a predominance of other life forms shown to be highly resistant to disturbance, such as hemicryptophytes and geophytes. These data contrast when compared with the total number of threatened species in Spain, where only hemicryptophytes are well represented (25.2%, $n = 149$), whereas therophytes and geophytes only represent ~9% ($n = 52$) and ~8% of the total number of species.

Vegetation capability to disperse and colonize new habitats are also important plant traits for survival in anthropogenic habitats, such as agricultural areas. For example, the reproductive success of plants depends initially on their pollination capacity. As the main pollination vectors of the taxa identified in this study are insects, as is also the case for the total number of threatened species in Spain (88%, $n = 488$), the loss of pollinators or their efficiency is one of the negative consequences detected in agricultural systems [33,40,97–99]. A clear example is the global commercialization of pollinators for use in crops, due to the absence of wild pollinators [100,101]. Small pollen loads can reduce fruit and seed formation, affecting seed viability, recruitment, progeny and vigor, and the genetic diversity of their populations [102–108]. In addition, identified threatened taxa are also characterized by low numbers and geographically restricted populations. The success of these populations living in fragmented landscapes is strongly dependent on the dispersal rate or the availability of dispersal vectors, as it can be a limiting factor for demographic recruitment, population continuity, and genetic exchange [109,110]. Overall, long-distance dispersal capacity may be key to the survival of populations in fragmented environments [111]. The predominant dispersal strategy in the taxa studied is mainly abiotic with the exception of some taxa (*Centaurea ultreiae*, *Leucanthemum gallaecicum*, *Limonium soboliferum* and *Plantago notata*) whose dispersal is carried out by ants and *Marsilea strigose*, whose vector is unknown. This is consistent with the mode of dispersal of the total number of threatened taxa in Spain, as abiotic dispersal predominates (85.6%, $n = 459$). Seed dispersal distances, both abiotic and by ants, are small and usually reach shorter distances than when other animals disperse seeds by epi- or endozoochory [112].

The future of plant taxa threatened by agriculture depends on their capability to survive in areas under diverse types of changes related to agriculture, but also on the intensity and direction of land use changes. An overall analysis of crop extension shows a general decrease in the extension of areas under agricultural use during the last three decades (Figure 5). According to this, and taking into account the high level of legal protection of most of the identified taxa (more than 90% of identified taxa are included in official lists; Table 1), one may expect a good conservation status of all taxa threatened by agriculture in Iberian Spain. However, a deeper analysis of land use dynamics shows that there are important changes in the area occupied by the different crops (Figure 5b), which reflects an important rate of crop extension occurring in parallel with agriculture abandonment, and changes to more intensive practices (irrigated crops, rice fields and tree crops have increased, while rainfed crops and other types of crops have decreased; Figure 5b). This could be one of the main reasons explaining why most of the identified populations are endangered, even though they have a high level of legal protection. Thus, it is clear that, although there are already mechanisms to protect them, more effort is needed by policy managers, land owners, and the society in order to ensure biodiversity conservation of plant taxa in areas endangered by agriculture. For example, in the U.S., the U.S. Endangered Species Act (ESA) has succeeded in protecting hundreds of taxa from extinction and improving their recovery over time [113,114]. However, threats to endangered taxa in the U.S. are still persistent and it is estimated that increased funding and continued management will be needed in the future to ensure their survival [114].

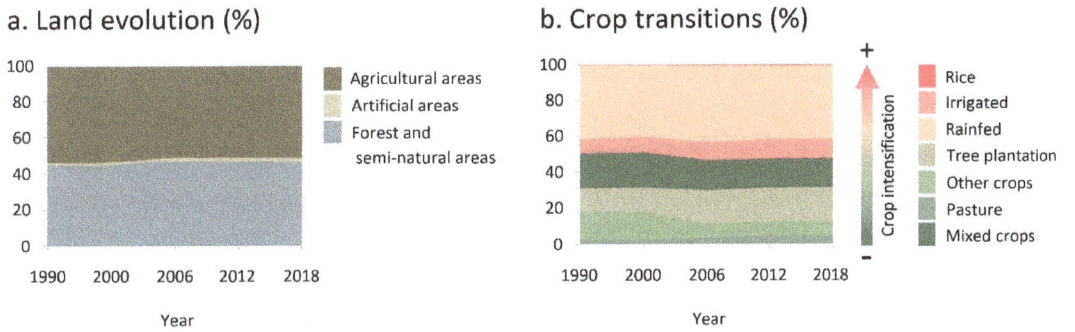

Figure 5. Evolution of changes in use in Spain between 1990 and 2018 according to CORINE. (**a**). Net change in natural and disturbed habitats. (**b**). Net evolution of the different types of agricultural use.

4.1. Crop Extension

The main impact for the plant taxa classified as threatened by agriculture in continental Spain is the loss of natural habitats due to increased agriculture. The extension of crops generates drastic changes in ecosystems in short periods, leading many taxa to immediate local extinction [115]. This also occurs when patches of natural habitat are maintained, because very frequently, they are small and threatened plants are permanently exposed to pressures from the surrounding areas [115]. Moreover, habitat extension reduction related to the expansion of agriculture reduced population size and has other indirect negative effects on plant population survival, such as the reduction of seed banks and the regenerative potential, both being essential for the survival of a large number of plant taxa [116]. All these together, implies that, even if taxa are still present in a favorable zone, local extinction is not avoided but postponed [115]. A clear example of the expansion of crops at the expense of the reduction of natural habitats is the expansion of greenhouses in the southeast of the peninsula [117] that affects, for example, *Androcymbium europaeum* [62].

Attending to the different life forms, there are examples of all of them in the list of species threatened by crop extension, the dominants being hemicryptophytes, phanerophytes, and geophytes. The predominance of these life forms within this category is probably due to their preference for natural areas, thus occupying remnants of natural vegetation close to agricultural fields. The only examples of phanerophytes (*Polygaloides balansae*, *Thymelaea lythroides* and *Vella pseudocytisus* subsp. *pseudocytisus*) and hydrophytes (*Marsilea batardae*, *Marsilea strigose*, and *Ranunculus lingua*) identified in this study are enclosed within this group. The negative effects of the extension of agriculture on phanerophytes are generally because trees and large plants included in this category are frequently removed during preliminary work to prepare the land for the installation of crops (i.e., clearing, leveling, etc., during the preliminary work to prepare the land for the installation of crops (clearing, leveling, etc.) [30]. Hydrophytes are linked to the margins of watercourses, lagoons, or temporary bodies of water. The expansion of crops can directly or indirectly imply the transformation, drainage or drying of the water point, which, together with the low ecological plasticity of some taxa, can cause their disappearance in the short term [62]. In this sense, Spain is one of the countries with the highest rates of groundwater depletion worldwide [118]. In addition, there are aquatic crops that can increase the likelihood of biological invasions. An example is the red swamp crayfish *Procambarus clarkii* (Girard, 1852), which is capable of spreading through rice crops and reaching high densities [119]. This taxon is common in Spain [120,121] and can have negative effects on crops and native biodiversity in invaded areas in a short time [122,123].

The predominant mode of dispersal in plants threatened by crop expansion is abiotic (anemochory, barochory, autochory, and hydrochory), with some exception in which ants (*Centaurea ultreiae*, *Leucanthemum gallaecicum*, *Limonium soboliferum*, and *Thymelaea lythroides*) facilitate seed dispersion. Therefore, the main handicap for this group is not the dispersal capacity, but the availability of suitable habitats for the dispersed seeds to germinate. As shown in Figure 2a, in many cases, the matrix in which threatened taxa are found is highly anthropogenic and remnants of natural vegetation are small and disconnected among them. As abiotic dispersal distances predominant in the taxa of this group are limited [112] even without increased crop cultivation, it is difficult for new populations to thrive. For example, in abiotic modes of dispersal, under optimal conditions (clear soil and morphologically adapted seeds), seeds at most reach distances of 500 m from the mother plant. In the case of dispersal by ants, they are also unable to disperse seeds over long distances, but they minimize predation and facilitate establishment [124].

4.2. Crop Intensification

Traditional farming systems, with low aggressive practices, harbor enormous biodiversity [125], and are key to the conservation of many threatened taxa. However, agricultural intensification is currently significantly decreasing the richness and functional diversity of different biological groups [30,126,127]. Agricultural intensification may cause dominant taxa to become more dominant and rare taxa to become extinct [128]; thus, having a more negative effect over the rare taxa [89,129]. For example, in England, between 1960 and 1997, the loss of rare taxa and the increase of more adaptable common taxa was detected as a consequence of agricultural intensification [129]. Furthermore, even if it is known that a threatened taxon is present in an intensively managed agricultural area, this information should be taken with caution. It is advisable to have good knowledge of the dynamics of its populations, as they may be faced with a gradual depletion of the seed bank [129]. Herbicide use and recurrent plowing have been identified as one of the main factors controlling the seed banks, which may accelerate local extinctions [130–132].

In our study, 14% of the threatened taxa are not affected by crop extension, but by crop intensification. Most of them, such as *Silene diclinis*, *Narcissus bujei*, and *Linaria nigricans*, are flexible taxa able to colonize and survive in some cultivated areas or in the borders of field crops under different levels of disturbance. As observed in Figure 1, the majority of taxa population included in this category are located on the east coast of the Iberian Peninsula, an area identified as a priority for threatened flora in Spain [80]. In most of the areas close to threatened populations, there are no significant net changes in the degree of intensification. However, in taxa, such as *Linaria nigricans*, there has been a greater increase in areas with more intensive agricultural management (Figure 3b). In this case, the fragmentation rate has been increasing over the last decades in some of the most important and largest populations, such as the population of *Linaria nigricans* located in Tabernas (Almeria) [133], where the irrigated olive grove area has increased from 400 ha in 1970 to 4336 ha in 2019 [29]. In addition, there has been a second process of intensification of existing crops [29].

Associated with this type of threat we have found three dominant life forms: hemicryptophytes, geophytes, and therophytes that may favor plant adaptation to survive in agricultural areas. For example, Druckenbrod and Dale [134] relate the increase of geophytes to disturbance by machinery in forested areas. Other authors, however, link the increase of therophytes to tillage, while indicating that geophytes and hemicryptophytes increased in undisturbed soils [66]. Similarly, Tarifa et al. [89] found that hemicryptophytes and therophyte life forms were favored by intensive management in olive orchards. These life forms have the ability to germinate from the seed banks or resprout when disturbances cease and suitable climatic conditions exist [90]. Consequently, they are able to survive and remain in transformed areas, such as agricultural fields. A persistence of seed bank viability has also been related to taxa that are annual or biennial [135], which favors the presence of therophytes. However, the intensification of agriculture and the massive use of agrochemicals may cause them to have adaptive disadvantages compared to other more generalist taxa, as described above. Therefore, all taxa we identified in this category that can colonize and survive in agricultural areas are now threatened by changes in management practices. This situation is aggravated for those taxa that depend on pollinators. As shown in Table 1, more than half of the identified taxa (*Astragalus nitidiflorus, Linaria nigricans, Narcissus bujei,* and *Silene diclinis)* have generalist entomophilous pollination, which will face an additional threat from agricultural intensification (for example see, Tarifa et al. [89]). This occurs mainly because crop intensification threatens the persistence of wild bee communities and pollination services [99], with important negative implications on the reproductive success of plants. Sometimes what happens is not that the number of bees or dominant taxa decreases, but that intensification reduces foraging success [95,136]. In woody crops, it has been shown that the structure of the pollinator network remains more or less stable under different management regimes (organic and intensive), but the most unique interactions do vary [136]. The risk of extinction of specialized and rare pollinators also affects certain endemic shrubland plants, because the quantity or quality of pollen and the reproductive output may be reduced in the absence of co-evolved pollinators [95,137].

Agricultural intensification also hinders seed dispersal, as it leads to a system characterized by fewer and less interconnected patches of optimal habitats for the threatened taxa. Within crops, at first, the removal of vegetation and the creation of open areas as a consequence of tough plowing, the use of livestock or herbicides, could favor abiotic dispersal plants, such as most of the taxa included in this category (Table 1) [138]. However, this is not usually the case when taking into account soil roughness and slope, factors that are also important for dispersal, as well as for germination and seedling establishment [139]. Recurrent plowing is common in some intensive crops and results in rough soils, which, under certain conditions, can improve the germination capacity of plants [140]. Nevertheless, roughness also increases resistance to movement and decreases seed dispersal distance, preventing colonization of other adjacent favorable agricultural areas. Agricultural intensification has also led to increased soil erosion [141], especially in areas with steep slopes. Soil erosion not only leads to nutrient impoverishment, but also accelerates desiccation and increases the burial depth of seeds [139]. This negatively affects seedling propagation, growth, and survival [139]. Moreover, taxa included in this category are small, which is an additional limitation for wind dispersal (e.g., Watkinson [142]).

4.3. Crop Abandonment

Europe is a continent that has been historically transformed and much of its land area is cultivated. For some threatened taxa, this has meant the loss of their primary habitats and has made their survival almost entirely dependent on the secondary agricultural habitats to which they have adapted [128]. A clear example is the flora and birds of the European steppes [143,144]. As these species have evolved with cultivation, when their preferred habitat (agricultural system) disappears, they are negatively affected [63]. Thus, the abandonment of crops is one of the main threats to most of the taxa included in this group, such as *Allium scaberrimum*, *Isatis aptera*, *Malvella sherardiana*, and *Verbascum fontqueri* (Table 1). Similar results have been observed in other well-studied groups that depend on the agricultural areas they inhabit, such as farmland birds [69,145].

Life forms of the four taxa identified as taxa threatened by land abandonment are hemicryptophytes, therophytes and geophytes. Although it is a very small number of species to draw clear conclusions about trait adaptation, it has been demonstrated that all of these life forms withstand disturbances, can live in crops, and are only displaced by other species when the crops are abandoned. This occurs because land abandonment often leads to interspecific competition for endangered taxa, which, in the end, may promote the increase in the richness and diversity of other more generalist plant species that may sometimes have adaptive advantages over threatened species [146].

Dispersal of taxa included in this group is mainly abiotic. Thus, it seems that revegetation after cultivation could minimize their chances of dispersal as the dispersal rate in open areas should be longer than in more densely vegetated areas [147]. However, as previously stated the number of species is very low to draw clear conclusions about it.

4.4. Conservation Implications

There is growing concern about how to reduce the impact of agricultural use on biodiversity and the scientific community considers the application of biodiversity conservation measures in these areas a key step to achieve effective biodiversity conservation at a global level [44]. For this reason, agri-environmental plans have been implemented in many regions to improve biodiversity in these areas. Some examples are, the Agri-Environment Schemes (AES) of the European Union (EU) Common Agricultural Policy (CAP). However, the measures have not been very effective [148–150] and sound scientific evaluations of the conservation status of taxa and the existing knowledge gaps are needed in order to support policy decisions and to prioritize conservation actions focused on the most threatened taxa [115]. By performing an overall evaluation of the state and potential evolution of the plant taxa threatened by changes in agricultural practices in Spain, we have found that there is an overall decrease in the extension of agricultural areas during the last three decades (Figure 5a). According to this, and considering the high level of legal protection of most of the identified taxa (more than 90% of identified taxa are included in official lists; Table 1), one may expect a good conservation status of all identified taxa. However, a deeper analysis of land use dynamics showed that there are important changes in the area occupied by the different crops (Figure 5b), which reflect an important rate of crop extension occurring in parallel with agriculture abandonment and changes to more intensive practices (irrigated crops, rice fields, and tree crops have increased, while rainfed crops and other types of crops have decreased; Figure 5b), all of these actions having important negative impacts on the plants considered in this study, as well as in all other plants that may not be well recognized as threatened taxa. Thus, although legal mechanisms do exist to protect them, more effort is needed by policy managers, landowners, and society to promote biodiversity conservation of plant taxa in areas endangered by agriculture.

Traditionally, there are two main approaches when facing the difficult and challenging task of reconciling biodiversity conservation with agriculture: (i) to implement measures to achieve sustainable and wildlife-friendly agriculture [91]; and (ii) to increase agriculture intensification in some areas and to minimize new conversions of natural habitats to cultivated areas in others [91]. The first approach proposes the implementation of measures to enhance biodiversity in already existing crops and mainly favors taxa threatened by crop intensification and abandonment. The main problems for its implementation may be the over-cost of the measures and a decrease in crop yields, which could imply an increase in natural habitat conversion rates, being detrimental to taxa affected by crop expansion. Increased intensification, on the other hand is expected to reduce pressure for taxa threatened by crop expansion and to avoid new taxa being included in this category due to the expansion of agriculture in non-altered territories. Nevertheless, it does increase pressure for plants that coexist in agro-ecosystems.

Most of the taxa identified as threatened by agricultural use in continental Spain are threatened by agriculture extension, as there are many plants unable to adapt to any type of agricultural management [59]. For these taxa, respectful and less productive agriculture that implies a greater conversion to cultivation may suppose an additional risk and a sustainable and well-managed intensification, in which natural habitats are conserved and with regulated abandonment of some areas with a proper plan for restoration, could be appropriate [91]. The proposed solution for taxa threatened by crop extension may be to the detriment of those threatened for other reasons (i.e., crop intensification and abandonment). In these cases, it is necessary to implement measures aimed at improving biodiversity in intensified landscapes or in areas where abandonment of cultivation is a threat to plants. For intensified crops, some of the measures to promote biodiversity proposed in scientific literature are: the reduction of the intensification level [151], to promote complexity and heterogeneity of the area by diversifying the agricultural landscape [27,152,153], to increase crop heterogeneity [154], to conserve remnants of natural vegetation [155], to preserve the margins of cultivated fields [156], to conserve riparian vegetation [157], to maintain or create ecological corridors [109], to perform actions to maintain and to improve vegetation cover and diversity within the crops [27], to reduce the use of agrochemicals [130], to identify and conserve key taxa and ecosystem functions [136], and to create green infrastructures such as ponds, hedges or buffer strips [128,158,159]. In the case of those taxa whose threat is crop abandonment [160], general measures could be the identification and maintenance of agricultural landscapes with a high conservation value.

All of the listed measures can benefit threatened taxa, but sometimes they are not sufficient, and additional specific actions are needed [128,151]. Spanish legislation makes the development of recovery plans for endangered taxa mandatory that include measures designed for threatened taxa. However, these plans have rarely been implemented and in others they are developed too late [161]. Thus, more effort is needed in order to implement long-term monitoring programs and warning systems able to detect new impacts, the rarefaction of populations or to evaluate the conservation measures implemented at an early stage. In extreme cases (very small and isolated populations, under great pressure), it is also necessary to develop ex situ conservation programs [162]. With these programs, rescue populations can be established, with which to reintroduce or reinforce populations in the future and conserve genetic viability [162]. Scientific collections preserved in natural history museums and academic institutions play an important role in their ex situ conservation programs for threatened taxa [163] and are responsible for preserving specimens and seeds. Herbaria have been documented as useful resources for improving the genetic diversity of threatened flora as they contain viable seeds and sometimes unique alleles not present in current taxa [164]. In addition, historical records can be obtained almost exclusively from specimens preserved in herbaria, so herbaria are important when making extinction risk assessments of plants [165–167]. However, despite their usefulness, their contributions are widely underestimated by both society and administrations [168] and are in crisis due to the reduction of resources [169]. As an additional recommendation, seeds of threatened species need to be conserved in germplasm banks and natural history collections should continue to be supported with funds and personnel.

In summary, conservation measures exist to promote biodiversity in agricultural landscapes, although few are specific for threatened flora. Moreover, it has been demonstrated that in most situations the adoption of these sustainable practices by farmers depends on incentives that provide a short-term economic benefit [170], which signifies a big effort for the different administrations and frequently only retard biodiversity loss [171]. Indeed, despite all global efforts for preserving global biodiversity, the sustainability gap is growing rather than closing [172], and many new species are threatened every year by the increase in agricultural land to guarantee food security for the global [173] population. Paradoxically, only two-thirds of the food produced in the world is consumed, and 14% of the losses occur in the post-harvest stages [174]. An illustrative example is that 114 kt of fruits and vegetables were discarded in Spain in 2009 [175]. Therefore, if biodiversity conservation, responsible consuming and the achievement of a sustainable production system is the goal, it is timely to promote a deep transformation of our social–ecological systems.

One such transformative shift could come through the reconnection with nature [176]. In recent years, there has been a significant increase in research that supports the need to strengthen human–nature connections (HNC) in agroecosystems to foster environmental and socio-cultural sustainability in agricultural landscapes [177–179]. This promotes the establishment of belonging, stewardship, and connections to nature [179]; thus, providing the social support that is needed to make agriculture and the protection of endangered flora compatible. Indeed, it has been demonstrated that links between nature and people may be more important for biodiversity conservation than indirect links based on incentive payments [143]. Even so, there is a general problem: at the societal level, little empathy has been detected for plants in relation to other biological groups, such as animals, a phenomenon known as "plant blindness" [180]. According to the leverage point hypothesis, the HNC can be approached from five dimensions [181]: material connections, experiential connections, cognitive connections, emotional attachments and philosophical perspectives. Most previous experiences in this line are focused on providing extra income to farmers and in to increase experience of population in agroecosystems, mainly achieving material and experiential connections. However, in order to achieve a real transformation to improve the emotional attachments, and the perspective that society has about what nature is, why it matters, and how humans ought to interact with it (philosophical perspective) would be more efficient. To deepen these connections, environmental education can be an important tool [182]. With environmental education, society can be made aware of the threatened taxa present in agricultural landscapes, their importance, and their threats. With experiences such as agrotourism, supported by environmental education, it is also possible to deepen the emotional and philosophical reconnection, and get consumers to decide to pay a little more for products grown in production systems that respect the environment and threatened plant species [175].

Regardless of the type of measure that we can implement for biodiversity conservation in agricultural areas, it should be a priority for society to be aware of the added value of biodiversity and the presence of endangered species in agricultural environments, and to promote their conservation. Therefore, reconnecting society with nature through agriculture is a challenge today and can be an effective tool to achieve better protection of threatened taxa in cultivated landscapes. This reconversion process must be accompanied by conservation support from the competent administrations and institutions. Moreover these institutions should promote the application of transdisciplinary and collaborative processes in which science, policy making, and society should work together to promote evidence-based biodiversity conservation practices [183]. For example, when developing land use policies, it is advisable to carry out exploratory studies involving different social actors working together in order to discuss potential solutions for the biodiversity crisis and to contribute toward improving the efficiency of policy instruments that will be reflected in later phases [184].

5. Conclusions

Agriculture-related activity causes negative impacts on threatened flora in continental Spain, mainly due to the crops extension, but also to the crop intensification or crop abandonment.

In Spain, the global extension of crops shows a generalized decrease during the last three decades. Nevertheless, when studied in detail, there are significant changes in the areas occupied by the different crops, which reflects an important pace of crop extension that occurs in parallel to the abandonment of agriculture and the shift towards more intensive practices.

The agricultural use of the territory and the biodiversity conservation are possible. For these, it is necessary to reduce and change consumption habits, to carry out rational land planning in which natural habitats are maintained, and to achieve a sustainable production system, in which specific measures for endangered flora are applied. These measurements may benefit from data within scientific collections, as these allow for the assessment of the loss of populations of threatened plant taxa and, in turn, facilitate the sustainable planning of the territory in which they are found.

Finally, to favor the conservation of flora threatened by agricultural use, it is necessary to promote a profound transformation of our socio-ecological systems. The most effective way to achieve it is the human-nature reconnection.

Supplementary Materials: The following are available online at https://www.mdpi.com/article/10.3390/agriculture11111097/s1, Table S1. Reclassification of land uses from CORINE land cover.

Author Contributions: Conceptualization, J.L.M.-P., E.G.-L., E.R.-C. and M.C.; Validation, Formal Analysis and Investigation, J.L.M.-P., E.G.-L. and E.R.-C.; Resources, E.G.-L.; Writing, J.L.M.-P.; Visualization, J.L.M.-P., E.G.-L., E.R.-C., M.C., P.B., M.S.-R. and A.L.-A.; Supervision, E.G.-L. and E.R.-C. All authors have read and agreed to the published version of the manuscript.

Funding: Not applicable.

Conflicts of Interest: The authors declare no conflict of interest.

References

1. Ellis, E.C. Anthropogenic transformation of the terrestrial biosphere. *Philos. Trans. R. Soc. A Math. Phys. Eng. Sci.* **2011**, *369*, 1010–1035. [CrossRef]
2. Foley, J.A.; DeFries, R.; Asner, G.P.; Barford, C.; Bonan, G.; Carpenter, S.R.; Chapin, F.S.; Coe, M.T.; Daily, G.C.; Gibbs, H.K.; et al. Global Consequences of Land Use. *Science* **2005**, *309*, 570–574. [CrossRef]
3. Ellis, E.C. Anthromes. In *Encyclopedia of the World's Biomes*; Elsevier: Amsterdam, The Netherlands, 2020; pp. 5–11.
4. Tilman, D.; Fargione, J.; Wolff, B.; D'Antonio, C.; Dobson, A.; Howarth, R.; Schindler, D.; Schlesinger, W.H.; Simberloff, D.; Swackhamer, D. Forecasting Agriculturally Driven Global Environmental Change. *Science* **2001**, *292*, 281–284. [CrossRef]
5. Steffen, W.; Richardson, K.; Rockström, J.; Cornell, S.E.; Fetzer, I.; Bennett, E.M.; Biggs, R.; Carpenter, S.R.; de Vries, W.; Wit, C.A.; et al. Planetary boundaries: Guiding human development on a changing planet. *Science* **2015**, *347*. [CrossRef]
6. Ramankutty, N.; Evan, A.T.; Monfreda, C.; Foley, J.A. Farming the planet: 1. Geographic distribution of global agricultural lands in the year 2000. *Global Biogeochem. Cycles* **2008**, *22*, 1003. [CrossRef]
7. Foley, J.A.; Ramankutty, N.; Brauman, K.A.; Cassidy, E.S.; Gerber, J.S.; Johnston, M.; Mueller, N.D.; O'Connell, C.; Ray, D.K.; West, P.C.; et al. Solutions for a cultivated planet. *Nature* **2011**, *478*, 337–342. [CrossRef] [PubMed]
8. Vitousek, P.M.; Mooney, H.A.; Lubchenco, J.; Melillo, J.M. Human Domination of Earth's Ecosystems. *Science* **1997**, *277*, 494–499. [CrossRef]
9. Liu, X.B.; Zhang, X.Y.; Wang, Y.X.; Suil, Y.Y.; Zhangl, S.L.; Herbert, S.; Ding, G. Soil degradation: A problem threatening the sustainable development of agriculture in Northeast China. *Plant Soil Environ.* **2010**, *56*, 87–97. [CrossRef]
10. Rodríguez, D. Greenhouse gas emissions of agriculture: A comparative analysis. In *Environmental Chemistry and Recent Pollution Control Approaches*; Saldarriaga-Noreña, H., Murillo-Tovar, M., Farooq, R., Dongre, R., Riaz, S., Eds.; IntechOpen: London, UK, 2019; pp. 21–40, ISBN 978-1-83968-063-2.
11. Moss, B. Water pollution by agriculture. *Philos. Trans. R. Soc. B Biol. Sci.* **2007**, *363*, 659–666. [CrossRef]
12. Peacock, E.; Haag, W.R.; Warren, M.L. Prehistoric Decline in Freshwater Mussels Coincident with the Advent of Maize Agriculture. *Conserv. Biol.* **2005**, *19*, 547–551. [CrossRef]
13. Böhlke, J.-K. Groundwater recharge and agricultural contamination. *Hydrogeol. J.* **2002**, *10*, 153–179. [CrossRef]
14. Carvalho, F.P. Agriculture, pesticides, food security and food safety. *Environ. Sci. Policy* **2006**, *9*, 685–692. [CrossRef]
15. Kyrikou, I.; Briassoulis, D. Biodegradation of Agricultural Plastic Films: A Critical Review. *J. Polym. Environ.* **2007**, *15*, 125–150. [CrossRef]
16. Baldock, D. *Agriculture and Habitat Loss in Europe*; WWF International: Gland, Switzerland, 1990; ISBN 2880850347.
17. Campbell, B.M.; Beare, D.J.; Bennett, E.M.; Hall-Spencer, J.M.; Ingram, J.S.; Jaramillo, F. Agriculture production as a major driver of the Earth system exceeding planetary boundaries. *Ecol. Soc.* **2017**, *22*. [CrossRef]
18. Burney, J.A.; Davis, S.J.; Lobell, D.B. Greenhouse gas mitigation by agricultural intensification. *Proc. Natl. Acad. Sci. USA* **2010**, *107*, 12052–12057. [CrossRef]
19. Tilman, D.; Balzer, C.; Hill, J.; Befort, B.L. Global food demand and the sustainable intensification of agriculture. *Proc. Natl. Acad. Sci. USA* **2011**, *108*, 20260–20264. [CrossRef] [PubMed]
20. Ut, T.T.; Kajisa, K. The impact of green revolution on rice production in Vietnam. *Dev. Econ.* **2006**, *44*, 167–189. [CrossRef]
21. Evenson, R.E.; Gollin, D. Assessing the Impact of the Green Revolution, 1960 to 2000. *Science* **2003**, *300*, 758–762. [CrossRef]

22. FAO. *Construyendo una Visión Común Para la Agricultura y Alimentación Sostenibles. Principios y Enfoques*; FAO: Rome, Italy, 2015.
23. Matson, P.A.; Parton, W.J.; Power, A.G.; Swift, M.J. Agricultural Intensification and Ecosystem Properties. *Science* **1997**, *277*, 504–509. [CrossRef] [PubMed]
24. Pingali, P.L. Green Revolution: Impacts, limits, and the path ahead. *Proc. Natl. Acad. Sci. USA* **2012**, *109*, 12302–12308. [CrossRef]
25. Eisenstein, M. Natural solutions for agricultural productivity. *Nature* **2020**, *588*, S58–S59. [CrossRef] [PubMed]
26. McNeely, J. How traditional agro-ecosystems can contribute to conserving biodiversity. In *Conserving Biodiversity Outside Protected Area the Role of Traditional Agro-Ecosystems*; Halladay, P., Gilmour, D.A., Eds.; IUCN Publishing Unit: Cambridge, UK, 1995.
27. Rey, P.J.; Manzaneda, A.J.; Valera, F.; Alcántara, J.M.; Tarifa, R.; Isla, J.; Molina-Pardo, J.L.; Calvo, G.; Salido, T.; Gutiérrez, J.E.; et al. Landscape-moderated biodiversity effects of ground herb cover in olive groves: Implications for regional biodiversity conservation. *Agric. Ecosyst. Environ.* **2019**, *277*, 61–73. [CrossRef]
28. Krebs, J.R.; Wilson, J.D.; Bradbury, R.B.; Siriwardena, G.M. The second Silent Spring? *Nature* **1999**, *400*, 611–612. [CrossRef]
29. Martínez-Valderrama, J.; Guirado, E.; Maestre, F.T. Unraveling Misunderstandings about Desertification: The Paradoxical Case of the Tabernas-Sorbas Basin in Southeast Spain. *Land* **2020**, *9*, 269. [CrossRef]
30. Flynn, D.F.B.; Gogol-Prokurat, M.; Nogeire, T.; Molinari, N.; Richers, B.T.; Lin, B.B.; Simpson, N.; Mayfield, M.M.; DeClerck, F. Loss of functional diversity under land use intensification across multiple taxa. *Ecol. Lett.* **2009**, *12*, 22–33. [CrossRef]
31. Wagner, D.L.; Grames, E.M.; Forister, M.L.; Berenbaum, M.R.; Stopak, D. Insect decline in the Anthropocene: Death by a thousand cuts. *Proc. Natl. Acad. Sci. USA* **2021**, *118*. [CrossRef]
32. Bennett, E.M.; Baird, J.; Baulch, H.; Chaplin-Kramer, R.; Fraser, E.; Loring, P.; Morrison, P.; Parrott, L.; Sherren, K.; Winkler, K.J.; et al. Ecosystem services and the resilience of agricultural landscapes. *Adv. Ecol. Res.* **2021**, *64*, 1–43. [CrossRef]
33. Martínez-Núñez, C.; Manzaneda, A.J.; Isla, J.; Tarifa, R.; Calvo, G.; Molina, J.L.; Salido, T.; Ruiz, C.; Gutiérrez, J.E.; Rey, P.J. Low-intensity management benefits solitary bees in olive groves. *J. Appl. Ecol.* **2020**, *57*, 111–120. [CrossRef]
34. Martínez-Núñez, C.; Rey, P.J.; Manzaneda, A.J.; García, D.; Tarifa, R.; Molina, J.L. Insectivorous birds are not effective pest control agents in olive groves. *Basic Appl. Ecol.* **2021**, *56*, 270–280. [CrossRef]
35. Dainese, M.; Martin, E.A.; Aizen, M.A.; Albrecht, M.; Bartomeus, I.; Bommarco, R.; Carvalheiro, L.G.; Chaplin-Kramer, R.; Gagic, V.; Garibaldi, L.A.; et al. A global synthesis reveals biodiversity-mediated benefits for crop production. *Sci. Adv.* **2019**, *5*, eaax0121. [CrossRef]
36. Millard, J.; Outhwaite, C.L.; Kinnersley, R.; Freeman, R.; Gregory, R.D.; Adedoja, O.; Gavini, S.; Kioko, E.; Kuhlmann, M.; Ollerton, J.; et al. Global effects of land-use intensity on local pollinator biodiversity. *Nat. Commun.* **2021**, *12*, 1–11. [CrossRef] [PubMed]
37. McKechnie, I.M.; Sargent, R.D. Do plant traits influence a species' response to habitat disturbance? A meta-analysis. *Biol. Conserv.* **2013**, *168*, 69–77. [CrossRef]
38. Martínez-Núñez, C.; Rey, P.J.; Salido, T.; Manzaneda, A.J.; Camacho, F.M.; Isla, J. Ant community potential for pest control in olive groves: Management and landscape effects. *Agric. Ecosyst. Environ.* **2021**, *305*, 107185. [CrossRef]
39. Potts, S.G.; Imperatriz-Fonseca, V.; Ngo, H.T.; Aizen, M.A.; Biesmeijer, J.C.; Breeze, T.D.; Dicks, L.V.; Garibaldi, L.A.; Hill, R.; Settele, J.; et al. Safeguarding pollinators and their values to human well-being. *Nature* **2016**, *540*, 220–229. [CrossRef] [PubMed]
40. Díaz, S.; Settele, J.; Brondízio, E.S.; Ngo, H.T.; Agard, J.; Arneth, A.; Balvanera, P.; Brauman, K.A.; Butchart, S.H.M.; Chan, K.M.A.; et al. Pervasive human-driven decline of life on Earth points to the need for transformative change. *Science* **2019**, *366*. [CrossRef]
41. Chamberlain, D.E.; Fuller, R.J.; Bunce, R.G.H.; Duckworth, J.C.; Shrubb, M. Changes in the abundance of farmland birds in relation to the timing of agricultural intensification in England and Wales. *J. Appl. Ecol.* **2000**, *37*, 771–788. [CrossRef]
42. Donald, P.F.; Green, R.E.; Heath, M.F. Agricultural intensification and the collapse of Europe's farmland bird populations. *Proc. R. Soc. London. Ser. B Biol. Sci.* **2001**, *268*, 25–29. [CrossRef]
43. Emmerson, M.; Morales, M.B.; Oñate, J.J.; Batáry, P.; Berendse, F.; Liira, J.; Aavik, T.; Guerrero, I.; Bommarco, R.; Eggers, S.; et al. How Agricultural Intensification Affects Biodiversity and Ecosystem Services. *Adv. Ecol. Res.* **2016**, *55*, 43–97. [CrossRef]
44. Wright, H.L.; Lake, I.R.; Dolman, P.M. Agriculture—A key element for conservation in the developing world. *Conserv. Lett.* **2012**, *5*, 11–19. [CrossRef]
45. Hobbs, P.R.; Sayre, K.; Gupta, R. The role of conservation agriculture in sustainable agriculture. *Philos. Trans. R. Soc. B Biol. Sci.* **2007**, *363*, 543–555. [CrossRef]
46. Altieri, M. *Agroecology: The Science of Sustainable Agriculture*; CRC Press: Boca Raton, FL, USA, 2018.
47. Thrupp, L.A. Linking Agricultural Biodiversity and Food Security: The Valuable Role of Agrobiodiversity for Sustainable Agriculture. *Int. Aff.* **2000**, *76*, 265–281. [CrossRef]
48. Diacono, M.; Trinchera, A.; Montemurro, F. An Overview on Agroecology and Organic Agriculture Strategies for Sustainable Crop Production. *Agron* **2021**, *11*, 223. [CrossRef]
49. Batáry, P.; Dicks, L.V.; Kleijn, D.; Sutherland, W.J. The role of agri-environment schemes in conservation and environmental management. *Conserv. Biol.* **2015**, *29*, 1006–1016. [CrossRef]
50. Wade, M.R.; Gurr, G.M.; Wratten, S.D. Ecological restoration of farmland: Progress and prospects. *Philos. Trans. R. Soc. B Biol. Sci.* **2007**, *363*, 831–847. [CrossRef]
51. Perrings, C.; Jackson, L.; Bawa, K.; Brussaard, L.; Brush, S.; Gavin, T.; Papa, R.; Pascual, U.; Ruiter, P. De Biodiversity in Agricultural Landscapes: Saving Natural Capital without Losing Interest. *Conserv. Biol.* **2006**, *20*, 263–264. [CrossRef] [PubMed]

52. Fischer, J.; Lindenmayer, D.B.; Manning, A.D. Biodiversity, ecosystem function, and resilience: Ten guiding principles for commodity production landscapes. *Front. Ecol. Environ.* **2006**, *4*, 80–86. [CrossRef]
53. Nabel, M.; Selig, C.; Gundlach, J.; Decken, H.; Klein, M. Biodiversity in agricultural used soils: Threats and options for its conservation in Germany and Europe. *Soil Org.* **2021**, *93*, 1–11. [CrossRef]
54. Santos, J.S.; Dodonov, P.; Oshima, J.E.F.; Martello, F.; Santos de Jesus, A.; Eduardo Ferreira, M.; Silva-Neto, C.M.; Ribeiro, M.C.; Collevatti, R.G. Landscape ecology in the Anthropocene: An overview for integrating agroecosystems and biodiversity conservation. *Perspect. Ecol. Conserv.* **2021**, *19*, 21–32. [CrossRef]
55. Singh, M.; Kumar, A.; Sharma, M. Conservation of Plant Diversity in Agroforestry Systems in a Biodiversity Hotspot Region of Northeast India. *Agric. Res.* **2021**, 1–13. [CrossRef]
56. Košulič, O.; Michalko, R.; Hula, V. Recent artificial vineyard terraces as a refuge for rare and endangered spiders in a modern agricultural landscape. *Ecol. Eng.* **2014**, *68*, 133–142. [CrossRef]
57. Blanca, G.; Cueto, M.; Fuentes, J.; Sáez, L.; Tarifa, R. *Linaria qartobensis* sp. nov. (Plantaginaceae) from the southern Iberian Peninsula. *J. Bot.* **2018**, *36*, e01914. [CrossRef]
58. Storkey, J.; Meyer, S.; Still, K.S.; Leuschner, C. The impact of agricultural intensification and land-use change on the European arable flora. *Proc. R. Soc. B Biol. Sci.* **2012**, *279*, 1421–1429. [CrossRef]
59. Lenzen, M.; Lane, A.; Widmer-Cooper, A.; Williams, M. Effects of Land Use on Threatened Species. *Conserv. Biol.* **2009**, *23*, 294–306. [CrossRef]
60. Richner, N.; Holderegger, R.; Linder, H.P.; Walter, T. Reviewing change in the arable flora of Europe: A meta-analysis. *Weed Res.* **2015**, *55*, 1–13. [CrossRef]
61. Osawa, T.; Kohyama, K.; Mitsuhashi, H. Areas of Increasing Agricultural Abandonment Overlap the Distribution of Previously Common, Currently Threatened Plant Species. *PLoS ONE* **2013**, *8*, e79978. [CrossRef]
62. Bañares, A.; Blanca, G.; Guemes, J.; Moreno, J.; Ortiz, S. *Atlas y Libro Rojo de la Flora Vascular Amenazada de España*; Dirección General de Conservación de la Naturaleza: Madrid, Spain, 2004; ISBN 84-8014-521-8.
63. Stoate, C.; Báldi, A.; Beja, P.; Boatman, N.D.; Herzon, I.; van Doorn, A.; de Snoo, G.R.; Rakosy, L.; Ramwell, C. Ecological impacts of early 21st century agricultural change in Europe–A review. *J. Environ. Manag.* **2009**, *91*, 22–46. [CrossRef]
64. Alignier, A.; Solé-Senan, X.O.; Robleño, I.; Baraibar, B.; Fahrig, L.; Giralt, D.; Gross, N.; Martin, J.-L.; Recasens, J.; Sirami, C.; et al. Configurational crop heterogeneity increases within-field plant diversity. *J. Appl. Ecol.* **2020**, *57*, 654–663. [CrossRef]
65. Fried, G.; Kazakou, E.; Gaba, S. Trajectories of weed communities explained by traits associated with species' response to management practices. *Agric. Ecosyst. Environ.* **2012**, *158*, 147–155. [CrossRef]
66. Zanin, G.; Otto, S.; Riello, L. Ecological interpretation of weed flora dynamics under different tillage systems. *Agric. Ecosyst. Environ.* **1997**, *66*, 77–188. [CrossRef]
67. Cornelissen, J.H.C.; Lavorel, S.; Garnier, E.; Díaz, S.; Buchmann, N.; Gurvich, D.E.; Reich, P.B.; Ter Steege, H.; Morgan, H.D.; Van Der Heijden, M.G.A.; et al. A handbook of protocols for standardised and easy measurement of plant functional traits worldwide. *Aust. J. Bot.* **2003**, *51*, 335–380. [CrossRef]
68. Renwick, A.; Jansson, T.; Verburg, P.H.; Revoredo-Giha, C.; Britz, W.; Gocht, A.; McCracken, D. Policy reform and agricultural land abandonment in the EU. *Land Use Policy* **2013**, *30*, 446–457. [CrossRef]
69. Traba, J.; Morales, M.B. The decline of farmland birds in Spain is strongly associated to the loss of fallowland. *Sci. Rep.* **2019**, *9*, 1–6. [CrossRef] [PubMed]
70. Moreno-Saiz, J.C.; Albertos, B.; Ruiz-Molero, E.; Mateo, R.G. The European Union can afford greater ambition in the conservation of its threatened plants. *Biol. Conserv.* **2021**, *261*, 109231. [CrossRef]
71. Bañares, A.; Blanca, G.; Güemes, J.; Moreno, C.; Ortiz, S. *Atlas y Libro Rojo de la Flora Vascular Amenazada de España-Adenda 2006*; Dirección General para la Biodiversidad-Sociedad Española de Biología de la Conservación de Plantas: Madrid, Spain, 2006; ISBN 978-84-8014-706-4.
72. Bañares, A.; Blanca, G.; Güemes, J.; Moreno, C.; Ortiz, S. *Atlas y Libro Rojo de la Flora Vascular Amenazada de España-Adenda 2008*; Dirección General de Medio Natural y Política Forestal (Ministerio de Medio Ambiente, y Medio Rural y Marino)-Sociedad Española de Biología de la Conservación de Plantas: Madrid, Spain, 2009; ISBN 978-84-8014-741-5.
73. Bañares, A.; Blanca, G.; Güemes, J.; Moreno, C.; Ortiz, S. *Atlas y Libro Rojo de la Flora Vascular Amenazada de España-Adenda 2010*; Dirección General de Medio Natural y Política Forestal (Ministerio de Medio Ambiente, y Medio Rural y Marino)-Sociedad Española de Biología de la Conservación de Plantas: Madrid, Spain, 2010.
74. Moreno, J.C.; Iriondo, J.M.; Martínez, F.; Martínez, J.; Salazar, C. *Atlas y Libro Rojo de la Flora Vascular Amenazada de España-Adenda 2017*; Ministerio para la Transición Ecológica-Sociedad Española de Biología de la Conservación de Plantas: Madrid, Spain, 2017.
75. Martínez, I.; Carreño, F.; Escudero, A.; Rubio, A. Are threatened lichen species well-protected in Spain? Effectiveness of a protected areas network. *Biol. Conserv.* **2006**, *133*, 500–511. [CrossRef]
76. Médail, F.; Quézel, P. Hot-spots analysis for conservation of plant biodiversity in the Mediterranean Basin. *Ann. Missouri Bot. Gard.* **1997**, *84*, 112–127. [CrossRef]
77. Myers, N.; Mittermeier, R.A.; Mittermeier, C.G.; da Fonseca, G.A.B.; Kent, J. Biodiversity hotspots for conservation priorities. *Nature* **2000**, *403*, 853–858. [CrossRef]
78. Benayas, R.; José, M.; Scheiner, S.M. Plant diversity, biogeography and environment in Iberia: Patterns and possible causal factors. *J. Veg. Sci.* **2002**, *13*, 245–258. [CrossRef]

79. Underwood, E.C.; Viers, J.H.; Klausmeyer, K.R.; Cox, R.L.; Shaw, M.R. Threats and biodiversity in the mediterranean biome. *Divers. Distrib.* **2009**, *15*, 188–197. [CrossRef]
80. Fernández-González, F.; Loidi, J.; Moreno, C.J.; Del Arco, M.; Fernández-Cancio, A. Impactos sobre la biodiversidad vegetal. In *Evaluación Preliminar de los Impactos en España por Efecto del Cambio Climático*; Centro de Publicaciones. Secretaría General Técnica, Ministerio de Medio Ambiente: Madrid, Spain, 2005; pp. 18–247, ISBN 84-8320-303-0.
81. Moreno Saiz, J.C. La diversidad florística vascular española. In *Biodiversidad: Aproximación a la Diversidad Botánica y Zoológica de España. Memorias de la Real Sociedad Española de Historia Natural*; Real Sociedad Española de Historia Natural: Madrid, Spain, 2011; Volume 9, pp. 75–107, ISBN 978-84-936677-6-4.
82. Moreno, J.C. *Lista Roja 2008 de la Flora Vascular Española*; Dirección General de Medio Natural y Política Forestal (Ministerio de Medio Ambiente, y Medio Rural y Marino, y Sociedad Española de Biología de la Conservación de Plantas): Madrid, Spain, 2008; ISBN 978-84-691-7375-6.
83. Algarra, J.; Blanca, G.; Cueto, M.; Phytotaxa, J.F.-U. New data on daffodils of the Narcissus nevadensis complex (Amaryllidaceae) in SE Spain: N. nevadensis subsp. herrerae subsp. nov., and N. nevadensis. *Phytotaxa* **2018**, *371*, 133–139. [CrossRef]
84. Molina, J.; Michaud, H.; Tison, J.; Fernandez Zamudio, R.; Véla, E. Allium scaberrimum. The IUCN Red List of Threatened Species. 2018. Available online: https://dx.doi.org/10.2305/IUCN.UK.2018-1.RLTS.T110805790A87775132.en (accessed on 27 August 2021).
85. Gutiérrez-Larruscain, D.; Santos-Vicente, M.; Anderberg, A.A.; Rico, E.; Martínez-Ortega, M.M. Phylogeny of the Inula group (Asteraceae: Inuleae): Evidence from nuclear and plastid genomes and a recircumscription of Pentanema. *Taxon* **2018**, *67*, 149–164. [CrossRef]
86. Pelser, P.B.; Veldkamp, J.F.; Van Der Meijden, R. New combinations in Jacobaea Mill. (Asteraceae, Senecioneae). *Compo News* **2006**, *44*, 1–11.
87. Al-Shehbaz, I.A. A generic and tribal synopsis of the Brassicaceae (Cruciferae). *Taxon* **2012**, *61*, 931–954. [CrossRef]
88. Maldonado, C.; Molina, C.I.; Zizka, A.; Persson, C.; Taylor, C.M.; Albán, J.; Chilquillo, E.; Rønsted, N.; Antonelli, A. Estimating species diversity and distribution in the era of Big Data: To what extent can we trust public databases? *Glob. Ecol. Biogeogr.* **2015**, *24*, 973–984. [CrossRef] [PubMed]
89. Tarifa, R.; Martínez-Núñez, C.; Valera, F.; González-Varo, J.P.; Salido, T.; Rey, P.J. Agricultural intensification erodes taxonomic and functional diversity in Mediterranean olive groves by filtering out rare species. *J. Appl. Ecol.* **2021**, *58*, 2266–2276. [CrossRef]
90. Raunkiaer, C. *The Life Forms of Plants and Statistical Plant Geography; Being the Collected Papers of C. raunkiaer*; Claredon Press: Oxford, UK, 1934.
91. Green, R.E.; Cornell, S.J.; Scharlemann, J.P.W.; Balmford, A. Farming and the Fate of Wild Nature. *Science* **2005**, *307*, 550–555. [CrossRef] [PubMed]
92. Pinke, G.; Király, G.; Barina, Z.; Mesterházy, A.; Balogh, L.; Csiky, J.; Schmotzer, A.; Molnár, A.V.; Pál, R.W. Assessment of endangered synanthropic plants of Hungary with special attention to arable weeds. *Plant Biosyst.* **2011**, *145*, 426–435. [CrossRef]
93. Bergmeier, E.; Strid, A. Regional diversity, population trends and threat assessment of the weeds of traditional agriculture in Greece. *Bot. J. Linn. Soc.* **2014**, *175*, 607–623. [CrossRef]
94. Lozano, F.D.; Atkins, K.J.; Moreno Sáiz, J.C.; Sims, A.E.; Dixon, K. The nature of threat category changes in three Mediterranean biodiversity hotspots. *Biol. Conserv.* **2013**, *157*, 21–30. [CrossRef]
95. Carman, K.; Jenkins, D.G. Comparing diversity to flower-bee interaction networks reveals unsuccessful foraging of native bees in disturbed habitats. *Biol. Conserv.* **2016**, *202*, 110–118. [CrossRef]
96. Erfanian, M.B.; Ejtehadi, H.; Vaezi, J.; Moazzeni, H. Plant community responses to multiple disturbances in an arid region of northeast Iran. *L. Degrad. Dev.* **2019**, *30*, 1554–1563. [CrossRef]
97. Basu, P.; Bhattacharya, R.; Ianetta, P. A decline in pollinator dependent vegetable crop productivity in India indicates pollination limitation and consequent agro-economic crises. *Nat. Preced.* **2011**. [CrossRef]
98. Gallai, N.; Salles, J.M.; Settele, J.; Vaissière, B.E. Economic valuation of the vulnerability of world agriculture confronted with pollinator decline. *Ecol. Econ.* **2009**, *68*, 810–821. [CrossRef]
99. Klein, A.-M.; Vaissire, B.E.; Cane, J.H.; Steffan-Dewenter, I.; Cunningham, S.A.; Kremen, C.; Tscharntke, T. Importance of pollinators in changing landscapes for world crops. *Proc. R. Soc. B Biol. Sci.* **2006**, *274*, 303–313. [CrossRef]
100. Velthuis, H.H.W.; Doorn, A. van A century of advances in bumblebee domestication and the economic and environmental aspects of its commercialization for pollination. *Apidologie* **2006**, *37*, 421–451. [CrossRef]
101. Carreck, N.L.; Williams, I.H.; Little, D.J. The movement of honey bee colonies for crop pollination and honey production by beekeepers in Great Britain. *Bee World* **2015**, *78*, 67–77. [CrossRef]
102. Németh, M.B.; Smith-Huerta, N.L. Pollen Deposition, Pollen Tube Growth, Seed Production, and Seedling Performance in Natural Populations of *Clarkia unguiculata* (Onagraceae). *Int. J. Plant Sci.* **2015**, *164*, 153–164. [CrossRef]
103. Kalla, S.E.; Ashman, T.L. The effects of pollen competition on progeny vigor in Fragaria virginiana (Rosaceae) depend on progeny growth environment. *Int. J. Plant Sci.* **2002**, *163*, 335–340. [CrossRef]
104. Brown, E.; Kephart, S. Variability in pollen load: Implications for reproduction and seedling vigor in a rare plant, Silene douglasii var. Oraria. *Int. J. Plant Sci.* **1999**, *160*, 1145–1152. [CrossRef] [PubMed]
105. Bertin, R.I. Effects of pollination intensity in Campsis radicans. *Am. J. Bot.* **1990**, *77*, 178–187. [CrossRef] [PubMed]

106. Chautá-Mellizo, A.; Campbell, S.A.; Bonilla, M.A.; Thaler, J.S.; Poveda, K. Effects of natural and artificial pollination on fruit and offspring quality. *Basic Appl. Ecol.* **2012**, *13*, 524–532. [CrossRef]
107. Brittain, C.; Potts, S.G. The potential impacts of insecticides on the life-history traits of bees and the consequences for pollination. *Basic Appl. Ecol.* **2011**, *12*, 321–331. [CrossRef]
108. Franzon, R.C.; Castro, C.M.; Raseira, M. do C.B. Genetic variability in surinam cherry populations originated from self-pollination and cross pollination, estimated by AFLP. *Rev. Bras. Frutic.* **2010**, *32*, 240–250. [CrossRef]
109. Poschlod, P.; Bakker, J.; Bonn, S.; Fischer, S. Dispersal of Plants in Fragmented Landscapes. In *Species Survival in Fragmented Landscapes*; Settele, C.J., Margules, P., Poschlod, K., Eds.; Henle: Dordrecht, The Netherlands, 1996; pp. 123–127, ISBN 978-94-009-0343-2.
110. Poschlod, P.; Bonn, S. Changing dispersal processes in the central European landscape since the last ice age: An explanation for the actual decrease of plant species richness in different habitats? *Acta Bot. Neerl.* **1998**, *47*, 27–44.
111. Cain, M.L.; Milligan, B.G.; Strand, A.E. Long-distance seed dispersal in plant populations. *Am. J. Bot.* **2000**, *87*, 1217–1227. [CrossRef]
112. Vittoz, P.; Engler, R. Seed dispersal distances: A typology based on dispersal modes and plant traits. *Bot. Helv.* **2007**, *117*, 109–124. [CrossRef]
113. Valdivia, A.; Wolf, S.; Suckling, K. Marine mammals and sea turtles listed under the U.S. Endangered Species Act are recovering. *PLoS ONE* **2019**, *14*, e0210164. [CrossRef]
114. Evans, D.; Che-Castaldo, J.; Crouse, D.; Davis, F.; Epanchin-Niell, R.; Flather, C.; Frohlich, R.; Goble, D.; Li, Y.; Male, T.; et al. Species recovery in the united states: Increasing the effectiveness of the Endangered Species Act. *Issues Ecol.* **2016**, *20*, 1–28.
115. Antonelli, A.; Smith, R.J.; Fry, C.; Simmonds, M.S.J.; Kersey, P.J.; Pritchard, H.W.; Abbo, M.S.; Acedo, C.; Adams, J. *State of the World's Plants and Fung 2020*; Royal Botanic Gardens: London, UK, 2020.
116. Wilson, P.; Aebischer, N.J. The distribution of dicotyledonous arable weeds in relation to distance from the field edge. *J. Appl. Ecol.* **1995**, 295–310. [CrossRef]
117. Mota, J.F.; Peñas, J.; Castro, H.; Cabello, J.; Guirado, J.S. Agricultural development vs biodiversity conservation: The Mediterranean semiarid vegetation in El Ejido (Almería, southeastern Spain). *Biodivers. Conserv.* **1996**, *5*, 1597–1617. [CrossRef]
118. Wada, Y.; Beek, L.P.H.; Bierkens, M.F.P. Nonsustainable groundwater sustaining irrigation: A global assessment. *Water Resour. Res.* **2012**, *48*, 6. [CrossRef]
119. Gherardi, F. Crayfish invading Europe: The case study of *Procambarus clarkii*. *Mar. Freshw. Behav. Physiol.* **2007**, *39*, 175–191. [CrossRef]
120. Cruz, M.J.; Rebelo, R. Colonization of freshwater habitats by an introduced crayfish, Procambarus clarkii, in Southwest Iberian Peninsula. *Hydrobiology* **2006**, *575*, 191–201. [CrossRef]
121. Gherardi, F.; Barbaresi, S. Invasive crayfish: Activity patterns of Procambarus clarkii in the rice fields of the Lower Guadalquivir (Spain). *Arch. Hydrobiol.* **2000**, 153–168. [CrossRef]
122. Gherardi, F.; Acquistapace, P. Invasive crayfish in Europe: The impact of Procambarus clarkii on the littoral community of a Mediterranean lake. *Freshw. Biol.* **2007**, *52*, 1249–1259. [CrossRef]
123. Cruz, M.J.; Rebelo, R.; Crespo, E.G. Effects of an introduced crayfish, Procambarus clarkii, on the distribution of south-western Iberian amphibians in their breeding habitats. *Ecography (Cop.)* **2006**, *29*, 329–338. [CrossRef]
124. Kelemen, K.; Kriván, A.; Standovár, T. Effects of land-use history and current management on ancient woodland herbs in Western Hungary. *J. Veg. Sci.* **2014**, *25*, 172–183. [CrossRef]
125. Gilmour, D.; Halladay, P. *Conserving Biodiversity Outside Protected Areas. The Role of Traditional Agro-Ecosystems*; Gilmour, D., Halladay, P., Eds.; IUCN: Gland, Switzerland; Cambridge, UK, 1995.
126. Tissier, M.L.; Kletty, F.; Handrich, Y.; Habold, C. Monocultural sowing in mesocosms decreases the species richness of weeds and invertebrates and critically reduces the fitness of the endangered European hamster. *Oecologia* **2017**, *186*, 589–599. [CrossRef]
127. Marshall, E.J.P.; Brown, V.K.; Boatman, N.D.; Lutman, P.J.W.; Squire, G.R.; Ward, L.K. The role of weeds in supporting biological diversity within crop fields. *Weed Res.* **2003**, *43*, 77–89. [CrossRef]
128. Kleijn, D.; Rundlöf, M.; Scheper, J.; Smith, H.G.; Tscharntke, T. Does conservation on farmland contribute to halting the biodiversity decline? *Trends Ecol. Evol.* **2011**, *26*, 474–481. [CrossRef]
129. Sutcliffe, O.L.; Kay, Q.O.N. Changes in the arable flora of central southern England since the 1960's. *Biol. Conserv.* **2000**, *93*, 1–8. [CrossRef]
130. José-María, L.; Research, F.S. Weed seedbanks in arable fields: Effects of management practices and surrounding landscape. *Weed Res.* **2011**, *51*, 631–640. [CrossRef]
131. Chauvel, B.; Gasquez, J.; Darmency, H. Changes of weed seed bank parameters according to species, time and environment. *Weed Res.* **1989**, *29*, 213–219. [CrossRef]
132. Levassor, C.; Ortega, M.; Peco, B. Seed bank dynamics of Mediterranean pastures subjected to mechanical disturbance. *J. Veg. Sci.* **1990**, *1*, 339–344. [CrossRef]
133. Peñas, J.; Benito, B.; Lorite, J.; Ballesteros, M.; Cañadas, E.M.; Martinez-Ortega, M. Habitat Fragmentation in Arid Zones: A Case Study of Linaria nigricans Under Land Use Changes (SE Spain). *Environ. Manag.* **2011**, *48*, 168–176. [CrossRef] [PubMed]
134. Druckenbrod, D.L.; Dale, V.H. Experimental response of understory plants to mechanized disturbance in an oak-pine forest. *Ecol. Indic.* **2012**, *15*, 181–187. [CrossRef]

135. Thompson, K.; Bakker, J.P.; Bekker, R.M.; Hodgson, J.G. Ecological correlates of seed persistence in soil in the north-west European flora. *J. Ecol.* **1998**, *86*, 163–169. [CrossRef]
136. Martínez-Núñez, C.; Manzaneda, A.J.; Rey, P.J. Plant-solitary bee networks have stable cores but variable peripheries under differing agricultural management: Bioindicator nodes unveiled. *Ecol. Indic.* **2020**, *115*, 106422. [CrossRef]
137. Forup, M.L.; Henson, K.S.E.; Craze, P.G.; Memmott, J. The restoration of ecological interactions: Plant–pollinator networks on ancient and restored heathlands. *J. Appl. Ecol.* **2008**, *45*, 742–752. [CrossRef]
138. Redbo-Torstensson, P.; Telenius, A. Primary and secondary seed dispersal by wind and water in Spergularia salina. *Ecography* **1995**, *18*, 230–237. [CrossRef]
139. Jiao, J.; Zou, H.; Jia, Y.; Wang, N. Research progress on the effects of soil erosion on vegetation. *Acta Ecol. Sin.* **2009**, *29*, 85–91. [CrossRef]
140. Johnson, E.A.; Fryer, G.I. Physical Characterization of Seed Microsites—Movement on the Ground. *J. Ecol.* **1992**, *80*, 823. [CrossRef]
141. Foucher, A.; Salvador-Blanes, S.; Evrard, O.; Simonneau, A.; Chapron, E.; Courp, T.; Cerdan, O.; Lefèvre, I.; Adriaensen, H.; Lecompte, F.; et al. Increase in soil erosion after agricultural intensification: Evidence from a lowland basin in France. *Anthropocene* **2014**, *7*, 30–41. [CrossRef]
142. Watkinson, A.R. The Demography of a Sand Dune Annual: Vulpia Fasciculata: III. The Dispersal of Seeds. *J. Ecol.* **1978**, *66*, 483. [CrossRef]
143. Fischer, J.; Hartel, T.; Kuemmerle, T. Conservation policy in traditional farming landscapes. *Conserv. Lett.* **2012**, *5*, 167–175. [CrossRef]
144. Kleijn, D.; Baquero, R.; Clough, Y.; Díaz, M.; De Esteban, J.; Fernández, F.; Gabriel, D.; Herzog, F.; Holzschuh, A.; Jöhl, R.; et al. Mixed biodiversity benefits of agri-environment schemes in five European countries. *Ecol. Lett.* **2006**, *9*, 243–254. [CrossRef] [PubMed]
145. Radović, A.; Nikolov, S.C.; Tepić, N.; Mikulić, K.; Jelaska, S.D.; Budinski, I. The influence of land abandonment on farmland bird communities: A case study from a floodplain landscape in Continental Croatia. *J. Vertebr. Biol.* **2013**, *62*, 269–281. [CrossRef]
146. Carlesi, S.; Bocci, G.; Moonen, A.C.; Frumento, P.; Bàrberi, P. Urban sprawl and land abandonment affect the functional response traits of maize weed communities in a heterogeneous landscape. *Agric. Ecosyst. Environ.* **2013**, *166*, 76–85. [CrossRef]
147. Soons, M.B.; Heil, G.W.; Nathan, R.; Katul, G.G. Determinants of Long-Distance Seed Dispersal By Wind In Grasslands. *Ecology* **2004**, *85*, 3056–3068. [CrossRef]
148. Kaligarič, M.; Čuš, J.; Škornik, S.; Ivajnšič, D. The failure of agri-environment measures to promote and conserve grassland biodiversity in Slovenia. *Land Use Policy* **2019**, *80*, 127–134. [CrossRef]
149. Pe'er, G.; Dicks, L.V.; Visconti, P.; Arlettaz, R.; Báldi, A.; Benton, T.G.; Collins, S.; Dieterich, M.; Gregory, R.D.; Hartig, F.; et al. EU agricultural reform fails on biodiversity. *Science.* **2014**, *344*, 1090–1092. [CrossRef]
150. Baker, D.J.; Freeman, S.N.; Grice, P.V.; Siriwardena, G.M. Landscape-scale responses of birds to agri-environment management: A test of the English Environmental Stewardship scheme. *J. Appl. Ecol.* **2012**, *49*, 871–882. [CrossRef]
151. Concepción, E.; Díaz, M. Medidas agroambientales y conservación de la biodiversidad: Limitaciones y perspectivas de futuro. *Ecosistemas* **2013**, *22*, 44–49. [CrossRef]
152. Lázaro, A.; Ecosistemas, C.T. Los cambios de uso del suelo como responsables del declive de polinizadores. *Ecosistemas* **2018**, *27*, 23–33. [CrossRef]
153. Gonthier, D.J.; Ennis, K.K.; Farinas, S.; Hsieh, H.-Y.; Iverson, A.L.; Batáry, P.; Rudolphi, J.; Tscharntke, T.; Cardinale, B.J.; Perfecto, I. Biodiversity conservation in agriculture requires a multi-scale approach. *Proc. R. Soc. B Biol. Sci.* **2014**, *281*, 20141358. [CrossRef]
154. Sirami, C.; Gross, N.; Baillod, A.B.; Bertrand, C.; Carrié, R.; Hass, A.; Henckel, L.; Miguet, P.; Vuillot, C.; Alignier, A.; et al. Increasing crop heterogeneity enhances multitrophic diversity across agricultural regions. *Proc. Natl. Acad. Sci. USA* **2019**, *116*, 16442–16447. [CrossRef] [PubMed]
155. Pacheco, R.; Vasconcelos, H.L.; Groc, S.; Camacho, G.P.; Frizzo, T.L.M. The importance of remnants of natural vegetation for maintaining ant diversity in Brazilian agricultural landscapes. *Biodivers. Conserv.* **2013**, *22*, 983–997. [CrossRef]
156. Wuczyński, A.; Dajdok, Z.; Wierzcholska, S.; Kujawa, K. Applying red lists to the evaluation of agricultural habitat: Regular occurrence of threatened birds, vascular plants, and bryophytes in field margins of Poland. *Biodivers. Conserv.* **2014**, *23*, 999–1017. [CrossRef]
157. Cole, L.J.; Brocklehurst, S.; Robertson, D.; Harrison, W.; McCracken, D.I. Exploring the interactions between resource availability and the utilisation of semi-natural habitats by insect pollinators in an intensive agricultural landscape. *Agric. Ecosyst. Environ.* **2017**, *246*, 157–167. [CrossRef]
158. Thiere, G.; Milenkovski, S.; Lindgren, P.E.; Sahlén, G.; Berglund, O.; Weisner, S.E.B. Wetland creation in agricultural landscapes: Biodiversity benefits on local and regional scales. *Biol. Conserv.* **2009**, *142*, 964–973. [CrossRef]
159. Hobbs, R.J. Can revegetation assist in the conservation of biodiversity in agricultural areas? *Pacific Conserv. Biol.* **1994**, *1*, 29–38. [CrossRef]
160. Suárez-Seoane, S.; Osborne, P.E.; Baudry, J. Responses of birds of different biogeographic origins and habitat requirements to agricultural land abandonment in northern Spain. *Biol. Conserv.* **2002**, *105*, 333–344. [CrossRef]
161. Moreno Saiz, J.C.; Domínguez Lozano, F.; Sainz Ollero, H. Recent progress in conservation of threatened Spanish vascular flora: A critical review. *Biol. Conserv.* **2003**, *113*, 419–431. [CrossRef]

162. Directrices de Uso de la Gestión ex Situ para la Conservación de Especies de la Comisión de Supervivencia de Especies de la UICN | IUCN Library System. Available online: https://portals.iucn.org/library/node/45186 (accessed on 7 September 2021).
163. Bowles, M.L.; Betz, R.F.; DeMauro, M.M. Propagation of Rare Plants from Historic Seed Collections: Implications for Species Restoration and Herbarium Management. *Restor. Ecol.* **1993**, *1*, 101–106. [CrossRef]
164. Nakahama, N.; Hirasawa, Y.; Minato, T.; Hasegawa, M.; Isagi, Y.; Shiga, T. Recovery of genetic diversity in threatened plants through use of germinated seeds from herbarium specimens. *Plant Ecol.* **2015**, *216*, 1635–1647. [CrossRef]
165. Willis, F.; Moat, J.; Paton, A. Defining a role for herbarium data in Red List assessments: A case study of Plectranthus from eastern and southern tropical Africa. *Biodivers. Conserv.* **2003**, *12*, 1537–1552. [CrossRef]
166. Lughadha, E.N.; Walker, B.E.; Canteiro, C.; Chadburn, H.; Davis, A.P.; Hargreaves, S.; Lucas, E.J.; Schuiteman, A.; Williams, E.; Bachman, S.P.; et al. The use and misuse of herbarium specimens in evaluating plant extinction risks. *Philos. Trans. R. Soc. B* **2019**, *374*, 20170402. [CrossRef]
167. Nualart, N.; Ibáñez, N.; Luque, P.; Pedrol, J.; Vilar, L.; Guàrdia, R. Dataset of herbarium specimens of threatened vascular plants in Catalonia. *PhytoKeys* **2017**, *77*, 41. [CrossRef]
168. Suarez, A.; Tsutsui, N. The Value of Museum Collections for Research and Society. *Bioscience* **2004**, *54*, 66–74. [CrossRef]
169. Dalton, R. Natural history collections in crisis as funding is slashed. *Nature* **2003**, *423*, 575. [CrossRef]
170. Piñeiro, V.; Arias, J.; Dürr, J.; Elverdin, P.; Ibáñez, A.M.; Kinengyere, A.; Opazo, C.M.; Owoo, N.; Page, J.R.; Prager, S.D.; et al. A scoping review on incentives for adoption of sustainable agricultural practices and their outcomes. *Nat. Sustain.* **2020**, *3*, 809–820. [CrossRef]
171. Phalan, B.; Balmford, A.; Green, R.E.; Scharlemann, J.P.W. Minimising the harm to biodiversity of producing more food globally. *Food Policy* **2011**, *36*, S62–S71. [CrossRef]
172. Fischer, J.; Manning, A.D.; Steffen, W.; Rose, D.B.; Daniell, K.; Felton, A.; Garnett, S.; Gilna, B.; Heinsohn, R.; Lindenmayer, D.B.; et al. Mind the sustainability gap. *Trends Ecol. Evol.* **2007**, *22*, 621–624. [CrossRef]
173. Williams, D.R.; Clark, M.; Buchanan, G.M.; Ficetola, G.F.; Rondinini, C.; Tilman, D. Proactive conservation to prevent habitat losses to agricultural expansion. *Nat. Sustain.* **2020**, *4*, 314–322. [CrossRef]
174. FAO. *Global Food Losses and Food Waste: Extent, Causes and Prevention*; FAO: Rome, Italy, 2011.
175. Martínez-Valderrama, J.; Guirado, E.; Maestre, F.T. Discarded food and resource depletion. *Nat. Food* **2020**, *1*, 660–662. [CrossRef]
176. Abson, D.J.; Fischer, J.; Leventon, J.; Newig, J.; Schomerus, T.; Vilsmaier, U.; von Wehrden, H.; Abernethy, P.; Ives, C.D.; Jager, N.W.; et al. Leverage points for sustainability transformation. *Ambio* **2016**, *46*, 30–39. [CrossRef] [PubMed]
177. Folke, C.; Jansson, Å.; Rockström, J.; Olsson, P.; Carpenter, S.R.; Chapin, F.S.; Crépin, A.-S.; Daily, G.; Danell, K.; Ebbesson, J.; et al. Reconnecting to the Biosphere. *Ambio* **2011**, *40*, 719–738. [CrossRef] [PubMed]
178. Zylstra, M.J.; Knight, A.T.; Esler, K.J.; Le Grange, L.L.L. Connectedness as a Core Conservation Concern: An Interdisciplinary Review of Theory and a Call for Practice. *Springer Sci. Rev.* **2014**, *2*, 119–143. [CrossRef]
179. Pérez-Ramírez, I.; García-Llorente, M.; Portilla, C.S.; Benito, A.; Castro, A.J. Participatory collective farming as a leverage point for fostering human-nature connectedness. *Ecosyst. People* **2021**, *17*, 222–234. [CrossRef]
180. Wandersee, J.H.; Schussler, E.E. Preventing Plant Blindness. *Am. Biol. Teach.* **1999**, *61*, 82–86. [CrossRef]
181. Ives, C.D.; Abson, D.J.; von Wehrden, H.; Dorninger, C.; Klaniecki, K.; Fischer, J. Reconnecting with nature for sustainability. *Sustain. Sci.* **2018**, *13*, 1389–1397. [CrossRef]
182. Jose, S.B.; Wu, C.-H.; Kamoun, S. Overcoming plant blindness in science, education, and society. *Plants People Planet* **2019**, *1*, 169–172. [CrossRef]
183. López-Rodríguez, M.D.; Castro, A.J.; Castro, H.; Jorreto, S.; Cabello, J. Science–policy interface for addressing environmental problems in arid Spain. *Environ. Sci. Policy* **2015**, *50*, 1–14. [CrossRef]
184. Van Ittersum, M.K.; Rabbinge, R.; Van Latesteijn, H.C. Exploratory land use studies and their role in strategic policy making. *Agric. Syst.* **1998**, *58*, 309–330. [CrossRef]

agriculture

MDPI

Article

Do Traditional Livestock Systems Fit into Contemporary Landscapes? Integrating Social Perceptions and Values on Landscape Change

Pablo Acebes [1,2,*], Zuriñe Iglesias-González [1] and Francisco J. Muñoz-Galvez [1]

1 Terrestrial Ecology Group (TEG-UAM), Departamento de Ecología, Universidad Autónoma de Madrid, 28049 Madrid, Spain; vincapervincalila@gmail.com (Z.I.-G.); fmunozgalvez@gmail.com (F.J.M.-G.)
2 Centro de Investigación en Biodiversidad y Cambio Global, Universidad Autónoma de Madrid, 28049 Madrid, Spain
* Correspondence: pablo.acebes@uam.es

Citation: Acebes, P.; Iglesias-González, Z.; Muñoz-Galvez, F.J. Do Traditional Livestock Systems Fit into Contemporary Landscapes? Integrating Social Perceptions and Values on Landscape Change. *Agriculture* **2021**, *11*, 1107. https://doi.org/10.3390/agriculture11111107

Academic Editors: José Luis Vicente-Vicente, Cristina Quintas-Soriano and María D. López-Rodríguez

Received: 4 October 2021
Accepted: 4 November 2021
Published: 6 November 2021

Abstract: European traditional cultural landscapes are increasingly modified by rural abandonment and urban growth processes. Acknowledged as of High Nature Value for providing multiple ecosystem services while contributing to human well-being, the future of these social-ecological systems is uncertain. Here we aim to (1) explore dominant land use and cover (LULC) changes linked to extensive livestock farming across an urban-rural gradient defined by a large city (Madrid) over the last three decades; (2) identify and classify the main driving forces shaping these landscape trajectories and; (3) acknowledge the main landscape values for promoting landscape stewardship under participatory governance frameworks. For doing so, we combine mapping analyses (CORINE Land cover) with stakeholder perceptions and positions. Our results show a dual process of progressive abandonment of agroecosystems linked to traditional livestock farming and an ever-increasing urban growth over the last three decades as the most important driving forces. The growing urban sprawl in areas close to Madrid begins to be perceived as problematic for interviewees. The decline of extensive livestock farming in detriment of tourism, particularly evident in rural areas far from Madrid, is perceived as a threat to the cultural heritage and traditions of rural people. This decline is also perceived as a worrying increase of wildfire risk. Stakeholders stressed the need of valuing extensive livestock farming to prevent rural-urban migration, dynamizing rural economies, conserving landscapes and traditions while producing food-quality products. Interviewees advocated for science-based, stakeholder-inclusive and participatory landscape planning and co-management, leading to more context-specific, regionalized policymaking.

Keywords: cultural landscapes; drivers of change; landscape planning; landscape stewardship; mixed methods; participatory governance; rural abandonment; stakeholder inclusion; urban growth

1. Introduction

Urbanization is one of the fundamental characteristics of European civilization. This growing process is producing a polarization of the territory between urban and rural areas, deeply impacting landscape dynamics [1]. While urban areas are facing explosive growth [2], remote rural areas are undergoing steep land abandonment [3,4]. As a consequence, there is an increasing societal demand in some of these regions to limit the rate of landscape change and to direct it towards more desirable pathways [1]. This change is particularly concerning in European traditional cultural landscapes. These landscapes are the result of 7000 years of human–nature coevolution [5], and are characterized by harboring high biodiversity, as well as providing multiple and varied ecosystem services, thus contributing to human well-being [6]. In fact, the concept of High Nature Value Farming—HNV is linked to these social-ecological systems and arises from the need to identify those forms of agricultural and livestock production characterized by low intensity, low inputs and reduced environmental impact. HNV systems are further threatened by depopulation and abandonment of traditional management models [7–9]. In parallel, many of these social-ecological systems are facing urban growth, resulting in new peri-urban landscapes shaping a heterogeneous mosaic of urban, rural and natural systems [2]. In these peri-urban landscapes, the co-occurring of farming, urban growth and outdoor leisure activities remains challenging.

Within this context, landscape change has become an emerging field of research [10]. Thus, landscape researchers argue for the need of understanding land use and land cover (LULC) changes to analyze landscape trajectories [10,11]. Land use is defined by the human use of the territory (e.g., agriculture, forestry, residential or industrial), while land cover refers to the physical and biological surface cover of the land (e.g., arable land, forest, pasture, water or artificial structures) [12]. A fruitful approach for understanding and analyzing the causes, processes and outcomes of landscape change is the concept of 'driving forces' [6,13]. This concept distinguishes between 'proximate drivers' and 'underlying drivers'. Proximate drivers refer to the human activities concerning land use that result in landscape changes (e.g., land abandonment, agricultural intensification, urban development). Underlying drivers represent the cultural, political, economic, technological and ecological factors such as agricultural policies, markets, or attitudes and beliefs that trigger those human actions [6,14]. This analytical framework has proven to be valuable for preventing and reducing tensions between conflicting land uses, even for predicting future scenarios [15,16]. It is also useful for developing strategies to achieve more desirable futures, and designing adequate policies [10,11]. However, the framework of proximate and underlying drivers has not yet been applied to extensive livestock systems. Furthermore, it could be of great value in discussing how these social-ecological systems can be enhanced, given their declining trend [7–9]. This approach fits with the idea of landscape stewardship, which has not received enough attention in landscape research [10]. 'Landscape stewardship' has been defined as all 'efforts to create, nurture and enable responsibility in landowners and resource users to manage and protect land and its natural and cultural heritage' [17]. Similarly, [18] believe that land management must consider the individual and societal values of landscapes, which emphasizes responsibility, collaboration, participation and communication in the planning and management of land resources. A core concept then linked to landscape stewardship is 'landscape value', that is, the place-based preferences of people associated with different biophysical and cultural landscape characteristics and elements [14]. Particularly, the idea of 'relational value', that is, the preferences, principles, and virtues associated with relationships, both interpersonal and as articulated by policies and social norms is gaining momentum in social-ecological systems [19,20]. Thus, the relational value is a relatively new social-ecological framing to conceptualize how people relate to and obtain value from their relationship with nature [21].

One way of addressing such multidimensional frameworks is through transdisciplinary approaches based on natural and social sciences. These approaches have proven to be efficient in understanding landscape trajectories, identifying the driving forces and contributing to a desired landscape stewardship [10,22]. Particularly, emerging methodologies and theories such as the Research and Innovation Approach [23] or the Theory of Change [24] seek to actively involve citizens, stakeholders, scientists, and policy makers in situations considered challenging. In this regard, there is a growing scientific interest and social demand for these approaches when looking for more inclusive and participatory governance frameworks to better understand landscape/ecosystem dynamics, human–nature interactions, or land management. Through the application of participatory governance frameworks, we can understand the willingness of people to act on multiple landscape functions that they perceive as crucial for their own well-being [10,25]. Governance can be understood as the structures and processes by which social systems manage their public affairs and generate and implement collective decisions to enhance societal well-being [26].

Following this epistemological approach, we conducted a case study of traditional cultural landscapes based on extensive livestock farming recognized as HNV in mountain and foothill areas of Madrid region (central Spain). This region is considered one of the European hotspots of urban sprawl [11]. Here, the coexistence of traditional farming practices with the growing urbanization processes is challenging for landscape planning and sustainable policymaking. We combined qualitative and quantitative methods to: (i) explore the most relevant landscape changes through the CORINE Land Cover project in an urban-rural gradient within Sierra de Guadarrama (Madrid region), and how these changes are perceived by different social actors in the territory; (ii) identify the proximate and underlying driving forces perceived as responsible for those landscape changes; (iii) examine which landscape values linked to these social-ecological systems are recognized by stakeholders and; (iv) study which forms of land stewardship they consider important to achieve multifunctional landscapes that enable sustainable landscape planning and management, based on participatory governance. Multifunctional landscapes provide food security, livelihood opportunities, maintenance of species and ecological functions as well as fulfil cultural, aesthetic and recreational needs [27]. However, the shift in land use practices, including land abandonment, has altered these landscapes and the benefits that flow from them, particularly extensive livestock systems [28,29]. With this study, we seek to contribute to land use science in addressing complex multifaceted "real-world problems" [10,30].

2. Materials and Methods

2.1. Study Areas

The study was conducted in two areas of central Spain with a long tradition of extensive livestock farming (Figure 1): Colmenar Viejo and Tres Cantos (hereafter CV), and Sierra del Rincón and surroundings (SR). CV corresponds to two contiguous peri-urban municipalities north of Madrid city located in the foothills and floodplains of the Guadarrama Mountains (mean distance from Madrid: 30 km; mean altitude: 900 m above sea level (m.a.s.l.)). The major land covers are pastures, wooded grasslands, shrubs and scattered Holm oak forests (*Quercus ilex*), and artificial (urban) covers. Although historically rural, with strong economic dependence on livestock, the proximity of Madrid city in an increasingly well-connected landscape of transport corridors and commercial centers has transformed the area in the last 40 years. SR comprises 9 municipalities located in the Guadarrama Mountains in the northeastern region of Madrid (mean distance from Madrid, 90 km; mean altitude, 1176 m.a.s.l.). It is a mountainous area dominated by Scots pine (*Pinus sylvestris*), oak forests (*Quercus pyrenaica*, *Q. petraea*), scrublands (mainly colonizing abandoned agricultural fields), and pastureland (including mesotrophic pastures developed on former agricultural fields). Currently, main land uses are extensive cattle farming and leisure activities. While CV has undergone an intense urban population growth in the last 40 years (1981: 21,159 inhabitants; 2020: 100,264 inhabitants; total area: 220 km^2), SR has suffered a depopulation process since the 1950s (1950: 2560 inhabitants; 2020: 1069 inhabitants; total area: 222 km^2) [31]. The Madrid region (8022 km^2) has undergone unprecedented transformation since the end of the 1980s, characterized by rapid and dispersed growth of transport networks and urban areas, accompanied by a steady decline in productive agricultural land [32,33].

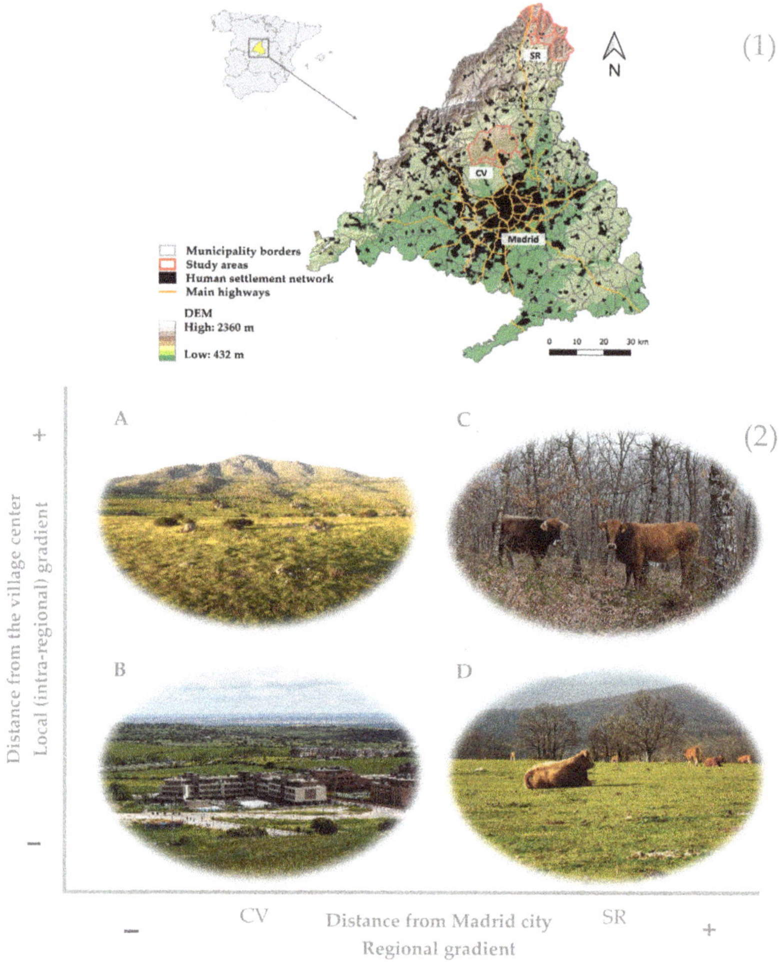

Figure 1. (**1**) Location of the study areas in Madrid region, Spain; (**2**) Pictures of the study areas within a double gradient: regional urban-rural gradient determined by Madrid city (x-axis), and a local gradient of close to/far from the village center (y-axis). DEM: Digital Elevation Model; CV: municipalities of Colmenar Viejo and Tres Cantos (photos **A**,**B**); SR: municipalities included in Sierra del Rincón (photos **C**,**D**).

2.2. Methodological Approach

Our methodology developed a multidisciplinary approach based on mixed methods that blended qualitative and quantitative research [34], by combining approaches from social and natural sciences for the in-depth understanding of socio-ecological realities linked to the territory [10]. We sought to understand perceptions and attitudes regarding LULC changes and the associated landscape values perceived by stakeholders for the last 30 years.

To reinforce the accuracy of the data provided by the participants, we used (i) triangulation strategies between subjects (i.e., asking the same questions to all stakeholder profiles) and (ii) crossed methods (i.e., by conducting a mixed sampling than combined semi-structured interviews with questionnaires). We further combine this approach with GIS mapping analyses aiming to characterize and analyze LULC changes in the study areas.

2.2.1. Land Cover Data and Land Cover Changes Analyses

We analyzed land use and land cover changes (LULC) in two areas located in the northern region of Madrid for the last 30 years, a period that has relevance to those interviewed, and of sufficient duration to identify landscape changes. The beginning of this period corresponds with the incorporation of Spain into the European Economic Union in 1986. As a result, the Common Agricultural Policy (CAP) came into force in the country. The CAP is the agricultural policy of the European Union (EU) that implements a system of subsidies for farmers and other programs (including rural development and environmental protection). The CAP impacts on agricultural landscapes, farmers and citizens across the continent and beyond [35].

To map landscape changes, we used CORINE Land Cover (CLC), a pan-European land coverage map for the entire EU territory (1:100,000 and Minimum Mapping Unit of 25 hectares) [36]. The CLC inventory was initiated in 1985 (reference year 1990). Updates, so far, have been produced in 2000, 2006, 2012, and 2018. We used 1990 and 2018 datasets to enable the identification of change over a 30-year time period. The spatial and temporal consistency of CLC layers makes it particularly appropriate for this type of analysis. We wanted to identify and characterize (i) land cover changes linked to agricultural uses (crop production and extensive livestock grazing), (ii) land cover changes due to urban growth (artificial areas) and (iii) land cover changes triggered by agricultural abandonment and the subsequent forest and shrubland expansion (forest and seminatural areas). Based on the CLC classes we defined four categories that allowed us to distinguish between (i) urban areas; (ii) crops; (iii) pastures; pasture-shrubland areas and agroforestry systems (extensive livestock grazing); and (iv) forest and shrubland areas (without livestock grazing). The description of these new categories is as follows:

- Artificial surfaces. Areas mainly used for dwellings, leisure urban parks, institutional buildings, industrial, commercial and transport networks, but also mines, dump areas or construction sites.
- Crop production areas. Areas used for cropping, which in our study areas are mainly rain-fed cereals and small orchards.
- Livestock grazing areas. Areas mainly covered by herbaceous vegetation and sparse shrubs in some cases, and agroforestry systems with clear signs of livestock grazing, such as water points, paths, stone walls, etc.
- Forest and shrubland areas. Areas covered with (semi-)natural woody vegetation such as forests (*Pinus* and *Quercus* species) and shrublands without signs of livestock presence and/or livestock grazing.

CLC in Spain changed the methodology in 2006 for mapping the territory [37,38]. As a result, there were overestimations in some CLC categories. In particular, transitional woodlands-shrub and sclerophyllous vegetation have been identified in areas corresponding to natural grasslands and forests. This meant that comparisons between CLC1990 and CLC2018 required additional interpretation to avoid misleading comparisons [37,38]. Thus, we validated the data by comparing all land cover categories in both datasets against high resolution aerial images and cross-checking with either expert knowledge or field work. Where required, we reclassified land cover categories (see Supplementary Materials for the reclassifying process). We combined the temporal analysis (1990–2018) with the assessment of two spatial gradients: (1) an urban-rural regional gradient defined by distance to/from Madrid city, in order to study the potential influence of a large city; it is expected to be more pronounced closer to it (i.e., in CV), whereas land abandonment is expected to be more intense far away from it (i.e., in SR); (2) a local intra-regional gradient of use within each area (CV and SR), assuming that urban growth and/or the maintenance of agricultural activities will be more marked around villages than away from them, where land abandonment will be more evident.

In order to analyze spatial gradients and temporal changes together, we conducted a sampling design that mapped a portion of the territory and not the entire study areas (Figure 2). Thus, we analyzed LULC in two sets of circular plots (hereinafter plots): one set including the urban/village center and close surroundings ('plot around village'), and another set of plots in areas more or less away from the village center ('plot away from village center'). This design allowed us to focus on specific land covers and uses and to effectively analyze the spatial gradients. With the plots around villages, we wanted to quantify the urban growth and how this affected the very near surrounding covers/uses. The aim of the plots away from village centers was to analyze LULC changes in non-urbanized areas, but with different intensity of human use (e.g., livestock grazing, forestry or abandoned/natural). We randomly placed plots away from village center, avoiding pine plantations in the case of SR (if possible). These covers planted in the 1960s, only had marginal forestry use and could mask some changes. The size (1 km radius) and number of plots ($n = 9$) away from village center were equal in both areas (CV and SR), as well as the size and number of plots around village center in SR, which coincided with the number of villages there ($n = 9$). In CV, we placed 2 plots of 3 km radius around the two urban centers (Colmenar Viejo and Tres Cantos). This size was modified and fitted to their urban size (Figure 2). Mean urban area in 2018 was of 0.054 km^2 in SR vs. 9.22 km^2 in CV. Due to the small size of villages in SR, that were below the Minimum Mapping Unit of CLC, we mapped these villages from aerial images from 1991 and 2018 to analyze possible changes in urban area between periods.

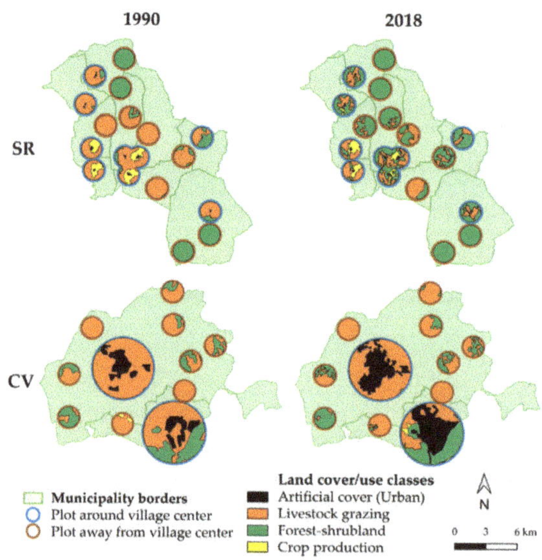

Figure 2. Mapping design and analysis based on two sets of plots placed on a portion of the territory. The map shows the land cover/use classes reclassified from CORINE Land Cover datasets of 1990 and 2018 for both study areas (CV and SR).

2.2.2. Key Informant Selection

We first designed a 'key informant sampling', that is, a sampling method based on the identification of informants with recognized knowledge and/or a relevant position in the territory. This approach was combined with a 'snowball sampling', i.e., asking each interviewee for more potential informants. We chose key informants holding different linkages with the territory who may show divergent positions and classified them into three groups (Table 1).

Table 1. Number (N) and type of stakeholders interviewed arranged in three social groups and by study area. Code: identification of stakeholders for recognizing the quotes in the text. All officers work for the local/regional government on environmental projects. CV: Colmenar Viejo and Tres Cantos; SR: Sierra del Rincón.

Social Group	Stakeholders	Inter-Territorial		CV		SR		
		N	Code	N	Code	N	Code	Ntotal
Ecologist	Academia	8	AC					8
	Environmental NGO	2	EN	1	Encv			3
Rural	Livestock farmer association	1	LFA					1
	Farmer/shepherd			5	Fcv	5	Fsr	10
	Beekeeper			1	BKcv	1	BKsr	2
	Hunter			1	HTcv	1	HTsr	2
Officer	Administration			2	Adcv	2	Adsr	4
	Environmental technician			1	Eocv	2	Eosr	3
	Forest ranger			3	FRcv	4	FRsr	7
	Veterinarian			2	VTcv	1	VTsr	3

2.2.3. Interview Design and Analyses

We conducted 43 semi-structured interviews between February and July 2020; nearly all were done via videoconference (e.g., Skype, Zoom) due to COVID-19 confinement, and lasted 70–90 min. Before starting the interview, we explained in detail the purpose of the research project, the expected duration, the further use of the data and the expected ways of dissemination. We requested an informed consent.

We proposed the following conversation topics: (1) LULC changes in the territory; (2) conflicts and synergies between land uses; (3) drivers of change of LULC; (4) role, challenges and threats of extensive livestock farming; (5) proposals to promote sustainable rural development and nature conservation, and; (6) public policies and societal demands. We designed a pilot interview with initial questions to check for the length, language suitability and potential sources of bias. The pilot interview questions were tested with four interviewees, and slightly adapted to ensure that interviewees clearly understood all the questions [39]. These pilot interviews were also included in the subsequent analysis. We voice recorded interviews and we transcribed them verbatim with the InqScribe 2.2.4 software for further analysis. Interview transcriptions were analyzed through a directed content analysis [40]. This type of analysis consists of sorting the responses (by researchers) into predetermined categories at the beginning of the analysis, which are modified and enriched as the transcript progresses. Thus, we first created coding categories based on the research questions. These categories were then modified and continuously reviewed based on the responses of the interviewees throughout the interview coding process. We used Atlas.ti 7.5.4 software (ATLAS.ti Scientific Software Development GmbH, Berlin, Germany) for this analysis.

Furthermore, we classified the final coding categories (i.e., once the direct content analysis was finished) according to the framework on LULC drivers of change given by [10]: proximate drivers (human actions that have a direct effect on landscape changes), and underlying drivers (the cultural, political, economic, technological and ecological factors that trigger those human actions) [6]. These driving forces were not given to the informants, but they were extracted from their discourses. We further classified underlying drivers based on the expertise of the researchers according to their scale of impact (local, regional, national, international). We also identified the landscape values and planning actions from interviewee discourses for an inclusive stewardship. The added value of this approach is the specific analysis of extensive livestock farming through the prism of the proximate and underlying drivers, the landscape values and landscape stewardship perceived or desired by different social actors of the territory.

2.2.4. Questionnaire Design and Analyses

We performed 41 questionnaires at the end of interviews to contrast and complete informant perceptions regarding shrub expansion and socio-political opinions about CAP, thus deepening on issues of greater interest or that can be better characterized with a quantitative approach. The questionnaire was made up of open and closed-list questions (pre-established options), and a Likert scale, which assessed the level of agreement and disagreement regarding a series of statements based on scientific literature and media. The data were transferred to a spreadsheet from which descriptive statistics were extracted and presented graphically using the R 3.5.2 software (R Foundation for Statistical Computing, Vienna, Austria) with ggplot2 package [41].

3. Results

3.1. Land Cover Changes across Urban-Rural Gradient: Mapping Analyses vs. Stakeholder Perceptions

Mapping analyses showed LC changes in both spatial gradients over the study period (Table 2). First, forest-shrubland and urban areas have increased over the last 30 years (37.6% and 54.2%, respectively). In contrast, livestock grazing and crop production areas have decreased (32.4% and 18.5%, respectively). Second, SR had more forest-shrubland areas (60.1%) than CV (21.1%). In SR these covers have 2-fold increase in 30 years. This shift has occurred mostly at the expense of decreasing pastures and other covers for livestock grazing. Third, in CV pastures and agroforestry systems were the predominant cover (53.4%). These covers have decreased especially in the near surroundings of villages (34.9%) due to urban growth, which has increased 2.2-fold in the last 30 years. Forth, forest-shrubland areas have increased less in areas far away from villages than in areas around villages, which is contradictory to the expected local gradient of use (close to/far from the village). Being territories of similar area (CV: 220 km^2; SR: 222 km^2), the urban cover in CV was 24.9 larger than in SR in 1990. In 2018, urban cover was already 43.5 larger in CV.

Table 2. Land uses and covers in the study areas based on Corine Land Cover Project (CLC 1990 and 2018), distinguishing between areas 'around village center' and 'away from village'. Two gradients are illustrated: urban growth and naturalness/abandonment. Δ is the rate of change (%) of each land cover between the two periods. CLC 1990 and 2018 are measured in km^2. CV: Colmenar Viejo and Tres Cantos municipalities; SR: Sierra del Rincón. * Urban cover in SR was estimated from high-resolution aerial imagery for the same dates.

Regional	Intra-Regional (Local)	Category	CLC 1990	CLC 2018	Δ
CV	Around village center	Livestock grazing	38.63	25.16	−34.9
		Crop production	0.07	0.50	602.3
		Forest-shrubland	8.42	9.85	16.9
		Artificial cover (Urban)	9.52	21.13	121.9
	Away from village center	Livestock grazing	20.60	20.21	−1.9
		Crop production	0.36	0.00	−100
		Forest-shrubland	7.36	8.11	10.2
SR	Around village center	Livestock grazing	19.68	11.73	−40.4
		Crop production	4.80	3.76	−21.8
		Forest-shrubland	3.02	12.02	297.6
		Artificial cover (Urban *)	0.38	0.49	27
	Away from village center	Livestock grazing	14.54	6.10	−58.1
		Forest-shrubland	13.78	22.22	61.2

Regarding stakeholder perceptions, all social groups in SR perceived a more intense increase of shrub expansion and afforestation in detriment of pastures. Stakeholders further perceived a landscape homogenization as a result of land abandonment (N = 22; 53%). *"The real abrupt change has occurred in SR; there are places where sheep even goat cannot get into. It is becoming a forest landscape"* (Fsr). Some officers and environmental NGOs (N = 3; 7%) further indicated that shrub expansion was less intense in CV due to livestock farming that remained the most widespread land use in the territory out of the urban growth areas. All social groups from CV and inter-territories highlighted the intense urban growth and the increase of transport infrastructures (highway, railway) in CV due to the proximity of Madrid city (N = 17; 41%). *"CV has suffered a vertiginous urban growth at expense of rural land"* (AC).

3.2. Proximate Drivers of Land Cover Changes

Almost all interviewees identified land use changes in recent decades as responsible for the perceived and mapped land cover changes (N = 37; 90%). The main perceived change was the progressive decrease of the primary sector (livestock farming, crop production and forestry) in favor of economic activities linked to secondary and tertiary sectors (N = 15; 36%; Figure 3). According to interviewee comments, a noticeable increase in construction activities for housing, transport corridors, services and light industry has happened in CV in detriment of pastures over the last 30 years (Figure 3): *"this is pure brick, here most of those who left the agriculture have become bricklayers, plumbers, electricians ... "* (FRcv). In SR, recreational activities have increased, although this growth has not led to a significant urban expansion, but to the rebuilding of old constructions for tourism purposes. This shift resulted in new forms of employment for local people (Figure 3): *"So, there has been a clear tertiarization of the economy in SR: everyone has set up a bar, a restaurant, a rural house, or a campsite.... or they have dedicated to things related to tourism"* (VTsr). Interviewees further perceived an increase of sport activities (hiking, biking, etc.) in natural areas (Figure 3). Several stakeholders acknowledged a decline of crop production (N = 13; 31%). This decline was characterized by the abandonment of private orchards and cereal crops (oat, rye wheat and barley) and their conversion into pastures for livestock grazing. Several stakeholders further expressed a reduction of forestry in SR (N = 10; 24%), especially wood harvesting for local consumption.

Livestock farming was considered by interviewees as the main land use in the territory and the activity that has contributed most to shape these landscapes. The changes of livestock farming was explained as the progressive shift in terms of livestock type and management modes since the second half of the 20th century (Figure 3); interviewees belonging to all social groups mentioned (i) the decline of small ruminants (sheep and goats), which where dominant until the 1970s, in favor of cattle (N = 28; 68%): *"We have shifted from seasonal shepherds holding small livestock to free-range cattle all year round"* (EN); (ii) a shift from dairy (intensive) cattle to beef (extensive) cattle farming (N = 13; 31%) in 1980s; and (iii) the shift towards larger farms managed by fewer farmers: *"There used to be more farms, and with more farmers, each one took his cattle to a specific area so that the whole territory was grazed"* (ADsr). Ecologists, officers and rural actors considered that these changes were responsible for the landscape changes (N = 23; 56%); contrary to sheep and goats, cattle graze in a limited proportion of the territory usually around villages and receive supplementary fodder (Figure 3). Likewise, stakeholders stated the progressive abandonment of fields (grazing areas) further away from villages and in those areas difficult to access for farmers. Stakeholders also mentioned the loss of traditional management practices such as transhumance, the underuse of drove roads and the complementary use of crops and pastures: *"A cereal-legume and fallow crop rotation was done and where the sheep came in to eat the stubble"* (Fsr).

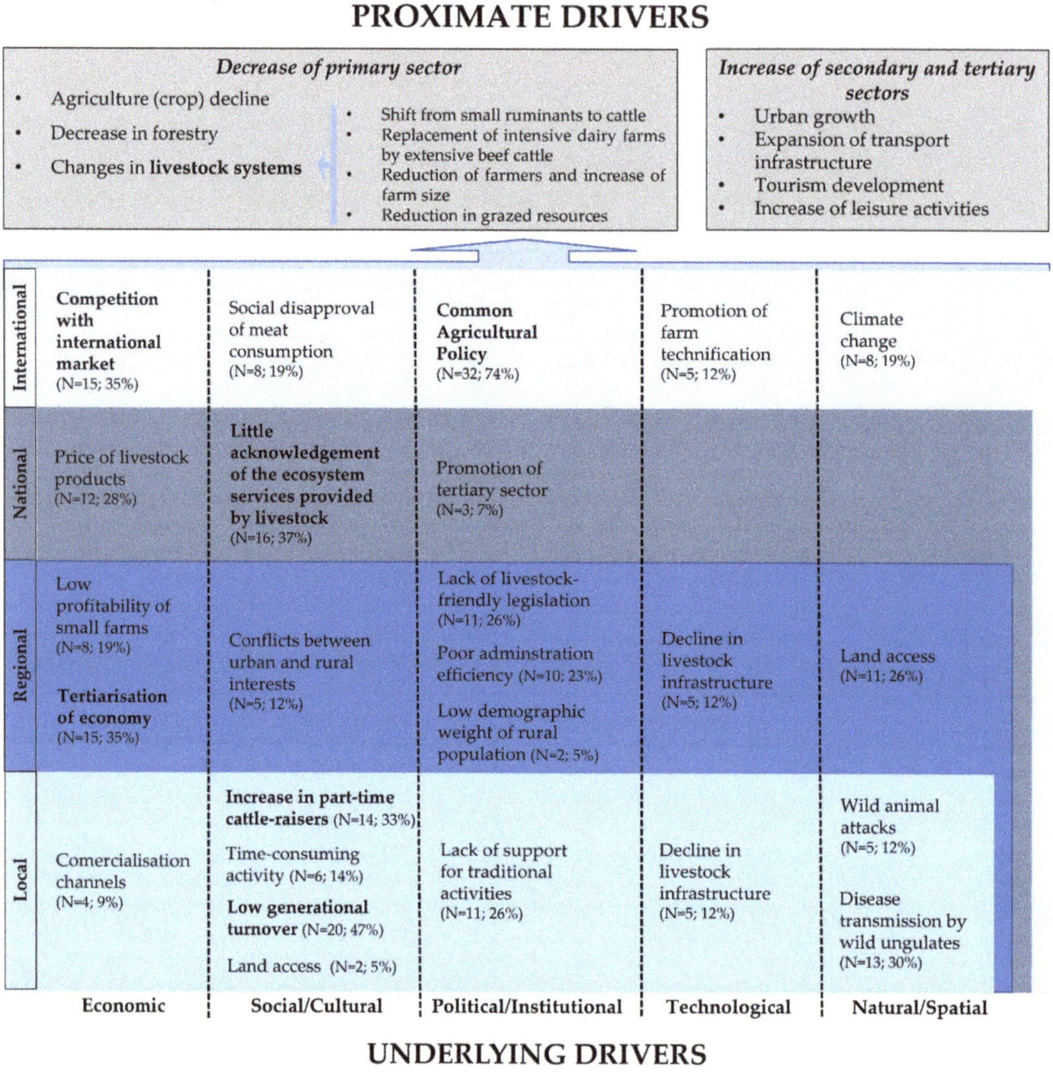

Figure 3. Proximate and underlying drivers of land use and land cover changes identified by stakeholders focusing on livestock-related factors in two study areas of Madrid region. The proximate drivers (top) identified by the stakeholders are shown according to the main topics. The underlying drivers (bottom) are adaptations of the topics mentioned by the interviewees, nested according to their scale of impact, and categorized by the typology suggested in [6]. The most frequently cited underlying drivers are highlighted in bold. Gaps of information are due to lack of mention by the interviewees.

3.3. Underlying Drivers of Land Cover Changes

Stakeholders identified several underlying drivers acting at different scales, from local to international. These drivers were related to economic, socio-cultural, political, technological and natural issues as responsible for LULC changes (Figure 3). Moreover, stakeholders related proximate drivers such as the aggregation of livestock on fewer farmers and their part-time dedication to underlying drivers such as low profitability of small farms, particularly those of sheep and goats; since beef cattle management is less time-consuming, farmers may have a second job. *"The main problem faced by farmers is the low farm profitability and low meat prices and other livestock-derived products*" (Fsr). This low profitability of farms together with very demanding work was explained as being responsible for the low generational turnover. In addition, the difficulty of land access for the incorporation of new farmers was emphasized as a major drawback (Figure 3); almost all farmers in the territory inherited land and/or livestock from their families.

Rural abandonment in SR was perceived as a lack of attention from administration to the rural claims and needs, where a feeling of abandonment was shared by local people (N = 3; 7%): "*a shift in the weight of farmers has occurred, who were a majority in the past, but currently are a minority from demographic and economic points of view*" (AC). Accordingly, several interviewees perceived that livestock farming was not a priority for administrations (N = 11; 26%), and that it was scarcely acknowledged for providing ecosystem services (Figure 3).

CAP subsidies were broadly perceived as an essential economic support of livestock farms: "If we were to remove the CAP, possibly 50–60% of livestock farming in Spain would disappear, since extensive livestock farming would not be competitive" (VTcv). In contrast, other interviewees argued that CAP produced an overdependence of livestock farmers on subsidies, thus favoring farming intensification. Interviewees generally considered that CAP 2023–2027 will bring benefits to extensive livestock farming (Figure 4).

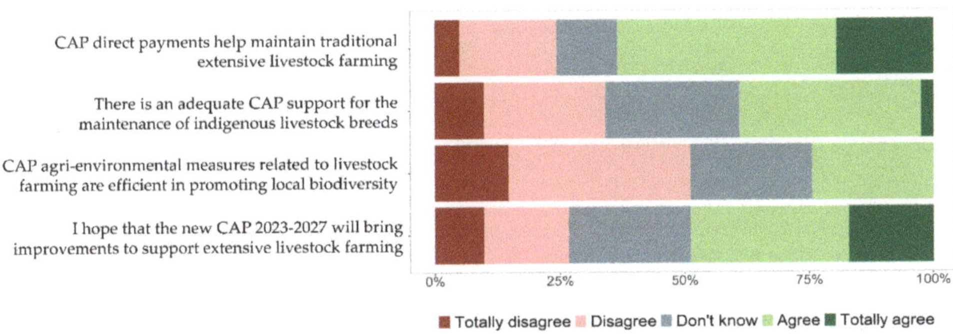

Figure 4. Positions of interviewed people regarding different issues related to the Common Agricultural Policy based on Likert scale responses from the questionnaires (N = 41). Values are expressed as the percentage of responses over the total number of people interviewed.

Alongside these economic, societal, cultural and political factors, technological factors were also identified as drivers of change in traditional livestock management. These drivers acted on local, regional (i.e., decline of drove roads, farming infrastructures, etc.) and international scales (technification and intensification of farms through policies such as the CAP). Regarding the natural drivers, interviewees mentioned those related to the expansion of wild large vertebrates as a result of rewilding processes linked to land abandonment; particularly, the insufficient response of the administration to the increasing attacks on livestock by large carnivores (grey wolf) and strict scavengers (vultures), as well as disease transmission by wild ungulates.

3.4. Landscape Values Perceived by Interviewees

We registered different opinions and attitudes amongst stakeholders regarding the landscape changes and values due to abandonment, the persistence of livestock farming and urban growth. Officers perceived shrub expansion as an opportunity (N = 7; 16%), while rural stakeholders and ecologists perceived it as a threat (N = 8 and N = 4, 19% and 9%, respectively). In CV, stakeholders tended to show polarized attitudes regarding this process. Even so, shrub expansion was perceived more as an opportunity in CV than in SR, where land abandonment was more evident (Figure 5a). Among the opportunities of shrub expansion, stakeholders considered this process would (i) benefit biodiversity,

(ii) prevent from soil erosion and (iii) provide food for goats and pollinators (Figure 5b). However, threats were numerically higher; the increase of wildfire risk was the most cited (see below). Some farmers were further concerned about the irreversible process of shrub encroachment and afforestation, leading to pasture loss (Figure 5b).

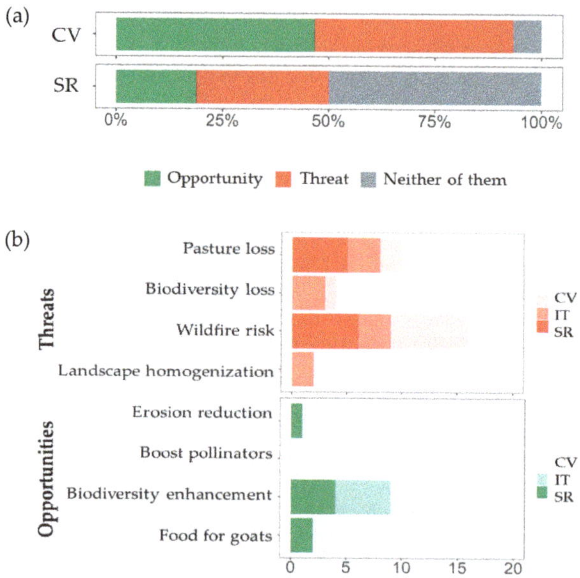

Figure 5. Stakeholder views on shrub expansion in the study areas (**a**) perceived as threat, opportunity or neither of them, shown as the percentage of responses over the total number of people interviewed within each study area, and (**b**) consequences of shrub expansion identified as threat or opportunity by interviewees represented by the number of citations. CV: interviewees from Colmenar Viejo and Tres Cantos; SR: interviewees from municipalities of Sierra del Rincón; IT: Interterritorial interviewees.

Focusing on livestock farming, the most pervasive positioning among stakeholders was the positive value given to extensive livestock farming in the territory such as: (i) a driver of employment and economic activity in rural areas (N = 17; 41%); (ii) the conservation of biocultural heritage comprising the uses and traditions of these systems (N = 5); (iii) the conservation of characteristic species (N = 29; 70%): "*there are many species of flora and fauna and EU Habitats of Community Interest that depend on these cultural landscapes*" (AC); and (iv) its role in wildfire prevention by removing vegetation that fuels wildfires (N = 19; 46%). This contrasted with the perception that the role of "fire brigades" could not be played by wild ungulates (N = 8; 19%): "*wild herbivorous animals are not going to fulfil the same functions (as livestock), because part of the functions depend on human practices and management of livestock*" (AC). Finally, (v) the promotion of locally produced, quality food systems.

Even so, almost half of the interviewees considered that the remaining livestock farms have undergone a process of intensification in recent years, negatively affecting pasture productivity (N = 19; 46%). Several interviewees further mentioned negative impacts of such practices on different (melliferous) plant and animal (i.e., edaphic invertebrate and amphibian) populations (N = 6; 14%). In addition, an academic and a livestock farmer warned on the impacts of climate change on grazing areas and stressed that current stocking rates would need to be adapted to pasture productivity. Several stakeholders from different social groups emphasized their concern about the increasing social rejection of livestock farming for its contribution to greenhouse gas emissions (N = 10; 24%).

Regarding the loss of primary sector in favor of secondary and tertiary sectors, a majority of actors believed that urban growth in CV was excessive, that urban growth threatened livestock farming and impacted on natural habitats and biodiversity (N = 17; 41%): *"Colmenar seems to want to pave the way so that agriculture is not a problem to go on (urban) growing"* (Fcv4). In contrast, local administration was favorable to this growth: *"I think that nowadays the general urban development plan has not expanded in a way that affects the rural world in such an aggressive way"* (ADcv).

There were divergent positions amongst interviewees on the promotion of tourism and recreational activities (tertiary sector): some stakeholders supported the opportunities brought by tourism as an alternative to declining livestock farming. Others argued that these landscapes would not be attractive for tourism without the livestock farming activity that preserves them: *"The SR is tourism, there is still some livestock farming, but in 15 years I don't know who wants to go and see a scrubland ... maybe what they wanted to watch were cattle, fields ... the demand for tourism is driving this landscape towards an inertia that is its own suicide"*. (AC).

3.5. Landscape Stewardship Desired by Interviewees

Half of the actors called for more inclusive participatory landscape planning in the medium and long term, bringing together administration, science and society (N = 22; 53%). The implementation of this planning was demanded as a process of integration at different levels: (i) policy interventions, between different management areas, e.g., rural development (agriculture, forestry, tourism) and biodiversity conservation; (ii) societal: between rural and urban perspectives; and (iii) livestock production: between traditional/conventional management practices and novel approaches. Likewise, there was also a plea for greater coordination between the social actors to achieve a sustainable use of the territory.

Furthermore, social actors requested context-specific solutions, since they perceived that policies were conceived in "offices" far away from rural areas (N = 17; 41%). *"Standard solutions are not valid, because each territory and each type of livestock has its own management"* (FRsr). Even actors from the administration recognized the scant flexibility and inefficiency of the administration. Some of the policy interventions most demanded by the interviewees, including farmers, officers and ecologists were (i) the engagement of livestock farmers in programs of fire control through grazing shrublands: (ii) the maintenance of HNV farming; and (iii) helping livestock farmers to carry out environmentally responsible practices (N = 9; 22%). A drawback highlighted by administration (and recognized by several farmers) to promoting and strengthening local/regional interventions was the lack of cohesion amongst farmers when dealing with administration: *"I believe that more unity is also needed in trying to ensure that there are representatives who truly represent the sector"* (VTcv). On the other hand, an important part of those context-specific policies should be focused on revitalizing rural areas such as SR. Here, interviewees demanded broadening and improving basic services such as health, education, transport or internet as a crucial step to prevent the emigration of rural population.

Stakeholders further recognized the crucial need for wider societal recognition of the economic, cultural and ecological roles of extensive livestock farming (N = 16; 39%): *"we would need education and consume local livestock products as the act of managing the territory and paying a fair price for it"* (AC). One of the instruments mentioned to recognize the added value of extensive livestock farming was distinctive names such as the Protected Geographical Indication or Protected Designation of Origin. In this vein, several actors, including some farmers, pointed out the relevance of recovering traditional management practices for: (i) increasing landscape sustainability and reducing land impact; (ii) removing intermediaries in product sales; and (iii) fostering local markets (direct sales to local consumers, restaurants, tourists).

4. Discussion

Our results show a dual process of progressive abandonment of HNV systems linked to traditional agricultural and livestock farming and an ever-increasing urban growth and touristification over the last three decades as the most important driving forces. This is explained by a generalized increase in the secondary and tertiary sectors driven by Madrid city. This pattern takes different paths according to the urban-rural gradient but shares the relegation of the primary sector to the background. Some authors suggest that the incorporation of Spain into the EEC (1986) led to important social, economic and technological effects that catalyzed these processes [42]. Particularly, land use decisions in Europe were strongly influenced by the CAP [35,43]. These results agree with previous Pan-European studies, mirroring the shared patterns and trends of landscape change in the European context [6,11,44–46], and particularly in the Iberian Peninsula (Portugal and Spain; [47,48]).

Our results show divergent landscape trajectories according to the regional urban-rural gradient in the most densely populated region of Spain (845.2 indiv./km^2; [49]): despite not being far from Madrid city (90 km), the more rural areas (SR) are undergoing a steady abandonment of traditional practices since the 1950s, whereas tourism and leisure activities are replacing traditional economies. In contrast, peri-urban areas with a rural history (CV) are facing an explosive urban sprawl, as described in other European areas [2]. Thus, urban society is exerting different pressures according to its needs for: (i) housing close to the workplace (CV); and (ii) leisure activities within distances easily accessed on weekends (SR).

Interviewees agreed with the land cover changes mapped with CORINE in both areas and within the urban-rural gradient. Interviewees also mentioned a more evident abandonment of land uses in areas far away from villages than the ones mapped in our analyses (particularly in SR). This may be due to the increased difficulty of mapping transitional covers corresponding to vegetation successional stages of abandonment, particularly the shrubland expansion [37,50]. An alternative explanation is that abandonment of areas away from villages was more intense before 1990. In any case, both mapping analyses and social perceptions seem to draw a good picture of landscape changes and trajectories in the study areas. Furthermore, interviewees could identify the patterns more difficult to capture at local scales with CORINE, thus highlighting the suitability of combining both approaches [10,45].

Stakeholders highlighted EU policies such as the earlier CAP and global food markets as high-impact underlying drivers responsible for the decline of pasture-based livestock farming at broad scales [35,45]. However, several interviewees recognized that recent CAP instruments have contributed to halting their abandonment [35,51]. Additionally, more regional or local underlying drivers may also be pushing these HNV systems to abandonment or conversion into urbanized areas. In this vein, interviewees pointed out societal, political and economic drivers such as the lack of generational turnover due to the hard life and low profitability of farms, as previously reported [52]. Local actors further stated the difficulty of becoming a shepherd/farmer due to limited or no access to land, despite fields being abandoned. This paradox was especially the case in SR, where land use for grazing do not compete with urban development. However, there is a social reluctance to offer lands for developing traditional activities to incoming people. This social driver could be regulated by the administration through legislation on land use and land tenure (i.e., regulation of prices and rent supply). On the other hand, the scarce support from administration was also highlighted by rural actors as an important political constraint. This perception produces tensions between rural stakeholders and the administration, leading to a lack of dialogue between the parties. Thus, the promotion of participatory governance processes is essential to unblock the situation [53–55].

Addressing and understanding people's perceptions and values with respect to landscape changes and trajectories is of great value in reflecting on and outlining modern policymaking [14,19,56,57]. Our results show how actors in the territory perceived and valued the processes of landscape change in a different manner. For people in rural areas, land abandonment means the permanent loss of pastures, the loss of biocultural heritage and their linkage with the territory [58]. For ecologists, land abandonment favors landscape homogenization [52]. Moreover, woody encroachment increases the risk of wildfire, which is especially worrying in Mediterranean ecosystems due to summer high temperatures and dry conditions in conjunction with global warming [59]. Some ecologists also emphasized that several habitats included in the EU Habitats Directive (92/43/EEC), depend on livestock use, so that its abandonment jeopardizes their conservation. Others argue that land abandonment prevents soil erosion, though recent studies have shown that abandonment of extensive livestock systems in central Spain decreases soil fertility, carbon sequestration [60] and microbial activity [61]. Some authors argue that land abandonment offers opportunities for biodiversity conservation, such as forest bird species [62] and large mammals [63,64]. Nevertheless, rewilding brings new threats and challenges for farmers, administration and society on how to coexist with wild fauna [65,66] which was underlined by stakeholders. Summarizing, we agree that the debate on agricultural abandonment should focus more on target-nuanced, context-dependent policy and management strategies [67]. What is more, to reach negotiated solutions in which multifunctional landscapes are preserved, participatory processes that encompass the diverse values and views of the stakeholders involved are urgently needed and demanded [68–70].The mounting urban sprawl in CV over the last decades is widely considered a threat for traditional livestock farming and cultural identity, causing air quality degradation (due to the massive use of private vehicles) and overcrowded use of the environment [2]. In general, interviewees showed divergent feelings and positions between desired landscape trajectories and those actually experienced, both in more rural areas and new urbanized areas. This poses the pressing need for building participatory governance models that incorporate people's values for landscape planning [71,72].

Several stakeholders considered that, by promoting extensive livestock farming, cultural landscapes can be preserved alongside the production of quality food products from farming systems that adhere to animal-welfare conditions not seen in the increasing industrial farming. Furthermore, by fostering local markets and initiatives such as the Km0 (eating food locally produced) or 'Farm to fork' (EU strategy that promotes fairer, healthier and environmentally-friendly food systems) [73] and valuing local products, we can contribute to preserving extensive livestock systems [74]. In addition, the disruptions caused by COVID-19 in several market supplies at global and regional scales highlight the need of strengthening local markets [71]. However, behind market regulations, the reduction in meat consumption is a pressing need worldwide, due to its major, negative consequences for land and water use and environmental change [75]. Extensive livestock farming certainly emerges as an alternative to industrial livestock production for its social, environmental and economic contribution in rural areas [76,77], where it leads to the maintenance of multifunctional landscapes [78]. Even more, by strengthening pasture-based livestock systems from the economic, ecological and technological dimensions, we can contribute to the increasing societal and scientific demand towards agroecological transitions of agricultural systems [79].

Thinking, valuing and planning the countryside is done mainly by urbanites while rural development is mainly focused upon the urban needs [1]. This was a common view among interviewees of the more rural areas (SR). To reverse this situation several actors advocated for science-based, stakeholder-inclusive and participatory landscape planning and co-management, leading to more context-specific, regionalized policymaking [11,80]. *"It seems as if the administration and the society do not know whether the use of cultural landscapes should be for production, conservation or tourism"* (AC). Thus, it is crucial to overcome sectoral approaches from different administrations (e.g., farming and rural development, nature conservation, tourism administration, and local, regional, national administration, etc.) which, to date, have promoted even opposing policy interventions regarding the countryside and rural development. The transition to more participatory and democratic governance models, in which different stakeholders get involved in landscape stewardship, was growing among social actors. Some of them went further and called for multifunctional landscape planning that integrates social, ecological, economic and cultural demands [56]. Multifunctional landscapes must ensure both ecological functions and ecosystem services [81]. They should be rooted in the idea of coexisting land uses based on traditional practices (agriculture, livestock, forestry), conservation aims and tourism activities, even urban growth, within sustainable planning management frameworks co-designed by stakeholders [82].

To conclude, our results show a shared pattern of landscape change with other European regions. This pattern is characterized by the progressive abandonment of traditional farming practices and their replacement by tourism in areas far from Madrid city (SR). In contrast, extensive livestock farming coexists with the growing urban sprawl in areas close to Madrid city (CV). If our aim as a society is to preserve the HNV landscapes, we must bring the needs of the rural stakeholders that support them to the forefront while we balance the impact of the urban population needs. Our results show how social actors do not feel satisfied with the current landscape trends. Stakeholders further advocate for participatory stakeholder-inclusive governance frameworks in search of more sustainable rural planning that lead to the coexistence of traditional practices with the increasing secondary and tertiary sectors. Finally, the identification of proximate and underlying drivers from interviewee discourses have proven to be valuable for analyzing landscape trajectories.

Supplementary Materials: The following are available online at https://www.mdpi.com/article/10.3390/agriculture11111107/s1, Information about the reclassification process. Table S1: The standard CORINE Land Cover nomenclature for the 3 level 1 classes identified in our study. Tables S2–S5: Description of the original CLC classes that make up the new categories for SR and CV in 1990 and 2018. Table S6: Land uses and covers at village-level based on Corine Land Cover Project (CLC 1990 and 2018).

Author Contributions: Conceptualization, P.A. and Z.I.-G.; methodology, P.A., F.J.M.-G. and Z.I.-G.; software, F.J.M.-G. and Z.I.-G.; validation, P.A.; investigation; P.A., F.J.M.-G., Z.I.-G.; writing—original draft preparation, P.A.; writing—review and editing, F.J.M.-G. and Z.I.-G. All authors have read and agreed to the published version of the manuscript.

Funding: This research received no external funding.

Institutional Review Board Statement: Ethical review and approval were waived for this study because the information used was completely anonymous; no personal information was compiled but we requested an informed consent of all interviewees before starting the interview (see "Interview design and analyses" section).

Informed Consent Statement: Informed consent was obtained from all subjects involved in the study.

Data Availability Statement: Not applicable.

Acknowledgments: Authors would like to thank three anonymous reviewers for contributions that improved the manuscript. Authors would especially like to thank all participants engaged in this study, and M. García-Llorente and D. López for contributing to the design of the interview/questionnaire. F.J.M.-G. hold a grant from Comunidad de Madrid–EU Social Fund (PEJD-2019-PRE/AMB-16775). Authors would also like to thank the REMEDINAL TE project (P2018/EMT-4338; Madrid Regional Government and EU Social Fund).

Conflicts of Interest: The authors declare no conflict of interest.

References

1. Antrop, M. Landscape Change and the Urbanization Process in Europe. *Landsc. Urban Plan.* **2004**, *67*, 9–26. [CrossRef]
2. Shaw, B.J.; van Vliet, J.; Verburg, P.H. The Peri-Urbanization of Europe: A Systematic Review of a Multifaceted Process. *Landsc. Urban Plan.* **2020**, *196*, 103733. [CrossRef]
3. MacDonald, D.; Crabtree, J.R.; Wiesinger, G.; Dax, T.; Stamou, N.; Fleury, P.; Gutierrez Lazpita, J.; Gibon, A. Agricultural Abandonment in Mountain Areas of Europe: Environmental Consequences and Policy Response. *J. Environ. Manag.* **2000**, *59*, 47–69. [CrossRef]
4. Van der Sluis, T.; Pedroli, B.; Frederiksen, P.; Kristensen, S.B.P.; Busck, A.G.; Pavlis, V.; Cosor, G.L. The Impact of European Landscape Transitions on the Provision of Landscape Services: An Explorative Study Using Six Cases of Rural Land Change. *Landsc. Ecol.* **2019**, *34*, 307–323. [CrossRef]

5. Montserrat, P. La fertilidad del paisaje. In *La Cultura que Hace el Paisaje*; La Fertilidad de la Tierra Ediciones: Navarra, Spain, 2009.
6. Plieninger, T.; Draux, H.; Fagerholm, N.; Bieling, C.; Bürgi, M.; Kizos, T.; Kuemmerle, T.; Primdahl, J.; Verburg, P.H. The Driving Forces of Landscape Change in Europe: A Systematic Review of the Evidence. *Land Use Policy* **2016**, *57*, 204–214. [CrossRef]
7. Oppermann, R.; Beaufoy, G.; Jones, G. *High Nature Value Farming in Europe*; Verlag Regionalkultur: Ubstadt-Weiher, Germany, 2012.
8. Lomba, A.; Guerra, C.; Alonso, J.; Honrado, J.P.; Jongman, R.; McCracken, D. Mapping and Monitoring High Nature Value Farmlands: Challenges in European Landscapes. *J. Environ. Manag.* **2014**, *143*, 140–150. [CrossRef]
9. Acebes, P.; Pereira, D.; Oñate, J.J. Towards the Identification and Assessment of HNV Dehesas: A Meso-Scale Approach. *Agrofor. Syst.* **2016**, *90*, 7–22. [CrossRef]
10. García-Martín, M.; Quintas-Soriano, C.; Torralba, M.; Wolpert, F.; Plieninger, T. Landscape Change in Europe. In *Sustainable Land Management in a European Context*; Springer International Publishing: Cham, Switzerland, 2021; pp. 17–37. ISBN 9783030508418.
11. Kuemmerle, T.; Levers, C.; Erb, K.; Estel, S.; Jepsen, M.R.; Müller, D.; Plutzar, C.; Stürck, J.; Verkerk, P.J.; Verburg, P.H.; et al. Hotspots of Land Use Change in Europe. *Environ. Res. Lett.* **2016**, *11*, 064020. [CrossRef]
12. Verburg, P.H.; Kok, K.; Pontius, R.G.; Veldkamp, A. Modeling land-use and land-cover change. In *Land-Use and Land-Cover Change*; Springer: Dordrecht, The Netherlands, 2006; pp. 117–135.
13. Klijn, J.A. Driving Forces behind Landscape Transformation in Europe, from a Conceptual Approach to Policy Options. In *The New Dimensions in European Landscapes*; Jongman, R.H.G., Ed.; Springer: Wageningen, The Netherlands, 2004; pp. 210–218.
14. Garcia-Martin, M.; Fagerholm, N.; Bieling, C.; Gounaridis, D.; Kizos, T.; Printsmann, A.; Müller, M.; Lieskovský, J.; Plieninger, T. Participatory Mapping of Landscape Values in a Pan-European Perspective. *Landsc. Ecol.* **2017**, *32*, 2133–2150. [CrossRef]
15. Munthali, M.G.; Davis, N.; Adeola, A.M.; Botai, J.O.; Kamwi, J.M.; Chisale, H.L.W.; Orimoogunje, O.O.I. Local Perception of Drivers of Land-Use and Land-Cover Change Dynamics across Dedza District, Central Malawi Region. *Sustainability* **2019**, *11*, 832. [CrossRef]
16. Gomes, L.C.; Bianchi, F.J.J.A.; Cardoso, I.M.; Schulte, R.P.O.; Arts, B.J.M.; Fernandes Filho, E.I. Land Use and Land Cover Scenarios: An Interdisciplinary Approach Integrating Local Conditions and the Global Shared Socioeconomic Pathways. *Land Use Policy* **2020**, *97*, 104723. [CrossRef]
17. Brown, J.; Mitchell, B. The Stewardship Approach and Its Relevance for Protected Landscapes. *Georg. Wright Forum* **2000**, *17*, 70–79.
18. Gundersen, V.; Mäkinen, K. Aldo Leopold and Stewardship: Lessons for Forest Planning and Management in the Nordic Countries? *Nor. Geogr. Tidsskr.* **2009**, *63*, 225–232. [CrossRef]
19. Chan, K.M.A.; Balvanera, P.; Benessaiah, K.; Chapman, M.; Díaz, S.; Gómez-Baggethun, E.; Gould, R.; Hannahs, N.; Jax, K.; Klain, S.; et al. Opinion: Why Protect Nature? Rethinking Values and the Environment. *Proc. Natl. Acad. Sci. USA* **2016**, *113*, 1462–1465. [CrossRef]
20. Arias-Arévalo, P.; Martín-López, B.; Gómez-Baggethun, E. Exploring Intrinsic, Instrumental, and Relational Values for Sustainable Management of Social-Ecological Systems. *Ecol. Soc.* **2017**, *22*, 43. [CrossRef]
21. Fischer, J.; Riechers, M.; Loos, J.; Martin-Lopez, B.; Temperton, V.M. Making the UN Decade on Ecosystem Restoration a Social-Ecological Endeavour. *Trends Ecol. Evol.* **2021**, *36*, 20–28. [CrossRef]
22. Kizos, T.; Verburg, P.H.; Bürgi, M.; Gounaridis, D.; Plieninger, T.; Bieling, C.; Balatsos, T. From Concepts to Practice: Combining Different Approaches to Understand Drivers of Landscape Change. *Ecol. Soc.* **2018**, *23*. [CrossRef]
23. Owen, R.; Macnaghten, P.; Stilgoe, J. Responsible Research and Innovation: From Science in Society to Science for Society, with Society. *Sci. Public Policy* **2012**, *39*, 751–760. [CrossRef]
24. Vogel, I. *Review of the Use of 'Theory of Change' in International Development*; 2012.
25. Enqvist, J.P.; West, S.; Masterson, V.A.; Haider, L.J.; Svedin, U.; Tengö, M. Stewardship as a Boundary Object for Sustainability Research: Linking Care, Knowledge and Agency. *Landsc. Urban Plan.* **2018**, *179*, 17–37. [CrossRef]
26. Folke, C.; Hahn, T.; Olsson, P.; Norberg, J. Adaptative governance of social-ecological systems. *Annu. Rev. Environ. Resour.* **2005**, *30*, 441–473. [CrossRef]
27. O'Farrell, P.J.; Anderson, P.M.L. Sustainable Multifunctional Landscapes: A Review to Implementation. *Curr. Opin. Environ. Sustain.* **2010**, *2*, 59–65. [CrossRef]
28. Cocca, G.; Sturaro, E.; Gallo, L.; Ramanzin, M. Is the Abandonment of Traditional Livestock Farming Systems the Main Driver of Mountain Landscape Change in Alpine Areas? *Land Use Policy* **2012**, *29*, 878–886. [CrossRef]
29. Delattre, L.; Debolini, M.; Paoli, J.C.; Napoleone, C.; Moulery, M.; Leonelli, L.; Santucci, P. Understanding the Relationships between Extensive Livestock Systems, Land-Cover Changes, and CAP Support in Less-Favored Mediterranean Areas. *Land* **2020**, *9*, 518. [CrossRef]
30. Zscheischler, J.; Rogga, S.; Busse, M. The Adoption and Implementation of Transdisciplinary Research in the Field of Land-Use Science—A Comparative Case Study. *Sustainability* **2017**, *9*, 1926. [CrossRef]
31. IECM (Instituto de Estadística de La Comunidad de Madrid). 2020. Available online: https://www.madrid.org/iestadis/ (accessed on 21 December 2020).
32. Hewitt, R.; Escobar, F. The Territorial Dynamics of Fast-Growing Regions: Unsustainable Land Use Change and Future Policy Challenges in Madrid, Spain. *Appl. Geogr.* **2011**, *31*, 650–667. [CrossRef]
33. Gallardo, M.; Martínez-Vega, J. Three Decades of Land-Use Changes in the Region of Madrid and How They Relate to Territorial Planning. *Eur. Plan. Stud.* **2016**, *24*, 1016–1033. [CrossRef]

34. Johnson, R.B.; Onwuegbuzie, A.J. Mixed Methods Research: A Research Paradigm Whose Time Has Come. *Educ. Res.* **2004**, *33*, 14–26. [CrossRef]
35. Pe'er, G.; Bonn, A.; Bruelheide, H.; Dieker, P.; Eisenhauer, N.; Feindt, P.H.; Hagedorn, G.; Hansjürgens, B.; Herzon, I.; Lomba, Â.; et al. Action Needed for the EU Common Agricultural Policy to Address Sustainability Challenges. *People Nat.* **2020**, *2*, 305–316. [CrossRef]
36. Copernicus CORINE Land Cover. Available online: https://land.copernicus.eu/pan-european/corine-land-cover2021 (accessed on 12 October 2021).
37. García-Álvarez, D.; Camacho Olmedo, M.T. Changes in the Methodology Used in the Production of the Spanish CORINE: Uncertainty Analysis of the New Maps. *Int. J. Appl. Earth Obs. Geoinf.* **2017**, *63*, 55–67. [CrossRef]
38. Soukup, T.; Büttner, G.; Feranec, J.; Hazeu, G.; Jaffrain, G.; Jindrova, M.; Kopecky, M.; Orlitova, E. CORINE Land Cover 2006–2012 changes: Analysis and assessment. In *European Landscape Dynamics*; CRC Press: Boca Raton, FL, USA, 2016; pp. 141–146, ISBN 131537286X.
39. Young, J.C.; Rose, D.C.; Mumby, H.S.; Benitez-Capistros, F.; Derrick, C.J.; Finch, T.; Garcia, C.; Home, C.; Marwaha, E.; Morgans, C.; et al. A Methodological Guide to Using and Reporting on Interviews in Conservation Science Research. *Methods Ecol. Evol.* **2018**, *9*, 10–19. [CrossRef]
40. Lune, H.; Berg, B.L. *Qualitative Research Methods for the Social Sciences*; Pearson: Harlow, UK, 2017.
41. Wickham, H. *Ggplot2: Elegant Graphics for Data Analysis*; Springer: New York, NY, USA, 2016; Available online: https://ggplot2.tidyverse.org (accessed on 19 May 2021).
42. Corbelle-Rico, E.; Butsic, V.; Enríquez-García, M.J.; Radeloff, V.C. Technology or Policy? Drivers of Land Cover Change in Northwestern Spain before and after the Accession to European Economic Community. *Land Use Policy* **2015**, *45*, 18–25. [CrossRef]
43. Lomba, A.; Moreira, F.; Klimek, S.; Jongman, R.H.; Sullivan, C.; Moran, J.; Poux, X.; Honrado, J.P.; Pinto-Correia, T.; Plieninger, T.; et al. Back to the Future: Rethinking Socioecological Systems Underlying High Nature Value Farmlands. *Front. Ecol. Environ.* **2020**, *18*, 36–42. [CrossRef]
44. Estel, S.; Kuemmerle, T.; Alcántara, C.; Levers, C.; Prishchepov, A.; Hostert, P. Mapping Farmland Abandonment and Recultivation across Europe Using MODIS NDVI Time Series. *Remote Sens. Environ.* **2015**, *163*, 312–325. [CrossRef]
45. Bürgi, M.; Bieling, C.; von Hackwitz, K.; Kizos, T.; Lieskovský, J.; Martín, M.G.; McCarthy, S.; Müller, M.; Palang, H.; Plieninger, T.; et al. Processes and Driving Forces in Changing Cultural Landscapes across Europe. *Landsc. Ecol.* **2017**, *32*, 2097–2112. [CrossRef]
46. Fuchs, R.; Herold, M.; Verburg, P.H.; Clevers, J.G.P.W.; Eberle, J. Gross Changes in Reconstructions of Historic Land Cover/Use for Europe between 1900 and 2010. *Glob. Chang. Biol.* **2015**, *21*, 299–313. [CrossRef] [PubMed]
47. Fernández-Nogueira, D.; Corbelle-Rico, E. Land Use Changes in Iberian Peninsula 1990–2012. *Land* **2018**, *7*, 99. [CrossRef]
48. Van der Zanden, E.H.; Carvalho-Ribeiro, S.M.; Verburg, P.H. Abandonment Landscapes: User Attitudes, Alternative Futures and Land Management in Castro Laboreiro, Portugal. *Reg. Environ. Chang.* **2018**, *18*, 1509–1520. [CrossRef] [PubMed]
49. INE Instituto Nacional de Estadística. 2020. Available online: https://www.ine.es/ (accessed on 3 November 2020).
50. Martínez-Fernández, J.; Ruiz-Benito, P.; Bonet, A.; Gómez, C. Methodological Variations in the Production of CORINE Land Cover and Consequences for Long-Term Land Cover Change Studies. The Case of Spain. *Int. J. Remote. Sens.* **2019**, *40*, 8914–8932. [CrossRef]
51. Lefebvre, M.; Espinosa, M.; Gomez y Paloma, S. *The Influence of the Common Agricultural Policy on Agricultural Landscapes. JS a. P. Report*; European Commission, Joint Research Center: Luxembourg, 2012; p. 83.
52. Bernués, A.; Ruiz, R.; Olaizola, A.; Villalba, D.; Casasús, I. Sustainability of Pasture-Based Livestock Farming Systems in the European Mediterranean Context: Synergies and Trade-Offs. *Livest. Sci.* **2011**, *139*, 44–57. [CrossRef]
53. Fischer, F. Participatory Governance: From Theory to Practice. In *The Oxford Handbook of Governance*; Oxford University Press: Oxford, UK, 2012. [CrossRef]
54. Newig, J.; Fritsch, O. Environmental Governance: Participatory, Multi-Level—And Effective? *Environ. Policy Gov.* **2009**, *19*, 197–214. [CrossRef]
55. Bermejo-Asensio, L.A.; Lobillo-Eguíbar, J.; Molina-Afonso, C. Aportes Del DRP (Diagnóstico Rural Participativo) a Las Metodologías Participativas y Aplicación a La Gestión de Los Recursos Naturales En La Gomera. *Práxis Particip. Desde Medio Rural.* **2003**, *6*, 71.
56. Verkerk, P.J.; Lindner, M.; Pérez-Soba, M.; Paterson, J.S.; Helming, J.; Verburg, P.H.; Kuemmerle, T.; Lotze-Campen, H.; Moiseyev, A.; Müller, D.; et al. Identifying Pathways to Visions of Future Land Use in Europe. *Reg. Environ. Chang.* **2018**, *18*, 817–830. [CrossRef]
57. Rolo, V.; Hartel, T.; Aviron, S.; Berg, S.; Crous-Duran, J.; Franca, A.; Mirck, J.; Palma, J.H.N.; Pantera, A.; Paulo, J.A.; et al. Challenges and Innovations for Improving the Sustainability of European Agroforestry Systems of High Nature and Cultural Value: Stakeholder Perspectives. *Sustain. Sci.* **2020**, *15*, 1301–1315. [CrossRef]
58. Quintas-Soriano, C.; Castro, A.J.; Castro, H.; García-Llorente, M. Impacts of Land Use Change on Ecosystem Services and Implications for Human Well-Being in Spanish Drylands. *Land Use Policy* **2016**, *54*, 534–548. [CrossRef]
59. Pausas, J.G.; Fernández-Muñoz, S. Fire Regime Changes in the Western Mediterranean Basin: From Fuel-Limited to Drought-Driven Fire Regime. *Clim. Chang.* **2012**, *110*, 215–226. [CrossRef]

60. Peco, B.; Navarro, E.; Carmona, C.P.; Medina, N.G.; Marques, M.J. Effects of Grazing Abandonment on Soil Multifunctionality: The Role of Plant Functional Traits. *Agric. Ecosyst. Environ.* **2017**, *249*, 215–225. [CrossRef]
61. Oggioni, S.D.; Ochoa-Hueso, R.; Peco, B. Livestock Grazing Abandonment Reduces Soil Microbial Activity and Carbon Storage in a Mediterranean Dehesa. *Appl. Soil Ecol.* **2020**, *153*. [CrossRef]
62. Sirami, C.; Brotons, L.; Burfield, I.; Fonderflick, J.; Martin, J.L. Is Land Abandonment Having an Impact on Biodiversity? A Meta-Analytical Approach to Bird Distribution Changes in the North-Western Mediterranean. *Biol. Conserv.* **2008**, *141*, 450–459. [CrossRef]
63. Carpio, A.J.; Apollonio, M.; Acevedo, P. Wild Ungulate Overabundance in Europe: Contexts, Causes, Monitoring and Management Recommendations. *Mammal Rev.* **2021**, *51*, 95–108. [CrossRef]
64. Cimatti, M.; Ranc, N.; Benítez-López, A.; Maiorano, L.; Boitani, L.; Cagnacci, F.; Čengić, M.; Ciucci, P.; Huijbregts, M.A.J.; Krofel, M.; et al. Large Carnivore Expansion in Europe Is Associated with Human Population Density and Land Cover Changes. *Divers. Distrib.* **2021**, *27*, 602–617. [CrossRef]
65. Carter, N.H.; Linnell, J.D.C. Co-Adaptation Is Key to Coexisting with Large Carnivores. *Trends Ecol. Evol.* **2016**, *31*, 575–578. [CrossRef]
66. Pozo, R.A.; Cusack, J.J.; Acebes, P.; Malo, J.E.; Traba, J.; Iranzo, E.C.; Morris-Trainor, Z.; Minderman, J.; Bunnefeld, N.; Radic-Schilling, S.; et al. Reconciling Livestock Production and Wild Herbivore Conservation: Challenges and Opportunities. *Trends Ecol. Evol.* **2021**, *36*, 750–761. [CrossRef] [PubMed]
67. Van der Zanden, E.H.; Verburg, P.H.; Schulp, C.J.E.; Verkerk, P.J. Trade-Offs of European Agricultural Abandonment. *Land Use Policy* **2017**, *62*, 290–301. [CrossRef]
68. Sinclair, K.; Curtis, A.; Freeman, P. Biosecurity in Multifunctional Landscapes: Challenges for Approaches Based on the Concept of 'Shared Responsibility'. *Prev. Vet. Med.* **2020**, *178*, 104682. [CrossRef] [PubMed]
69. Cockburn, J.; Cundill, G.; Shackleton, S.; Rouget, M.; Zwinkels, M.; Cornelius, S.A.; Metcalfe, L.; van den Broeck, D. Collaborative Stewardship in Multifunctional Landscapes: Toward Relational, Pluralistic Approaches. *Ecol. Soc.* **2019**, *24*. [CrossRef]
70. Vila-Subirós, J.; Rodríguez-Carreras, R.; Varga, D.; Ribas, A.; Úbeda, X.; Asperó, F.; Llausàs, A.; Outeiro, L. Stakeholder Perceptions of Landscape Changes in the Mediterranean Mountains of the North-Eastern Iberian Peninsula. *Land Degrad. Dev.* **2016**, *27*, 1354–1365. [CrossRef]
71. Esgalhado, C.; Guimarães, M.H.; Lardon, S.; Debolini, M.; Balzan, M.V.; Gennai-Schott, S.C.; Simón Rojo, M.; Mekki, I.; Bouchemal, S. Mediterranean Land System Dynamics and Their Underlying Drivers: Stakeholder Perception from Multiple Case Studies. *Landsc. Urban Plan.* **2021**, *213*, 104134. [CrossRef]
72. Herrera-Calvo, P.M.; García-Fernández, J. Dialogue, Participation and Social Innovation Initiatives for Improving Territorial Governance in Rural Environments. *Ciudades* **2021**, 45–64. [CrossRef]
73. Commission, E. A Farm to Fork Strategy for Fair, Healthy and Environmentally-Friendly Food System. In Proceedings of the Communication from the Commision to the European Parliament, the Council, the European Economic and Social Comittee and the Comitte of the Regions, Brussels, Belgium, 20 May 2020.
74. Herrera-Calvo, P.M.; Majadas-Andray, J.; Esteban, A.; Ramírez, N.; Rico, L. *La Ganadería Extensiva, Una Actividad Clave Para Nuestra Alimentación*; Fundación Entretantos: Valladolid, Spain, 2018.
75. Godfray, H.C.J.; Aveyard, P.; Garnett, T.; Hall, J.W.; Key, T.J.; Lorimer, J.; Pierrehumbert, R.T.; Scarborough, P.; Springmann, M.; Jebb, S.A. Meat Consumption, Health, and the Environment. *Science* **2018**, *361*. [CrossRef]
76. Manzano-Baena, P.; Salguero-Herrera, C. Mobile Pastoralism in the Mediterranean: Arguments and Evidence for Policy Reform and to Combat Climate Change. Available online: http://medconsortium.org/wp-content/uploads/2017/12/MobilePastoralismMotherDocument_December2017_ForWeb.pdf (accessed on 30 September 2021).
77. Manzano, P.; Burgas, D.; Cadahía, L.; Eronen, J.T.; Fernández-Llamazares, Á.; Bencherif, S.; Holand, Ø.; Seitsonen, O.; Byambaa, B.; Fortelius, M.; et al. Toward a Holistic Understanding of Pastoralism. *One Earth* **2021**, *4*, 651–665. [CrossRef]
78. Bruley, E.; Locatelli, B.; Lavorel, S. Nature's Contributions to People: Coproducing Quality of Life from Multifunctional Landscapes. *Ecol. Soc.* **2021**, *26*, 12. [CrossRef]
79. Ollivier, G.; Magda, D.; Mazé, A.; Plumecocq, G.; Lamine, C. Agroecological Transitions: What Can Sustainability Transition Frameworks Teach Us? An Ontological and Empirical Analysis. *Ecol. Soc.* **2018**, *23*. [CrossRef]
80. Fagerholm, N.; Eilola, S.; Kisanga, D.; Arki, V.; Käyhkö, N. Place-Based Landscape Services and Potential of Participatory Spatial Planning in Multifunctional Rural Landscapes in Southern Highlands, Tanzania. *Landsc. Ecol.* **2019**, *34*, 1769–1787. [CrossRef]
81. Holt, A.R.; Alix, A.; Thompson, A.; Maltby, L. Food Production, Ecosystem Services and Biodiversity: We Can't Have It All Everywhere. *Sci. Total. Environ.* **2016**, *573*, 1422–1429. [CrossRef]
82. Lavorel, S.; Grigulis, K.; Leitinger, G.; Kohler, M.; Schirpke, U.; Tappeiner, U. Historical Trajectories in Land Use Pattern and Grassland Ecosystem Services in Two European Alpine Landscapes. *Reg. Environ. Chang.* **2017**, *17*, 2251–2264. [CrossRef]

agriculture

Article

Participatory Mapping of Demand for Ecosystem Services in Agricultural Landscapes

Carmen Schwartz [1,2,*], Mostafa Shaaban [1], Sonoko Dorothea Bellingrath-Kimura [1,2] and Annette Piorr [1]

1 Leibniz Centre for Agricultural Landscape Research (ZALF e.V.), Eberswalder Straße 84, 15374 Müncheberg, Germany; mostafa.shaaban@zalf.de (M.S.); belks@zalf.de (S.D.B.-K.); apiorr@zalf.de (A.P.)
2 Thaer-Institute of Agricultural and Horticultural Sciences, Humboldt-University Berlin, Unter den Linden 6, 10117 Berlin, Germany
* Correspondence: carmen.schwartz@zalf.de

Abstract: Agricultural land use systems have been optimized for producing provisioning ecosystem services (ES) in the past few decades, often at the expense of regulating and cultural services. Research has focused mainly on the supply side of ES and related trade-offs, but the demand side for regulatory services remains largely neglected. The objective of this paper is to evaluate the usefulness of participatory geographic information system (PGIS) methods for demand assessment in larger rural and agrarian contexts by identifying spatially explicit demand patterns for ES, thereby enlarging the body of participatory approaches to ES-based land use management. Accordingly, we map, assess, and statistically and spatially analyze different demands for five ES by different stakeholder groups in agricultural landscapes in three case studies. The results are presented in a stakeholder workshop and prerequisites for collaborative ES management are discussed. Our results show that poor correlation exists between stakeholder groups and demands for ES; however, arable land constitutes the highest share of the mapped area of demands for the five ES. These results have been validated by both the survey and the stakeholder workshop. Our study concludes that PGIS represents a useful tool to link demand assessments and landscape management systematically, especially for decision support systems.

Keywords: participatory mapping; ecosystem services; demand; PGIS; agricultural landscapes

Citation: Schwartz, C.; Shaaban, M.; Bellingrath-Kimura, S.D.; Piorr, A. Participatory Mapping of Demand for Ecosystem Services in Agricultural Landscapes. *Agriculture* **2021**, *11*, 1193. https://doi.org/10.3390/agriculture11121193

Academic Editor: Francesco Caracciolo

Received: 13 October 2021
Accepted: 22 November 2021
Published: 26 November 2021

1. Introduction

Agricultural systems are genuinely social–ecological systems, with the possibility of producing a wide variety of provisioning ecosystem services (ES) and providing key ecological processes and regulatory services. The magnitude of the supply of agricultural ES is influenced by the interactions between the social and ecological systems, i.e., the farmer, current political regulations, consumer choices and the farming ecosystem [1,2].

Agricultural areas have been characterized by intensification, mechanization and a reduction in the labor force in the past few decades [3]. While this process has been considered essential for achieving food security, regulating ES have been mainly negatively affected by this process, such as pollination, agrobiodiversity, water cycling and clean air [4]. The continuous process of concentration of large parts of the land in the hands of few owners in the North-east of Germany has led to an increase in the average field size with increasing attention paid to maximizing the production of provisioning services, often reached through the reduction of landscape elements, such as tree rows and hedges, with negative consequences for regulating ES [5].

The transition to a more sustainable form of land use must fully account for the economic, ecological and social implications of agricultural productivity. The services and dis-services generated by these systems affect the stability of local and global ecosystems and, by extension, the people living in these systems [2].

Scientifically displaying the value of ES in agricultural landscapes has gained increased attention in the past few years. Efforts have been undertaken to display the biophysical, economic, environmental and social value of land use systems in monetary [6,7], non-monetary [8] and spatially explicit ways [9]. The Common International Classification of ES (CICES) [10] is a widely used assessment framework. It lists three main categories of ES: (1) provisioning, (2) regulation and maintenance and (3) cultural (see examples for each category in Table 1). These services are generated by underlying structures, processes and functions of the ecosystems. Biodiversity is the diversity of all living organisms and is considered to be both a function service that many other processes and services depend on and a service because it has direct benefits to human well-being. Our study focuses on one provisioning service (biomass yield) and four regulatory services (biodiversity, carbon sequestration, erosion control and water availability).

Table 1. Examples of provisioning, regulation and maintenance, and cultural ES modified according to CICES [10].

Section	Division	Group	Class	Class Type	Simple Descriptor
Provisioning	Biomass	Cultivated terrestrial plants for nutrition, materials or energy	Cultivated terrestrial plants (including fungi, algae) grown for nutritional purposes	Crops by amount, type (e.g., cereals, root crops, soft fruit)	Any crops and fruit grown by humans for food; food crops
	Biomass	Cultivated terrestrial plants for nutrition, materials or energy	Cultivated plants (including fungi, algae) grown as a source of energy	By amount, type, source	Plant materials used as a source of energy
Regulation and Maintenance	Regulation of physical, chemical and biological conditions	Regulation of baseline flows and extreme events	Hydrological cycle and water flow regulation (including flood control and coastal protection)	By depth/volumes	Regulating the flows of water in our environment
	Regulation of physical, chemical and biological conditions	Pest and disease control	Pest control (including invasive species)	By reduction in incidence, risk, area protected by type of living system	Controlling pests and invasive species
Cultural	Direct, in situ and outdoor interactions with living systems that depend on presence in the environmental setting	Physical and experiential interactions with natural environment	Characteristics of living systems that that enable activities promoting health, recuperation or enjoyment through active or immersive interactions	By type of living system or environmental setting	Using the environment for sport and recreation; using nature to help stay fit
	Direct, in situ and outdoor interactions with living systems that depend on presence in the environmental setting	Intellectual and representative interactions with the natural environment	Characteristics of living systems that are resonant in terms of culture or heritage	By type of living system or environmental setting	The things in nature that help people identify with the history or culture of where they live or come from

1.1. Mapping & PGIS Approaches

Participatory mapping of ES has gained increased attention in urban and rural contexts in the last few years [11–13]. Participatory geographic information system (PGIS) tools have been used to involve stakeholders in spatially explicit ES assessments by combining survey questions with a mapping component. They have proven useful for engaging people and their knowledge of landscapes in identifying and valuing ES in direct relationship to the landscape they originate from or are provided in. The PGIS tools have been used for highlighting the spatial heterogeneity of ES [14], perceived trade-offs and synergies [15], and flows of ES [16].

1.2. Demand and Supply Assessments and Trade-Offs

Identifying both an ecosystem's capacity to provide services (the supply side) and the social demand for those services (the demand side) remains a challenge in ES research [17]. Few studies combine assessments of both demand and supply of ES within the same study and region [18–20]; therefore, limited evidence exists on the demand for ES in relation to the supply within the same area. Spatially explicit knowledge about demand can show the connectivity between ecosystems and the beneficiaries of their services, and can predict competition over resources or possibilities for cooperation [21]. Combined approaches of supply and demand can capture the biophysical conditions for ES supply and societal needs that influence the actual supply that can be addressed via participatory mapping.

Geijzendorffer et al. [22] developed a scheme that encompasses five interlinked components along the supply demand continuum: interest, demand, match, managed supply and potential supply. An interest in ES becomes a demand only through the actual allocation of scarce resources, such as time or money, to fulfil this interest in a specific area and time. What follows from this definition is the identification of three types of mismatches between demand and supply: (i) actual uptake of ES is higher than the ecosystem can sustainably supply, (ii) managed supply leads to the production of certain services at the expense of others, trade-offs occur, and (iii) demand is unsatisfied due to insufficient supply. This scheme allows for the identification of trade-offs that a simple overlay of demand and supply maps might miss [23]. Some studies display trade-offs and synergies as a balance for the overall spatial entity under consideration [24], but studies that display trade-offs in a spatially explicit context are rare. Such information would be crucial for regionally optimized decision-making as different ES, due to biophysical and environmental processes and flows, are characterized by different spatial extensions. The trade-offs identified depend on the valuation method with which ES are assessed [25]. The dominance of biophysical and monetary supply assessments leads to a potential bias in the trade-offs identified. Therefore, including trade-offs in the demand assessments has the advantage of identifying trade-offs that would escape in a (biophysical or monetary) supply assessment. We visualize the preferences for ES in the demand assessment by different actors depending on the perceived utilities they expect from these ES. Trade-offs between demands are visualized when they are mapped in a spatial context, where the actors make decisions regarding allocating and specifying the areas of demands. Bringing the spatial ES supply demand trade-offs together helps one to identify the decision behaviors of the actors in response to the ecological behavior. It can help potential trade-offs to be foreseen that can arise if several ES are demanded but cannot be supplied in the same area. Furthermore, it includes a beneficiary perspective on trade-offs and can be a way of identifying potential bundles of ES that people demand in the same area with implications for management that reduces trade-offs between selected ES.

1.3. Assessment of Regulatory Services

A genuine problem with accounting for regulatory services is that they are nonmarketable goods and services and thus stay invisible if demand is assessed only for goods consumed. However, regulating ES are essential for the maintenance and perpetuation of the entire ecosystem. Regulating ES encompass a wide range of services closely linked to functions and processes that sustain the existence of the ecosystem itself and all living entities contained within it [26]. Regulating and cultural services are often traded off against the production of food, fiber and fuel [27]. Assessing regulating ES can be a preventive measure for avoiding trade-offs. Most demand assessments focus on provisioning services by assessing the marketed quantity (such as the yield of agricultural or forestry goods), while some on cultural services assess the willingness to pay or the distance travelled to visit natural environments (i.e., recreation). Regulatory services are more difficult to assess, and so are their benefits on human well-being that might occur with spatial or temporal delay or might not be perceivable at all. The more complex relationship between regulatory services and human well-being is one reason for a striking demand assessment gap in this field [28]. Furthermore, beneficiaries might not be aware of their demand for regulatory services, which constrains the possibility of assessing it [29]. Defining clear beneficiary groups in advance helps to avoid spatial mismatches between ES and their beneficiaries [21].

With this background from the literature, we were able to identify four main research gaps: (1) a lack of translation of trade-offs into the spatial dimension. ES assessments have become precise in identifying gaps of supply of singular and multiple ES. However, trade-offs between ES, especially regulating ES, become visible only if displayed in a spatially explicit context; (2) a lack of sufficient differentiation by beneficiaries of ES. Demands are directly related to the people benefiting from the supply. In the case of regulating ES, beneficiaries can be the people living directly in the surrounding area of the ES produced (e.g., water regulation) or people benefitting from the global effects of climate regulation worldwide (e.g., carbon sequestration); (3) a lack of studies spatially assessing the demand for regulating and provisioning ES. Most studies using PGIS assess preferences for cultural ES. Empirical assessment of the demand for and understanding of regulating ES is missing; and (4) the lack of an empirical grounding of Geijzendorffer's concept regarding demand.

In order to address these gaps, our study combines the assessment of demand for selected regulation ES with a digital mapping exercise. The goal of this approach is to assess demands differentiated by stakeholder group, ES and region. We want to find out whether demands differ by stakeholder group, and if demands for different, non-synergetic ES show spatial overlay and thus cause conflicts in land use decisions. We investigate supply perceived and demand stated, thereby analyzing the gap between the state perceived and the state desired. We choose a stakeholder-based approach to relate the demand formulated directly to specific groups of beneficiaries. In our study, we furthermore aim to include the stated interest in relation to the perceived current supply in a spatially explicit location for all ES individually. Thereby, we capture the interest formulated by individuals and relate it to the potential supply perceived by the participants.

We propose an approach of combining participatory mapping with a questionnaire that allows the evaluation of the perceived current supply and demand for several ES. The supply of ES in agricultural landscapes depends on the configuration of the site, and increases with extent [30]. Therefore, we combine the evaluation of ES with a mapping component that allows the mapping of large areas, sub regions and single plot areas. Furthermore, we discuss the areas highlighted in the mapping exercise in a stakeholder workshop. With this approach, we aim at obtaining information about possible agricultural areas of interest for ES management on a landscape level.

Our research questions are: Do demands for ES differ between stakeholder groups? Is there a relationship between demand for ES and current land use? Can we identify "hot spots" of demand for ES?

2. Materials and Methods

2.1. Study Regions

We selected three case study areas (CSAs) located in three administrative NUTS3 districts of Brandenburg, a federal state in the northeast of Germany. The districts were chosen according to different degrees of heterogeneity in soil composition, natural vegetation and different types of land use, displayed and visualized in Table 2 and Figure 1.

Table 2. Land use in all case study areas, 2019 (MOL = Märkisch-Oderland; OPR = Ostprignitz-Ruppin; UM = Uckermark) [27].

	MOL	OPR	UM
Total Area (km^2)	2158.5	2508.65	3058.35
Agricultural Area (km^2)	1255 (58.14% of total area)	1253 (49.95% of total area)	1766 (57.77% of total area)
Agricultural Area—Crops (km^2)	1160 (92.43% of agricultural area)	907 (72.39% of agricultural area)	1472 (83.35% of agricultural area)
Agricultural Area—Grassland (km^2)	91 (7.25% of agricultural area)	341 (27.21% of agricultural area)	293 (16.6% of agricultural area)
Perennial cultures and others (km^2)	4 (0.32% of agricultural area)	5 (0.32% of agricultural area)	1 (0.05% of agricultural area)
Forest Area (km^2)	510.27 (23.6% of total area)	813.76 (32.44% of total area)	748.11 (24.46% of total area)

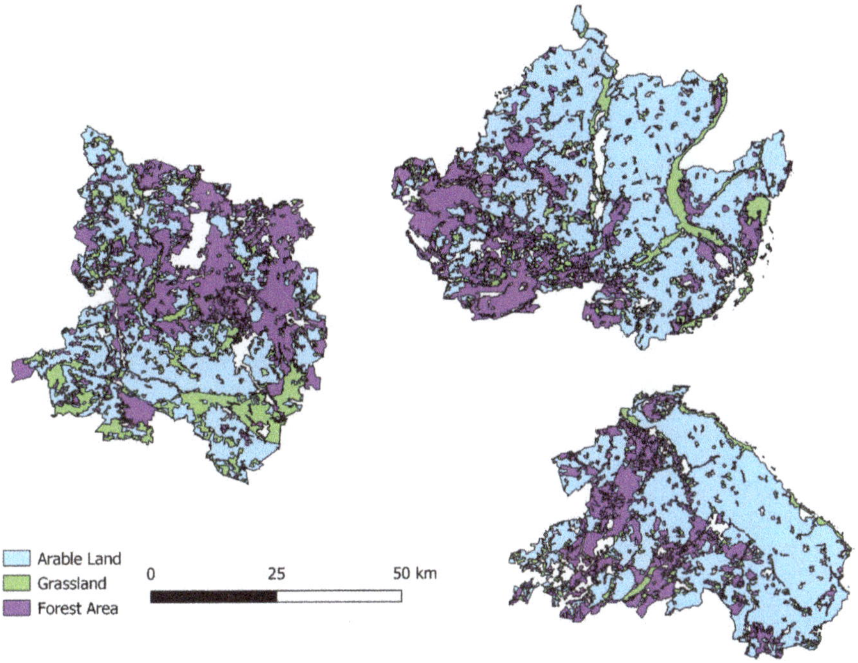

Figure 1. Land use in all case study areas.

Märkisch-Oderland (MOL) covers an area of 2158.67 km^2. The area used for agriculture in 2019 covered 1255 km^2 (58.14%) of the total area, of which 1160 km^2 were used for crops and 91 km^2 as permanent grasslands. Forest area comprises 510.27 km^2 or 23.6% of the total area [31].

Ostprignitz-Ruppin (OPR) covers 2509 km^2 and is located in the northwest of Brandenburg. It was founded in 1993, with the administrative center in Neuruppin. Agricultural and forest area cover 50 and 32%, respectively, of the total area. There are 18 nature reserves in OPR. Tourism, with possibilities for camping, hunting and water sports are important economic activities in addition to agriculture [32].

Uckermark (UM) covers an area of 3076.93 km^2 with 118,947 inhabitants [32]. In 2019, the agricultural area covered 1766 km^2 [27], of which 83% is used for crops, and 16.6% for grassland. Forest area covers 748 km^2 or 24% of the total area.

2.2. Questionnaire Design

We used Maptionnaire [33], a commercial PGIS tool that allows the active involvement of stakeholders in land use decision-making by capturing their perceptions. The tool has the advantage of integrating survey questions with spatial data. It allows participants to indicate areas and spots of interest on an interactive map and relate these spatial data to specific questions and attributes. The questionnaire runs on a web platform with a specific URL for each survey. It was originally developed for efficient interaction with stakeholders and better informed decision-making in urban contexts, but has proven useful for research in rural areas [12] and landscape planning [9].

After an introduction to the goal of the survey and the topic of ES, participants were able to self-assign to different stakeholder categories. Consequently, the survey was structured by the five ES assessed. We chose five ES for the demand assessment, especially services related to soil functioning and yield. These are: (1) biodiversity, (2) carbon sequestration, (3) erosion control, (4) water availability and (5) yield. Regulatory services and functions are essential for maintaining the economic viability and long-term functioning of ecosystems [26]. We based our definition of the respective ES on the CICES [10]. The survey followed the same structure for each ES.

The ES was explained and agricultural management practices that influence the supply of the respective ES were mentioned in a detailed description. Participants were asked to self-assess their knowledge of these ES based on the previous explanation. Participants were asked in a mapping exercise to map one to three areas they consider relevant for the ES (Appendix A, Figure A1). Participants were asked to estimate the current perceived supply levels of the ES ("How do you estimate current supply levels?") and to then state their demand of the same ES as a percentage of the optimum state ("how high should the supply be?") within the same areas mapped by them in pop-up windows with closed-end-questions. This procedure was similar for all five ES and was followed by a brief section collecting demographic and socioeconomic data from the participants. At the end of the survey, participants had the possibility to evaluate the length, quality and relevance of the questionnaire.

The questionnaire consists of 28 pages. The core questions regarding the demand for ES take up only 10 pages, whereas the rest includes an introduction to the survey, consent to data protection, demographic classifications of the participants and evaluation of the questionnaire. Some questions were followed by sub-questions in pop-up windows, leading to a variation in the number of questions to be answered between 26 and 30, depending on the stakeholder group. Depending on the number of areas mapped, participants were able to answer between 19 (one area mapped per ES) and 47 questions (three areas mapped per ES) regarding the evaluation of ES. This leads to a total number of questions to be answered varying between a minimum of 26 and a maximum of 58.

There are four main advantages of this way of proceeding: firstly, the data collected contains information on the ecosystem the service is provided by, and on the beneficiary, i.e., the person by whom the demand is formulated. Secondly, it contains information about the gap between the current state perceived and the state demanded. The underlying assumption is that the formulated demand depends on the perception of the current state. Thirdly, it shows possible trade-offs that can arise if several non-synergetic ES are demanded in one area by displaying them in a spatial manner. Fourthly, it follows the concept by Geijzendorffer of distinction between the demand for goods marketed and goods desired.

2.3. Survey Dissemination and Scope of Sampling

The questionnaire was open from 1 March to 30 November 2020. The reason for this long period was the sudden outbreak of the COVID-19 pandemic and its impact on the availability of stakeholders and the impossibility of face-to-face visits.

After a pretest with selected participants, a thorough search for stakeholders in the region was conducted based on a spatial raster with previously defined categories related to our chosen ES—agriculture, forestry, nature conservation, tourism, inhabitants and others. The target audience were potential multipliers in our CSAs, i.e., people with a sufficiently large network and the possibility to distribute the questionnaire further, from the different stakeholder groups in the areas identified. Management bodies of protected areas and environmental and agricultural associations were contacted in all three districts in Brandenburg, as well as organizations and action groups based in Berlin but working on a regional level. The questionnaire and the project were presented to them via email and telephone call. Furthermore, we distributed the questionnaire via social media channels and the homepage of the Leibniz Centre for Agricultural Landscape Research, and printed postcards with QR-codes directing those interested to the questionnaire. They were distributed in frequented places of the study regions during August 2020.

2.4. Data Analysis

The QGIS and R studio were used for the spatial and statistical analysis, respectively. Only completed questionnaires were included in the analysis. We conducted an overall analysis for the whole study region and analyzed subregions separately regarding interesting features and results.

2.4.1. Statistical Analysis

Using RStudio 1.3.959 [34] open source software, we estimate the correlation between stakeholder categories, supply perceived and demand stated for the five ES. The correlation between a nominal variable (stakeholder categories) and a continuous variable has been estimated through applying the function Intraclass Correlation Coefficient (ICCest) that uses the variance components from a one-way analysis of variance. The confidence interval is estimated by applying the type "THD", which is based upon the exact confidence limit equation in Searle [35] and can be used for unbalanced data (see Thomas and Hultquist [36]; Donner [37]). The aim of this analysis is to check whether actors from the same stakeholder category have a similar or close perception of the supply and demand or not. An example of the code used is shown in the text box below. This correlation has been further validated by coding the stakeholder categories with numerical values and implementing scatter plots between the respective coded stakeholder categories and supply perceived and demand stated.

```
Example of the code:
intraclasscc <- ICC::ICCest(Stakeholder_category,Water_Perceived_supply_average, data = NULL,
alpha = 0.05,CI.type = c("THD")) write.table(intraclasscc, 'cor.txt')
```

An average self-assessment of knowledge about the different ES was conducted for each CSA, and a ranking of the ES perceived as important common goods in the region. The relationship between the knowledge of ES and the perceived importance of ES was investigated.

The minimum, maximum and average values of the observations of the supply perceived S^k_{avg} and demand stated D^k_{avg} for each ES k in each CSA as well as for the three CSAs together were calculated and consequently the respective absolute values of the gap G^k_{avg} between supply perceived and demand stated was calculated according to Equation (1). We also created a boxplot based on the observations of the three CSAs regarding the same three variables to depict their dispersion across the median, 1st and 3rd quartile values. The boxplot has the advantage of showing outliers that also reflect poor knowledge and/or awareness that could otherwise be neglected consideration.

$$G^k_x = \left| S^k_x - D^k_x \right|, \ where\ x = min, maxor\ average \tag{1}$$

2.4.2. Spatial Analysis

The areas mapped contain information about the extent of the area that participants estimate relevant for ES supply, and information on their state of ES supply perceived and their demand for the mapped areas. We analyzed the mapped areas in relation to the attributes assigned to the areas by the participants, and in relation to the land use they are mapped upon.

In the first step, the different areas mapped for each ES were overlaid and graduated colors were distributed according to the indicated level of ES perceived and demanded in each area. For the same area, we calculated the gap between the supply and demand. We identified patterns of interest, and hotspots of demand and supply perceived in each CSA. We further focused on the visual identification of areas that were mapped for multiple ES, indicating potential trade-offs between ES supply.

In the second step, datasets on land use were integrated into the analysis. Data on arable land, grassland and forest from Corine Land Cover (CLC) 2018 [38] were used in order to assess the correlation between the areas mapped to different land uses. Specifically, the filtered categories were composed of (1) arable land: non-irrigated cropland (CLC code 211), (2) grassland: pasture and grassland (CLC code 231), and (3) forest: coniferous, deciduous and mixed forest (CLC codes 311, 312, 313, respectively). The areas mapped by the participants for each ES, were overlaid with the CLC 2018 maps. The overlaid areas were calculated in km^2 and % of the total mapped area.

2.5. Stakeholder Workshop

We discussed the results of the questionnaire with ten stakeholders from science, agriculture, entrepreneurship and regional management in an online stakeholder workshop. A focused open discussion emphasized the question regarding how participatory data can be used in regional management and landscape planning. We evaluated the criteria of successful participatory work and identified ways of implementing management strategies that can improve the supply of ES based on the assessment of demand.

3. Results

3.1. Respondents and Background

We collected a total of 53 complete questionnaires, of which 30 were collected in MOL, 14 in UM and 9 in OPR. The sample population was 40% male, 30% female, 2% diverse and 28% of the participants did not answer the gender question. Age distribution ranged between 21 and 80. Regarding affiliation with stakeholder groups, 34% selected farming and agriculture, 24% selected the stakeholder category of science, and 8% were related to civil society (Figure 2).

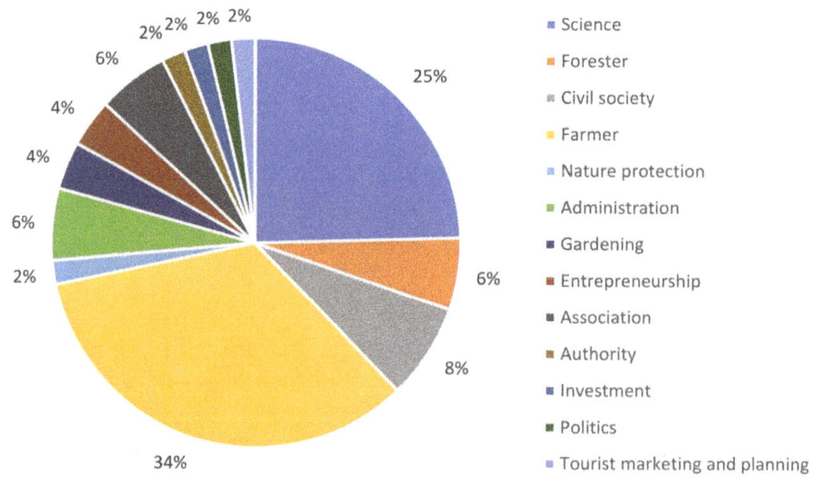

Figure 2. Percentage of stakeholder categories in the three case study areas.

Of the 18 farmers who completed the questionnaire, 10 reported working according to the guidelines of organic farming, 5 worked as conventional farmers, 1 was in transition to organic and 2 did not specify. Farm sizes varied between less than 20 ha (5 answers),

20–100 ha (7 answers), 100–500 ha (3 answers) and more than 500 ha (2 answers). The most frequent farm products are food crops, meat and vegetables (Appendix A, Figure A2).

3.2. Statistical Analysis

The intraclass correlation estimation showed a poor correlation between the variable stakeholder category against the average demand stated and the supply perceived of the different ES (Table 3). This be a consequence of the small sample size of actors from some stakeholder categories who took part in the survey. However, based on our results, no connection between the stakeholder group and the perception of current supply or demand for ES can be drawn. However, the highest intraclass correlation coefficient has been found with the supply of erosion control perceived, which has been checked via a scatter plot as shown in Figure 3 using numerical codes for stakeholder categories as shown in Table 4. It can be observed that farmers' supply of erosion control perceived falls to between 10 and 40%, whereas actors from science were split into two groups with high and low perception of the supply. Nevertheless, it ought to be noted that their mapped areas are not located in the same geographical location that justifies the variations in their perceptions, either.

Table 3. Intraclass correlation coefficients and confidence interval estimation results between stakeholder categories and supply perceived and demands for ES (CS = carbon sequestration; Bio = biodiversity; EC = erosion control).

Stakeholder_Category	ICC	LowerCI	UpperCI	*n*	k	varw	vara
CS_Perceived_supply_average	−0.069	−0.305	0.167	13	3.553	700.048	−45.100
CS_Demand_average	0.024	−0.173	0.378	13	3.553	1875.583	45.957
Bio_Perceived_supply_average	0.032	−0.168	0.387	13	3.553	1001.460	32.876
Bio_Demand_average	0.013	−0.180	0.365	13	3.553	2085.036	28.245
Yield_Perceived_supply_average	−0.112	−0.254	0.196	13	3.553	1228.897	−123.892
Yield_Demand_average	−0.009	−0.194	0.339	13	3.553	1844.923	−15.690
Water_Perceived_supply_average	−0.075	−0.233	0.250	13	3.553	652.103	−45.432
Water_Demand_average	−0.073	−0.232	0.253	13	3.553	1864.221	−126.847
EC_Perceived_supply_average	0.157	−0.084	0.518	13	3.553	555.733	103.313
EC_Demand_average	−0.001	−0.189	0.348	13	3.553	1573.915	−0.861

Abbreviation–Definition: ICC—intraclass correlation coefficient; LowerCI—lower confidence interval limit, where the confidence level is set by alpha; UpperCI—upper confidence interval limit, where the confidence level is set by alpha; *n*—total number of individuals or groups used in the analysis; k—number of measurements per individual or group. In an unbalanced design, k is always less than the mean number of measurements per individual/group and is calculated using the equation in Lessells and Boag [39]; varw—within individual or group variance; vara—among individual or group variance.

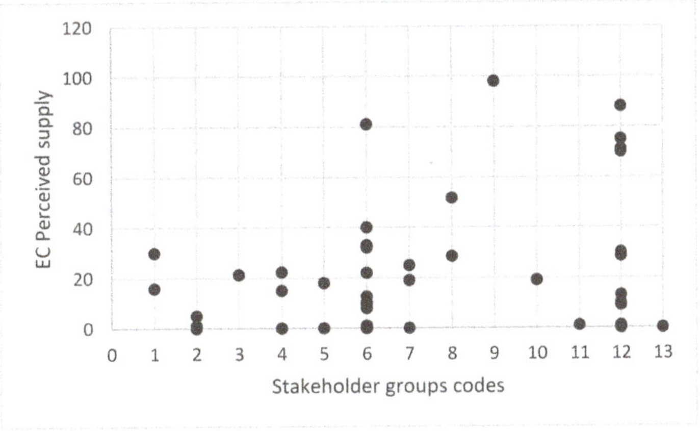

Figure 3. Scatter plot between the supply of erosion control (EC) perceived and coded stakeholder groups to display the correlation between the two variables.

Table 4. Codes of stakeholder groups used for the scatter chart in Figure 3.

Stakeholder Category	Code	Stakeholder Category	Code
Administration	1	Gardening	8
Association	2	Investment	9
Authority	3	Nature protection	10
Civil society	4	Politics	11
Entrepreneurship	5	Science	12
Farmer	6	Tourist marketing and planning	13
Forester	7	N.B. Zero value = not answered	

Average values of knowledge self-assessment for the different ES ranked between 60 and 70%, where biodiversity and water availability scored highest with 70%, and carbon sequestration lowest with 62.5%. Average awareness in UM and OPR was highest for erosion control with 79.9 and 76.8%, respectively, whereas water availability scored highest in MOL with 73%. Biodiversity and water availability were chosen most often in the ranking of ES as important public goods with 31 and 34 votes, respectively. Again, erosion control scored higher in OPR and UM than in MOL.

The average demand stated exceeded the average supply perceived for all ES in the test regions. Current supply levels were estimated to be the lowest on average for erosion control (24% of the optimum state), while biodiversity and yield were estimated to have higher supply levels with 47 and 55% of the optimum, respectively. Water availability scored relatively low in MOL (31%) and UM (32%), and relatively high in OPR (48%), leading to an average value of 33%, the same as carbon sequestration (33%). Average demand values stated were lowest for yield (79%), carbon sequestration (79%) and erosion control (80%), and highest for biodiversity (85%) and water availability (89%). The average supply demand gap was highest for erosion control and water availability (56%) and lowest for yield (24%) (Appendix A, Figures A3 and A4). The range of supply perceived is rather wide, from 24 to 55 average percentage points, whereas demands stated are at least 24% higher than supplies, and much narrower, ranging from 79 to 89%.

Figure 4 shows a boxplot comparing the supply perceived, demand stated and the gap for all ES values. The boxplot shows the dispersion of data for the three variables, which are stretched between 0 and 100 for almost all ES. However, the median value of the supply perceived is about 20% for carbon sequestration, water availability and erosion control, which is also very close to the 1st quartile value, indicating an agreement on a narrow value between 10 and 20%. Meanwhile, the demands for these ES comes in between 70 and 85%, which is justified by a gap ranging from 40 to 60%. On the contrary, it is more likely to lie between 60 and 75% for the supply of yield perceived with a close value of demand (75–80%), which makes the gap very small (about 20%). Surprisingly, the demand for biodiversity shows a high agreement between 90 and 100% and a low perceived supply (20–40%). The high dispersion between the median and the 1st or 3rd quartile or between the minimum or maximum value and the 1st or 3rd quartile, respectively, suggests some outlier data that might have originated from the low awareness of the participants about the status of the ES in the CSAs.

3.3. Spatial Analysis

3.3.1. Demand Area Index

The analysis of the areas indicated by respondents and ES was carried out for each district separately and aggregated over participants. We obtained information on the supply perceived and demand stated for the same area, and calculated the gap between supply and demand. Some areas or parts of an area were mapped several times by different participants or for different ES and were counted as often as they were mapped in this analysis. Only areas within the questionnaire boundaries were used for the spatial evaluation. The demand area index captures the total added area of all areas mapped. Table 5 shows these overall area values as an index for the three CSAs in km^2.

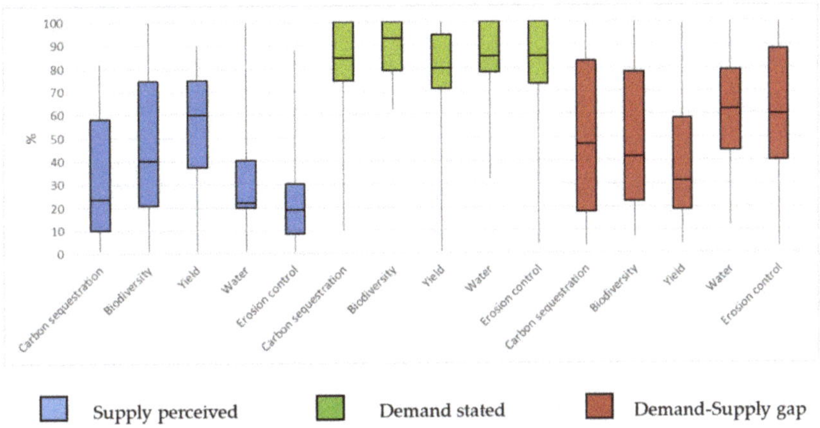

Figure 4. Average supply and demand assessment for ES in all three CSAs.

Table 5. Area mapped in km^2.

	MOL (km^2)	% of Total Area Mapped	OPR (km^2)	% of Total Area Mapped	UM (km^2)	% of Total Area Mapped
Biodiversity	21,551.5	13.67	10,212.7	16.19	8266.94	24.88
Carbon Sequestration	9710.42	6.16	25,911.9	41.08	2333.75	7.02
Erosion Control	41,159.3	26.10	6358.08	10.08	12234	36.82
Water Availability	69,534.8	44.09	15,032.2	23.83	7482.88	22.52
Yield	15,739	9.98	5559.28	8.81	2911.41	8.76
Total area mapped (km^2)	157,695.02		63,074.16		33,228.98	

The total area mapped was largest in MOL due to the high number of participants. The ES with the largest share mapped was water availability in MOL, with 44% of the total area mapped, followed by erosion control (26%) and biodiversity (14%). The largest area in OPR was mapped for carbon sequestration (41%), followed by water availability (24%) and biodiversity (16%). Erosion control (36%), biodiversity (25%) and water availability (22%) were the areas with the largest overall surface in UM. Yield scored below 10% of the total area mapped in all three districts.

Example Erosion Control in MOL

We use the results from the CSA MOL and the ES erosion control exemplarily to display relationships between the demand stated, supply perceived and the mapped areas.

The level of supply perceived indicated in MOL varied between 1 and 88 on a scale from 0 to 100% (Figure 5a). A total of 32 out of 45 participants (71%) rated current supply levels at 50 or lower, 4 participants (8%) rated supply levels higher than 50, and 9 participants (20%) only mapped an area without answering the question on supply levels.

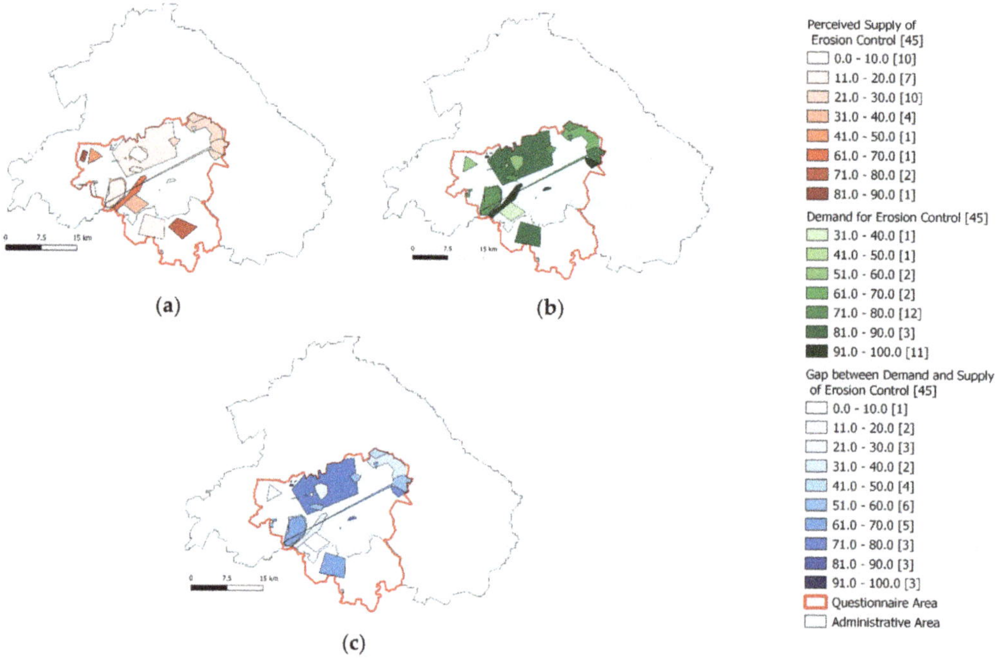

Figure 5. Supply perceived (**a**), demand stated (**b**) and the gap (**c**) for erosion control in MOL.

The levels of demand stated ranged from 40 to 100 (Figure 5b), where 30 out of 45 respondents (67%) rated their demand higher than 50, participants (4%) found their demand to be below 50, and 13 participants (29%) did not give an indication about their demand.

The demand levels indicated were in all cases higher than the supply levels perceived. The gap between the supply perceived and demand stated ranged from 7 to 98 (Figure 5c). The gap between the current state and the state demanded was higher than 50 in 20 of 45 cases (44%). Respondents found the discrepancy between demand and supply to be lower than 50 in 12 cases (27%). Thirteen respondents did not indicate either their demand or supply perceived, so the gap could not be calculated. Figure 5a–c show that the gap between demand and supply is highest where the supply perceived is rated rather low (<30%), and the demand rather high (>60%). Small gaps between demand and supply occur where both the state perceived and the demand are high or low. However, in a few areas the current state perceived equals the demand stated. The results do not show hot spots of demand for erosion control. However, in combination with demand maps for other ES, overlays of demand for multiple ES in almost all regions can be identified.

3.3.2. Mapped Area and Land Use

We again use the results from the CSA MOL, especially for erosion control, to demonstrate results for the analysis of the relationship between mapped areas and land use. Arable land covers 1160 km^2 or 53.7% of the total area in MOL. Results show that the largest areas mapped lies on arable land (Figure 6). This reflects the larger share of arable land in the landscape and is also an indication that the awareness of ES is higher in arable land than in forest and grassland areas. Table 6 shows the surface of the areas mapped by the participants in MOL for each ES, and its distribution by land use type. Areas mapped were only calculated once and values reflect the surface area mapped. Participants mapped proportionally more area on arable land than on forest or grassland sites. The share mapped on arable land for all ES was between 41 and 66% of the total area mapped. An average of between 5 and 14% were mapped on grassland, and between 18 and 39% on forest area. This reflects the general distribution of arable land, which is comparably higher than grassland and forest in all three CSAs (see Section 2.1), and even exceeds the proportion of arable land on the total area. The share mapped on arable land for some ES clearly exceeds the general share of arable land due to a coupled assignment of spatial entities to different ES, which can also be interpreted as an indication of the increased awareness of potential trade-offs between yield and other ES on arable land.

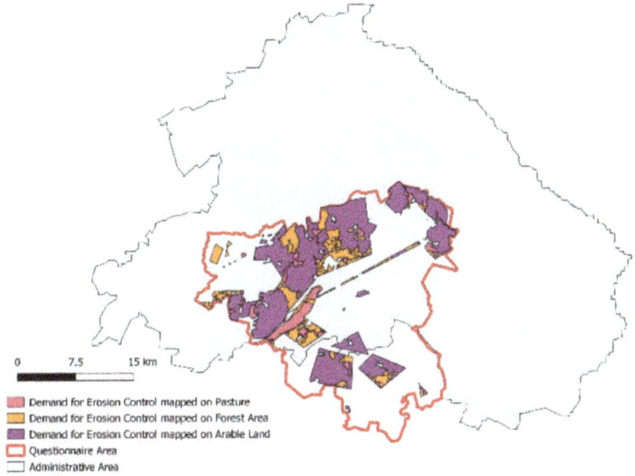

Figure 6. Demand for Erosion Control and Land Use in MOL.

Table 6. Mapped area for all ES per land use in Märkisch-Oderland.

	Biodiversity (km^2)	% of Total Surface Area Mapped	Carbon Seques-tration (km^2)	% of Total Surface Area Mapped	Erosion Control (km^2)	% of Total Surface Area Mapped	Water Avail-ability (km^2)	% of Total Surface Area Mapped	Yield (km^2)	% of Total Surface Area Mapped
Arable land	7294.76	40.80	5505.98	64.12	15,738.6	63.30	23,261.8	49.09	6970.03	65.60
Pasture	1836.81	10.27	1172.54	13.66	2223.18	8.94	3598.08	7.59	507.87	4.78
Forest area	6868.79	38.41	1526.91	17.78	5083.63	20.45	15,358.5	32.41	2794.77	26.30
Total surface area (km^2)	17,881		8586.82		24,864.8		47,388.1		10,625.2	

The largest areas in MOL are mapped for water availability, followed by erosion control and biodiversity.

3.3.3. Stakeholder Workshop

Our intention in conducting an online workshop with ten local representatives of the stakeholder groups addressed in the survey was to validate how far the results from this demand mapping exercise could be useful in regional and landscape management. Four general conclusions could be drawn from the stakeholder side. (1) A systemic approach combining the generation of knowledge on ES and the communication and dissemination of this knowledge is necessary. Participatory tools can be a helpful vehicle for improving the collection of perceptions, but the overarching goal has to be a more systemic perspective of management for a multitude of ES. (2) Trade-offs impede the achievement of measures to improve multiple ES. The PGIS can help to identify parts of these trade-offs. (3) The possibilities and limitations of participation should be clarified when inviting stakeholders to take part. Motivation decreases if people are asked for their opinion but their suggestions find no reflections in reality. (4) The more explicit the goal of participation, the more targeted a manner in which it can be carried out. The use of PGIS in workshops with technical guidance could be useful in regional development strategies or for landscape planning and decision-making in municipalities.

4. Discussion

In this discussion, we critically review our study results. In order to find out how this research could be applied in other areas, we then work out interfaces with other studies.

The results suggest that there is no difference between the supplies demanded and perceived according to the stakeholder group. Local people show similar understandings of the surrounding landscape, independent of their profession or affiliation. Furthermore, demands exceed the supply perceived of all ES in all CSAs. For all ES, participants suggest that the current levels do not match the state needed for an ecological equilibrium. The gap between demand stated and supply perceived is highest for erosion control and water availability (56%) and lowest for yield (24%). Participants mapped less than 10% of the total area mapped for yield in all three CSAs. Biodiversity and water availability are the two ES most demanded. Both are also the ES most recognized by the participants as important public goods. In addition, participants report having a generally high knowledge about these ES. Higher self-assessed knowledge can lead to a higher awareness of demand for the respective ES.

The comparison of areas mapped and current land use on the mapped areas indicates an increased awareness of ES on arable land in comparison to grassland and forest area. The share of areas mapped on arable land exceeds the share of arable land in the CSAs for most ES. This result is in line with the problem focus on erosion control and water availability, both features of arable land that particularly emerge with rather uniform cropping patterns and low diversification of landscape and crop rotations. This causal explanation has also been raised by the stakeholder side when presenting and discussing these results.

The focus of our empirical study is to attempt to locate and evaluate other categories of ES both spatially and explicitly using PGIS. Since a comparable direct subjective level of experience or perception, as with cultural dimensions (e.g., beauty, harmony, naturalness), cannot be assumed here, we allowed all respondents equal information access to the queried ES and functions—biodiversity, carbon sequestration, erosion control, water availability, yield—and to infer indications of their condition.

When interpreting the results, it is important to keep in mind that experience ratings and information may have interacted to varying degrees here. It cannot be excluded, for example, that the ES erosion control and water availability are mentioned primarily because their assessment can be more observation-based than, for instance, the more abstract knowledge-based ES carbon sequestration.

At the same time, at the very least, we felt it was important and necessary to contrast the distribution of mapped ES demand preferences and ES supply situations perceived for the purposes of consistency analysis according to Brown et al. [40]. These authors mention spatial accuracy and credibility (reputation, trustworthiness and motivations of the spatial data contributor) as key data quality features for the use of public participatory mapping in land use planning. We addressed spatial accuracy with the proof of logical consistency with land use classes distribution ('validity-as-accuracy'). Regarding 'validity-as-credibility,' we did not find any difference in the outcome depending on who did the assessment.

The use of participatory instruments can be a powerful tool for democratizing land use planning and bringing transparency into decisions. The PGIS can be a useful tool for investigating people's perception of ES in the landscape and capturing their desired state. However, possible pitfalls should be considered and avoided. In particular, the reason for participation should be clarified, and the possibilities and limitations should be made clear to the participants. The format of map-based identification of very specific landscape requires much knowledge of the landscape itself and elaborate digital skills. A more targeted way in larger rural areas might be to use this tool in workshops or in smaller groups of local experts under technical guidance. This would allow the collection of more background information on location-specific perceptions and notions of individuals, as well as discursive statements between experts. Such could provide a valuable tool for local land use governance processes.

PGIS and stakeholder analysis could in further research be combined with geostatistical or modeling approaches for obtaining robust and site-specific results. Simulation models or optimization models have the advantage of allowing quantification, upscaling and systemic assessments. However, leaving out stakeholder involvements in these assessments leads to limited practical utility for site-specific decision making [41]. The combination of both participatory mapping and analytical or modeling approaches can produce integrated and practical results for land-use decisions, both in trade-off analyses and in ES valuation.

The combination of participatory mapping of non-cultural ES with monetary valuations is another field of potential use. The reason why the empirical focus of social needs has been predominantly on cultural ES so far is that the preferences stated can also be expressed most clearly here, since they are based on individual evaluations of experience. Fagerholm and Käyhkö [42] describe social landscape evaluations as subjectively experienced and related to location as well as context. Monetary valuations of ES can give an idea about their value in relation to societies' gross domestic product but often miss out on functionalities and system dynamics that are not yet understood. Furthermore, they foster an idea of the replaceability of ES by financial means. The combination of mapping with monetary approaches could capture both numerical and spatial values that enable a closer approximation of the economic and social value of the ES. A study by Kenter [43] combines choice experiment with participatory systems modelling, participatory mapping and psychometric analysis. Results show that with a participatory component, participants were better able to include personal values, a public perspective or place-based values into the monetary valuation of ES and the analysis of trade-offs [43].

The use of decision support systems (DSS) play an increasing role in the choice of management strategies for the supply of multiple ES, climate change mitigation and biodiversity preservation in agriculture and forestry [44,45]. The challenge of balancing different demands for ES in agricultural landscapes creates a necessity for DSS that integrate demands for provisioning, regulating and cultural ES across spatial and temporal dimensions. The usefulness of results from ES demand and supply assessment for DSS requires understandable data representation. More research is needed on the question of how ES information can be integrated into DSS in a decision-supportive way [46]. Our research shows one possibility of representing demand in a spatially explicit way that could be integrated with biophysical data on supply and recommendations for management in DSS.

5. Conclusions

We used a map-based questionnaire to collect data on demand and supply perception of five ES, formulated by different stakeholder groups in three CSAs. We discussed the results of the survey with experts from the stakeholder groups in a stakeholder workshop. Our aim is to evaluate the usefulness of PGIS methods in larger rural and agrarian contexts and to contribute a new methodological approach for assessing spatially explicit demands for regulating and provisioning ES. The demands for ES play a growing role in the management of ES, especially in agricultural areas. A harmonization of different demands from different stakeholders can avoid trade-offs and alleviate the decision for management strategies to improve multiple ES supply.

Our study shows no significant differences in demands between stakeholder groups. Our results rather suggest the importance of including local knowledge on landscapes in land use decisions and give the first indication that people from different stakeholder categories have profound knowledge of their surrounding ecosystems.

These results are preliminary, and we encourage further systematic investigation into its procedural aspects. We recommend the methodology presented as a starting point for demand analyses in similar agrarian contexts to generate results for comparison. The issue of spatial trade-off mapping is a valuable investigation area, where we recommend a smaller scale and reduced number of ES with a rather distinct functional interrelation as a starting point. This research could be extended by complementary geostatistical or analytical approaches.

Author Contributions: Conceptualization, A.P., C.S. and M.S.; methodology: Section 2.1, C.S.; Section 2.2, C.S. and A.P.; Section 2.3, C.S. and A.P., Section 2.4.1, M.S.; Section 2.4.2, C.S. resources, A.P. and S.D.B.-K.; writing—original draft preparation, C.S.; writing—review and editing, A.P., M.S. and S.D.B.-K.; visualization, C.S. and M.S.; supervision, A.P and S.D.B.-K.; project administration, A.P. and S.D.B.-K.; funding acquisition, A.P. and S.D.B.-K. All authors have read and agreed to the published version of the manuscript.

Funding: This research was funded by the German Federal Ministry of Education and Research (BMBF) through the Digital Agriculture Knowledge and Information System (DAKIS) Project, grant number FKZ 031B0729A, and the APC was funded by the DAKIS Project.

Institutional Review Board Statement: Not applicable.

Informed Consent Statement: Not applicable.

Acknowledgments: We would like to thank all reviewers for their efforts and fruitful feedback on the article.

Conflicts of Interest: The authors declare no conflict of interest. The funders had no role in the design of the study; in the collection, analyses, or interpretation of data; in the writing of the manuscript; or in the decision to publish the results.

Appendix A

Appendix A.1. Ecosystem Services

In our study, we focus on the mapping of five regulating ES in the landscapes of Brandenburg. The selection of ES follows the typology of CICES and includes four regulating and one provisioning ES. The descriptions we gave in the questionnaire of the five ES chosen were as follows.

Appendix A.2. Erosion Risk and Erosion Control

Soil erosion describes the removal of fine-grained topsoil by wind or water leading to a deterioration in the soil quality. Erosion is a natural process but is often greatly enhanced or triggered by the use of soils. Negative impacts of erosion can range from a reduction in soil fertility, loss of important soil functions and crop failure to flooding and contamination of trails and roads. The erosion risk of sites is largely determined by natural factors, such as slope, rainfall intensity and soil characteristics. Land use by humans, particularly the geometry and size of cropland, selection of crop types and intensity of tillage, can influence erosion risk significantly. Appropriate erosion control measures include establishing diverse crop rotations, ensuring long soil cover through cover crops and undersowing, slope-parallel tillage, and conservation tillage, for example, mulch seeding. At the landscape level, windbreak plantings in the form of hedges and woody elements and the permanent greening of slope hollows and depth contours up to change of use in areas particularly at risk of erosion can be useful.

Appendix A.3. Water Availability

Almost 98% of agricultural land in Germany is fed by green water, i.e., rainwater. The amount of groundwater and surface water available to plants is influenced by meteorological and hydrological factors. Climate change is altering the dynamics of the temporal availability of rainwater so that periods of heavy rainfall and drought may occur. Rising temperatures also favor evaporation rates. At the same time, topography and soil characteristics affect a soil's ability to hold plant-available water. Appropriate land use measures can influence the average water availability positively—the soil's water-holding capacity can be improved by building up soil organic carbon. Crop and variety selection can be adapted using varieties with lower water requirements.

Appendix A.4. Carbon Sequestration

Soil organic matter is around 50% carbon and is an important feature of soil fertility. Soils with a high organic matter content can store and release more nutrients and water to plants than soils with less organic matter. In addition, carbon sequestration in soils is increasingly seen as a way of reducing atmospheric carbon levels as mitigation and adaptation strategies for global climate change. Increasing (anthropogenic) carbon sequestration in soils and plants can be achieved by:

- favoring biomass growth. Perennial crops and woody plants, in particular the removal of the greenhouse gas CO2 from the atmosphere by building up biomass, thus contributing to climate change mitigation.
- the development of organic matter-rich soil horizons by adding organic material (compost, crop residues).

Appendix A.5. Biodiversity

Biological diversity or biodiversity ensures the vital services provided by nature. Biodiversity encompasses all living things in their various habitats: in the soil, in water and on land—from animals and plants to fungi and bacteria. In addition to species diversity, biodiversity includes genetic diversity and the diversity of communities of organisms. Biodiversity provides key regulatory services to ecosystems, such as the pollination of crops, soil fertility, protection against environmental disasters, such as floods, landslides and avalanches, purification of water and air, decomposition of waste and pollutants, and natural pest control. Human interference with nature alters the food base and habitats of organisms and thus ecosystems. Urban sprawl, landscape fragmentation and increasing land use are negatively affecting the number of habitats. At the same time, land use holds great potential for implementing biodiversity-enhancing measures. These include the creation of large reservoirs of biodiversity of farm animals, cultivated plants, habitats and wild organisms adapted to them.

Appendix A.6. Yield

Yield is the biomass of crops and fodder plants harvested and marketed per year per hectare. On arable land, this refers to food crops, energy crops, fodder crops and industrial crops. On forestry land, yield refers to the biomass of wood.

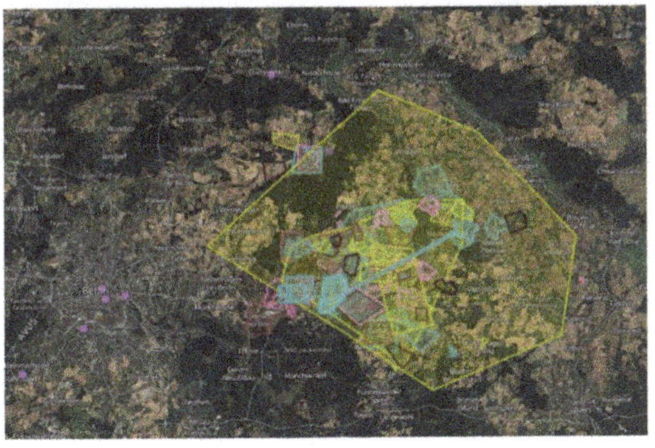

Figure A1. Screenshot of Maptionnaire survey surface.

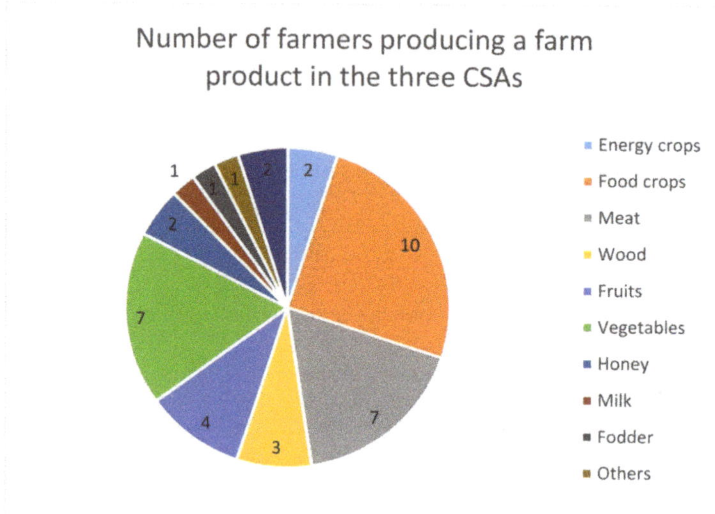

Figure A2. Variety of farm products produced in all three CSAs.

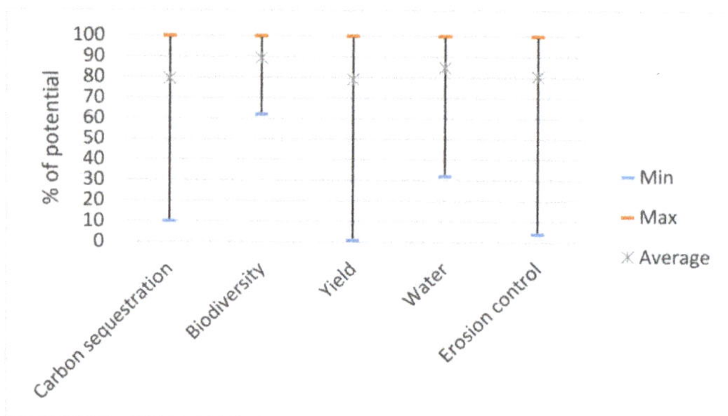

Figure A3. Demand stated for ES in all three CSAs.

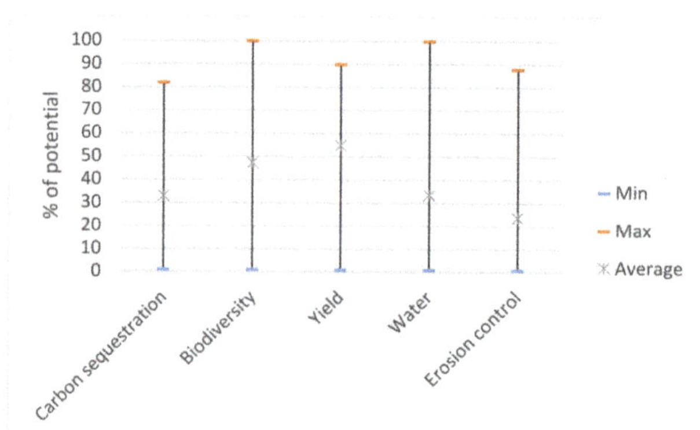

Figure A4. Supply perceived for ES in all three CSAs.

References

1. Power, A.G. Ecosystem services and agriculture: Tradeoffs and synergies. *Philos. Trans. R. Soc. B Biol. Sci.* **2010**, *365*, 2959–2971. [CrossRef] [PubMed]
2. Zhang, W.; Ricketts, T.H.; Kremen, C.; Carney, K.; Swinton, S.M. Ecosystem services and dis-services to agriculture. *Ecol. Econ.* **2007**, *64*, 253–260. [CrossRef]
3. Altieri, M.A. Sustainable agricultural development in Latin America: Exploring the possibilities. *Agric. Ecosyst. Environ.* **1992**, *39*, 1–21. [CrossRef]
4. Rasmussen, L.V.; Coolsaet, B.; Martin, A.; Mertz, O.; Pascual, U.; Corbera, E.; Dawson, N.; Fisher, J.A.; Franks, P.; Ryan, C.M. Social-ecological outcomes of agricultural intensification. *Nat. Sustain.* **2018**, *1*, 275–282. [CrossRef]
5. Ungaro, F.; Zasada, I.; Piorr, A. Turning points of ecological resilience: Geostatistical modelling of landscape change and bird habitat provision. *Landsc. Urban Plan.* **2017**, *157*, 297–308. [CrossRef]
6. Costanza, R.; d'Arge, R.; De Groot, R.; Farber, S.; Grasso, M.; Hannon, B.; Limburg, K.; Naeem, S.; O'Neill, R.V.; Paruelo, J.; et al. The value of the world's ecosystem services and natural capital. *Nature* **1997**, *387*, 253–260. [CrossRef]
7. Sandhu, H.S.; Wratten, S.; Cullen, R.; Case, B. The future of farming: The value of ecosystem services in conventional and organic arable land. An experimental approach. *Ecol. Econ.* **2008**, *64*, 835–848. [CrossRef]
8. Sherrouse, B.C.; Clement, J.M.; Semmens, D.J. A GIS application for assessing, mapping, and quantifying the social values of ecosystem services. *Appl. Geo.* **2011**, *31*, 748–760. [CrossRef]
9. Van Berkel, D.B.; Verburg, P.H. Spatial quantification and valuation of cultural ecosystem services in an agricultural landscape. *Ecol. Indic.* **2014**, *37*, 163–174. [CrossRef]
10. Haines-Young, R.; Potschin, M. Common International Classification of Ecosystem Services (CICES) V5.1 and Guidance on the Application of the Revised Structure. 2018. Available online: www.cices.eu (accessed on 24 November 2021).
11. Wolff, S.; Schulp, N.; Verburg, P. Mapping ecosystem services demand: A review of current research and future perspectives. *Ecol. Indic.* **2015**, *55*, 159–171. [CrossRef]
12. Garcia-Martin, M.; Fagerholm, N.; Bieling, C.; Gounaridis, D.; Kizos, T.; Printsmann, A.; Müller, M.; Lieskovský, J.; Plieninger, T. Participatory mapping of landscape values in a Pan-European perspective. *Landsc. Ecol.* **2017**, *32*, 2133–2150. [CrossRef]
13. García-Díez, V.; García-Llorente, M.; González, J.A. Participatory mapping of cultural ecosystem services in madrid: Insights for landscape planning. *Land* **2020**, *9*, 244. [CrossRef]
14. Fagerholm, N.; Torralba, M.; Moreno, G.; Girardello, M.; Herzog, F.; Aviron, S.; Burgess, P.; Crous-Duran, J.; Ferreiro-Domínguez, N.; Graves, A.; et al. Cross-site analysis of perceived ecosystem service benefits in multifunctional landscapes. *Glob. Environ. Chang.* **2019**, *56*, 134–147. [CrossRef]
15. Plieninger, T.; Torralba, M.; Hartel, T.; Fagerholm, N. Perceived ecosystem services synergies, trade-offs, and bundles in European high nature value farming landscapes. *Landsc. Ecol.* **2019**, *34*, 1565–1581. [CrossRef]
16. Palomo, I.; Martín-López, B.; Potschin, M.; Haines-Young, R.; Montes, C. National Parks, buffer zones and surrounding lands: Mapping ecosystem service flows. *Ecosyst. Serv.* **2013**, *4*, 104–116. [CrossRef]
17. Castro, A.J.; Verburg, P.H.; Martín-López, B.; Garcia, M.; Cabello, J.; Vaughn, C.C.; López, E. Ecosystem service trade-offs from supply to social demand: A landscape-scale spatial analysis. *Landsc. Urban Plan.* **2014**, *132*, 102–110. [CrossRef]
18. Schulp, C.; Lautenbach, S.; Verburg, P. Quantifying and mapping ecosystem services: Demand and supply of pollination in the European Union. *Ecol. Indic.* **2013**, *36*, 131–141. [CrossRef]
19. Nedkov, S.; Burkhard, B. Flood regulating ecosystem services—Mapping supply and demand, in the Etropole municipality, Bulgaria. *Ecol. Indic.* **2012**, *21*, 67–79. [CrossRef]
20. Kroll, F.; Müller, F.; Haase, D.; Fohrer, N. Rural–urban gradient analysis of ecosystem services supply and demand dynamics. *Land Use Policy* **2012**, *29*, 521–535. [CrossRef]
21. Bagstad, K.J.; Johnson, G.; Voigt, B.; Villa, F. Spatial dynamics of ecosystem service flows: A comprehensive approach to quantifying actual services. *Ecosyst. Serv.* **2013**, *4*, 117–125. [CrossRef]
22. Geijzendorffer, I.R.; Martin-Lopez, B.; Roche, P. Improving the identification of mismatches in ecosystem services assessments. *Ecol. Indic.* **2015**, *52*, 320–331. [CrossRef]
23. Lautenbach, S.; Mupepele, A.-C.; Dormann, C.F.; Lee, H.; Schmidt, S.; Scholte, S.S.K.; Seppelt, R.; Van Teeffelen, A.J.A.; Verhagen, W.; Volk, M. Blind spots in ecosystem services research and challenges for implementation. *Reg. Environ. Chang.* **2019**, *19*, 2151–2172. [CrossRef]
24. Burkhard, B.; Kroll, F.; Nedkov, S.; Müller, F. Mapping ecosystem service supply, demand and budgets. *Ecol. Indic.* **2012**, *21*, 17–29. [CrossRef]
25. Martin-Lopez, B.; Gómez-Baggethun, E.; García-Llorente, M.; Montes, C. Trade-offs across value-domains in ecosystem services assessment. *Ecol. Indic.* **2014**, *37*, 220–228. [CrossRef]
26. Smith, P.; Ashmore, M.R.; Black, H.I.J.; Burgess, P.J.; Evans, C.D.; Quine, T.A.; Thomson, A.M.; Hicks, K.; Orr, H.G. REVIEW: The role of ecosystems and their management in regulating climate, and soil, water and air quality. *J. Appl. Ecol.* **2013**, *50*, 812–829. [CrossRef]
27. Rodríguez, J.; Beard, T.D., Jr.; Bennett, E.; Cumming, G.; Cork, S.; Agard, J.; Dobson, A.P.; Peterson, G.D. Trade-offs across space, time, and ecosystem services. *Ecol. Soc.* **2006**, *11*, 1. [CrossRef]

28. Mengist, W.; Soromessa, T.; Feyisa, G.L. A global view of regulatory ecosystem services: Existed knowledge, trends, and research gaps. *Ecol. Process.* **2020**, *9*, 1–14. [CrossRef]
29. Syrbe, R.-U.; Grunewald, K. Ecosystem service supply and demand–the challenge to balance spatial mismatches. *Int. J. Biodivers. Sci. Ecosyst. Serv. Manag.* **2017**, *13*, 148–161. [CrossRef]
30. Goldman, R.L.; Thompson, B.H.; Daily, G.C. Institutional incentives for managing the landscape: Inducing cooperation for the production of ecosystem services. *Ecol. Econ.* **2007**, *64*, 333–343. [CrossRef]
31. Maerkisch-Oderland, L. Statistischer Jahresbericht 2020. 2021. Available online: https://www.maerkisch-oderland.de/datei/anzeigen/id/31785,1249/statistischer_jahresbericht_2020.pdf (accessed on 24 November 2021).
32. Berlin-Brandenburg AfS. Bodennutzung und Ernte. 2021. Available online: https://statistik-berlin-brandenburg.de/regionalstatistiken/r-gesamt_neu.asp?Ptyp=410&Sageb=41002&creg=BBB&anzwer=8 (accessed on 24 November 2021).
33. Maptionnaire 2021. Available online: https://maptionnaire.com/ (accessed on 24 November 2021).
34. RStudio Team. *RStudio: Integrated Development Environment for R*; RStudio—PBC: Boston, MA, USA, 2021; Available online: https://www.rstudio.com (accessed on 8 September 2021).
35. Searle, S.R.; Gruber, M.H. *Linear Models*; John Wiley & Sons: Hoboken, NJ, USA, 2016.
36. Thomas, J.D.; Hultquist, R.A. Interval Estimation for the Unbalanced Case of the One-Way Random Effects Model. *Ann. Stat.* **1978**, *6*, 582–587. [CrossRef]
37. Donner, A. The use of correlation and regression in the analysis of family resemblance. *Am. J. Epidemiol.* **1979**, *110*, 335–342. [CrossRef]
38. EEA. Corine Land Cover (CLC). 2018. Version 20b2. Available online: https://land.copernicus.eu/pan-european/corine-land-cover/clc2018 (accessed on 24 November 2021).
39. Lessells, C.; Boag, P.T. Unrepeatable repeatabilities: A common mistake. *Auk* **1987**, *104*, 116–121. [CrossRef]
40. Brown, G.; Sanders, S.; Reed, P. Using public participatory mapping to inform general land use planning and zoning. *Landsc. Urban Plan.* **2018**, *177*, 64–74. [CrossRef]
41. Klapwijk, C.; van Wijk, M.; Rosenstock, T.; van Asten, P.; Thornton, P.; Giller, K. Analysis of trade-offs in agricultural systems: Current status and way forward. *Curr. Opin. Environ. Sustain.* **2014**, *6*, 110–115. [CrossRef]
42. Fagerholm, N.; Käyhkö, N. Participatory mapping and geographical patterns of the social landscape values of rural communities in Zanzibar, Tanzania. *Fenn. Int. J. Geogr.* **2009**, *187*, 43–60.
43. Kenter, J.O. Integrating deliberative monetary valuation, systems modelling and participatory mapping to assess shared values of ecosystem services. *Ecosyst. Serv.* **2016**, *21*, 291–307. [CrossRef]
44. Grêt-Regamey, A.; Sirén, E.; Brunner, S.H.; Weibel, B. Review of decision support tools to operationalize the ecosystem services concept. *Ecosyst. Serv.* **2017**, *26*, 306–315. [CrossRef]
45. Nordström, E.-M.; Nieuwenhuis, M.; Başkent, E.Z.; Biber, P.; Black, K.; Borges, J.G.; Bugalho, M.N.; Corradini, G.; Corrigan, E.; Eriksson, L.O.; et al. Forest decision support systems for the analysis of ecosystem services provisioning at the landscape scale under global climate and market change scenarios. *Eur. J. For. Res.* **2019**, *138*, 561–581. [CrossRef]
46. Klein, T.; Celio, E.; Grêt-Regamey, A. Ecosystem services visualization and communication: A demand analysis approach for designing information and conceptualizing decision support systems. *Ecosyst. Serv.* **2015**, *13*, 173–183. [CrossRef]

agriculture

MDPI

Article

Reconnecting Farmers with Nature through Agroecological Transitions: Interacting Niches and Experimentation and the Role of Agricultural Knowledge and Innovation Systems

Cynthia Giagnocavo [1,*], Miguel de Cara-García [2], Mónica González [3,4], Melchor Juan [5], José Ignacio Marín-Guirao [2], Sepide Mehrabi [6], Estefanía Rodríguez [2], Jan van der Blom [7] and Eduardo Crisol-Martínez [7,8,9]

1. Cátedra COEXPHAL-UAL on Horticulture, Cooperative Studies and Sustainable Development, Department of Economics and Business, University of Almería, Carretera Sacramento s/n, 04120 La Cañada, Almeria, Spain
2. Department of Sustainable Crop Protection, Andalusian Institute of Agricultural and Fisheries Research Training (IFAPA) La Mojonera, Camino San Nicolás, 1, 04745 La Mojonera, Almeria, Spain; franciscom.cara@juntadeandalucia.es (M.d.C.-G.); josei.marin@juntadeandalucia.es (J.I.M.-G.); mestefania.rodriguez@juntadeandalucia.es (E.R.)
3. Experimental Station of Foundation Cajamar, 04710 Las Palmerillas, Almeria, Spain; mgonzalez@agrobio.es
4. Agrobio S.L. Ctra Nacional 340, Km 419, nº 81, 04746 La Mojonera, Almeria, Spain
5. Laboratory of Phytopathology, LABCOLOR-COEXPHAL (Laboratory of Association of Vegetable and Fruit Growers of Almeria), C\Esteban Murillo, 3, 04746 La Mojonera, Almeria, Spain; melchor@coexphal.es
6. Department of Economics and Business, University of Almería, Carretera Sacramento s/n, 04120 La Cañada, Almeria, Spain; sepidemehrabi@ual.es
7. Department of Crop Production Techniques, COEXPHAL (Association of Vegetable and Fruit Growers of Almeria), C\Esteban Murillo, 3, 04746 La Mojonera, Almeria, Spain; jvdblom@coexphal.es (J.v.d.B.); ecrisol@coexphal.es (E.C.-M.)
8. Escuela de Agronomía, Facultad de Ciencias Agrarias y Forestales, Universidad Católica del Maule, Casilla 7-D, Curico 3349001, Chile
9. EcoLaVerna Integral Restoration Ecology, Bridestown, Kildinan, T56CD39 Co. Cork, Ireland

* Correspondence: cgiagnocavo@ual.es

Citation: Giagnocavo, C.; de Cara-García, M.; González, M.; Juan, M.; Marín-Guirao, J.I.; Mehrabi, S.; Rodríguez, E.; van der Blom, J.; Crisol-Martínez, E. Reconnecting Farmers with Nature through Agroecological Transitions: Interacting Niches and Experimentation and the Role of Agricultural Knowledge and Innovation Systems. *Agriculture* **2022**, *12*, 137. https://doi.org/10.3390/agriculture12020137

Academic Editors: José Luis Vicente-Vicente, Cristina Quintas-Soriano and María D. López-Rodríguez

Received: 15 December 2021
Accepted: 14 January 2022
Published: 20 January 2022

Abstract: Sustainability transitions in agriculture are explored through an analysis of niche initiatives within a common production system, relying on sustainable transitions, multi-level perspectives, and agroecological frameworks, and involving multi-actor, agricultural knowledge, and innovation systems (AKIS). The article focuses on how experimental niches and sustainable activities affect farmers' relationships with nature, and the reconceptualisation of the production system in which they operate, particularly where this system is embedded in less sustainable conventional or dominant regimes and landscapes. The need for fundamental changes, in the way that humans interact with nature, is widely argued for in order to achieve sustainable development, and farmers occupy a central role through participation in complex networks of agri-food systems. They have also found themselves disconnected from nature through conventional agri-industrial production practices. Four niches (biological control, ecological restoration, soil health, and ecological pond management) within the greenhouse sector of Almeria (SE Spain) are explored in a case study. Our results indicate that a farmer's interaction with nature is functional, but through agroecological practices, a deeper understanding of the ecosystems in which greenhouse landscapes are embedded may be gained. As they become more connected to nature and benefit from ecosystem services, they can transition to more sustainable agricultural systems.

Keywords: human nature connectedness; sustainability transitions; agricultural innovations; multi-level perspective; agroecology; agricultural knowledge and innovation systems (AKIS); conservative agriculture practices; knowledge co-production; mediterranean horticulture; integrated pest management; greenhouses; soil health; biological control; pond naturalisation; collective action; socio-ecological systems

1. Introduction

In this paper, we explore processes of sustainability transitions in agriculture through the analysis of four niche initiatives in Almeria, South-East Spain, within a common intensive greenhouse production system, involving diverse multi-actor, agricultural knowledge and innovation systems (AKIS). A central focus of our research is to understand how involvement in niche experimental and sustainable activities affects a farmer's relationship with nature, and the reconceptualisation of the production system in which such niches operate, particularly where this system is heavily embedded in less sustainable conventional or dominant regimes and landscapes. The need for fundamental changes in the way that humans interact with nature is widely argued for in order to achieve sustainable development [1] and farmers; both, in their roles as humans and producers within the complex network of agri-food systems, occupy crucial roles. Through agricultural activity, a farmer's interaction with nature is functional and economic. However, through the practice of agriculture, a deeper understanding of the ecosystems in which farming activities are located may be gained by farmers, their organisations, and, in general, their AKIS. As they become more connected to nature through practice, they can transition to more sustainable agricultural systems.

Agricultural activity has a significant social, economic, and environmental impact, particularly in intensive systems. Recent studies based on Crippa et al. [2] and Rockstrom et al. [3], amongst others, demonstrate that dominant agricultural and food systems have led to serious and ongoing resource depletion and severe and inequitable environmental and social impacts. In addition to academic calls for transformation, social and political stakeholders are also calling for change in agri-food systems [4–7]. The United Nations Environment Programme (UNEP) [8] has made it abundantly clear that making the transformation of food and water systems more equitable and resilient is an urgent goal, citing the need to implement a wide range of agricultural management systems and approaches, including conservation agriculture, organic farming, agroecology, integrated pest and nutrient management, soil and water conservation, agroforestry, and irrigation management, to name a few (p. 121), applicable to various farming systems.

More specifically, the UNEP report refers to the need for a reduction in nitrogen and phosphorus imbalances to reduce pollution of freshwater, groundwater, and coastal zones. In addition, overuse of pesticides and fertilisers can produce several negative consequences, including damage to ecosystems, biota and human health, and environmental pollution, among others [9–13]. Biodiversity loss is taking place rapidly due to various human activities, agriculture being one of them [9,10,14,15], and for this reason, to ensure food security, as well as for other non-anthropocentric reasons, it is crucial to reconcile agricultural production and biodiversity conservation.

"Humanity is waging war on nature" is the more direct, non-academic language used by the Secretary General of the United Nations, Antonio Guterres, who stated bluntly "Making peace with nature is the defining task of the 21st century. It must be the top, top priority for everyone, everywhere" [16].

However, globally, agricultural business generally appear to have difficulties in calling a truce: market influences, supported by policy, have resulted in increased farm size and vertical and horizontal integration [17], as well as intense power concentration in the inputs and distributor ends of global supply chains [18], leaving little bargaining power for small and medium farmers. Agricultural liberal market policy measures in Europe (and elsewhere) have often favoured increased production and intensification at the expense of biodiversity and ecosystem services [19]. According to the European Environmental Bureau (EEB), Dupeux [20], and Poore and Nemecek [21], food systems are responsible for 26% of greenhouse gas emissions worldwide; they not only contribute to climate change, but also to the deterioration of ecosystems and unprecedented levels of species loss [22]. Natural resources in which farmers rely on are under such pronounced overexploitation that an extinction crisis is a threat [23]. The same EEB report observes that "European farmers are lurching from crisis to crisis with an ageing farming population which struggles to attract new young farmers".

All of these influences have had an impact on agri-food systems, from seed to fork. Farmers, their advisors, and their relevant AKIS, which include networks of individuals, research and education, bridging institutions, business and enterprise, and the enabling environment institutions and policies, have also been influenced by agri-food system pressures and the demand to increase production and optimise efficiencies, while at the same time, becoming aware that farming activity depends on ecological sustainability. AKIS is considered a key concept in identifying, analysing, and assessing the various actors in the agricultural sector, as well as their communication and interaction for innovation processes [24]. It should be noted that certain literature also refers to Agricultural Innovation Systems or "AIS", which are similar to AKIS, but with a more pronounced focus on system innovations [25]. For the purposes of this article, we will refer to AKIS, which is understood to include agricultural knowledge, and the various technical, organisational, social, and institutional organisations involved in innovation and transition processes.

Calls for changing agri-food systems are increasingly framed in the context of sustainability transitions [4,26–28], including in the agri-food sector [1,29–34]. The Sustainability Transitions: Policy and Practices report was published in 2019 by the Sustainability Transitions Research Network (STRN) [35], and studies a wide range of sectors, agriculture being recently added as an "official" group. Sustainability transitions are sociotechnical transitions that are associated with sustainability targets and that switch systems to more sustainable modes of production and consumption [28]. They deal with fundamental changes that are complex, multi-dimensional (technical, organisational, institutional, political, and sociocultural), and generally are long-term and uncertain [36,37]. They can be disruptive, contested—that is, they involve tradeoffs for different actors—and affect different parts of the value chain. Sustainability transitions and pathways are also highly dependent on the context of sectors, places, and social or technical maturity/readiness levels. These transitions are a multi-actor process, which co-evolve, navigating between both stability and change. The sustainability transitions literature does not ignore that power and politics also play a central role, although it is not reductionist.

Coupled with the framework of sustainability transitions, is the multi-level perspective (developed by Rip and Kemp 1998 [38]) with reference to climate change, and further developed by Smith et al. [37] and Geels [39,40], which develops three central concepts that will be utilised in this paper: niche, regime, and landscape. Niche (micro level) refers to small networks of actors that carry out innovative activities and by virtue of their experimental, limited nature, or shared "space" are protected from the dominant systems, whether the market or otherwise. Regime (meso level) refers to the dominant, incumbent social–technical system including formal and informal rules. The regime includes technologies, institutions, and actors, and they offer coherence, stability, and are not prone to radical transformation, but rather incremental adjustments. Landscape (macro level) refers to broad societal trends, macro economic trends, political developments, cultural, and societal values, and exogenous events, such as crises, demographic changes, climate change, etc. Changes in the landscape can open up opportunities for niches and put pressure on regimes to change [41].

Because of the context dependent nature of sustainability transitions and the various levels in which they operate, existing AKIS may enhance innovation niches to support sustainable transitions across the various systems implicated in agricultural activity through leveraging collective and integrated innovation from different levels of activity [25].

This article focuses mainly on farmers and the production phase of agriculture. While some criticism has been directed at scholars for not "adequately addressing food systems" [5,26,41], by involving the whole of the classic supply chain actors, such as consumers, distributors, and processors, it should be recognised that there are multiple systems at play in agri-food, which are not focused only on linear supply chain relationships, but also on agro-ecological, socio-ecological, sociotechnical, or nexus systems (i.e., food, water, energy), which are often implicated mainly in production activities. Analysis of the production stage, as this article does, is not necessarily based on a more limited approach to systems, but rather on various different interacting systems that may include non-market as well as market systems. In particular, the production system, which encompasses much more than agricultural production activity within farm boundaries, also implicates social–ecological or human–environment systems that describe human behaviour and the interaction with other systems, such as the water, biodiversity, and ecosystems, as well as sociotechnical systems, where ecological, biophysical, and geographical dependencies occur [25,30,42]. Necessarily, any study of transitions, even in the agri-food sector, implicates a wide range of disciplines [14,43–45].

Returning to the main theme then, of connecting farmers to nature, this paper engages with the imperative to "make peace with nature" from what may seem a rather surprising or unsuitable candidate for an argument for sustainability transition: an intensive agricultural system involving 32,000 ha of fruits and vegetables (plastic) greenhouse farming in a semi-arid region of Almeria, South-East Spain.

Part of the process of "making peace with nature" implies a deeper understanding of and relationship with nature. "Connectedness" to nature is proven to have a positive impact on an expanded sense of valuing non-human species and also leads to pro-environmental and conservation behaviour [46,47], particularly amongst farmers [48]. This connection has been defined as the extent to which individuals include nature in their emotional and cognitive perspectives [49,50]. In arguments for a transition to a more sustainable agriculture, much emphasis has been put on consumers and their power to reshape production through market demand [51,52]. In agroecological transitions, the consumer relationship to growers is noted as being transformative [53]. However, this begs the question of what a relationship with growers will achieve if the growers themselves are not connected to nature and engaging in sustainable agriculture.

As a consequence of industrial agri-food systems, coupled with demographic trends that have seen a significant shift of people from rural to urban areas, not only has there been a disconnection of a large proportion of the European population with agroecosystems, within which such agricultural activity is carried out, but farmers themselves have also found themselves disconnected from nature by virtue of their participation in conventional agri-industrial production [54,55]. A farmer's relationship with nature is a fundamental cornerstone of any attempt to transition to sustainable agricultural systems. Consequently, the relationship with nature is also influential in shaping other people's understanding of agriculture, including, but not limited to, local communities, consumers, and civil society. The question then is whether farmers themselves can reconnect with nature through the adoption of more sustainable agricultural practices, and more importantly, how, given the constraints of markets, path dependencies and lock-ins. Recent European initiatives, such as the Farm to Fork or the European Green Deal, have refocused on the importance of farmers in sustainable agricultural transitions, as well as their advisors and other AKIS [56], but putting this into practice at a system level is challenging.

This article addresses this challenge by combining several analytic approaches, which are then applied to four niche initiative case studies.

2. Material and Methods

This article employs mixed methods, triangulating desk research, experimental and project results, and is centred around four case studies representative of related niche activity within the Almeria intensive farming system. It is a result of a multi-actor authorship approach. We use several overlapping frameworks to explore these niches: sustainability transition frameworks [57]; multi-level perspective [39]; and Gliessman's five levels of food system change [53], coupled with multi-actor AKIS, as well as Geels [36] and Pereira et al.'s [58], analysis of multi-level transitions and the dangers of lock-ins [59], which inhibit transitions.

Gliessman's five levels of agroecology provide us with a conceptual link for classifying "levels" of agri-food system change based on the relationship of growers to other actors in food systems and also their role in bringing about sustainability transitions through incremental and transformational change [53] (Figure 1). Level 1 is concerned with optimising resource use efficiencies, level 2 with the substitution of conventional practices with agroecological practices, and level 3 with the redesign of agroecosystems, the latter being part of transformational change. The fourth and fifth levels go beyond the farm into the food system and the societies in which they are embedded. These five levels altogether represent a roadmap to outline a process for transforming a food system into an agroecological one.

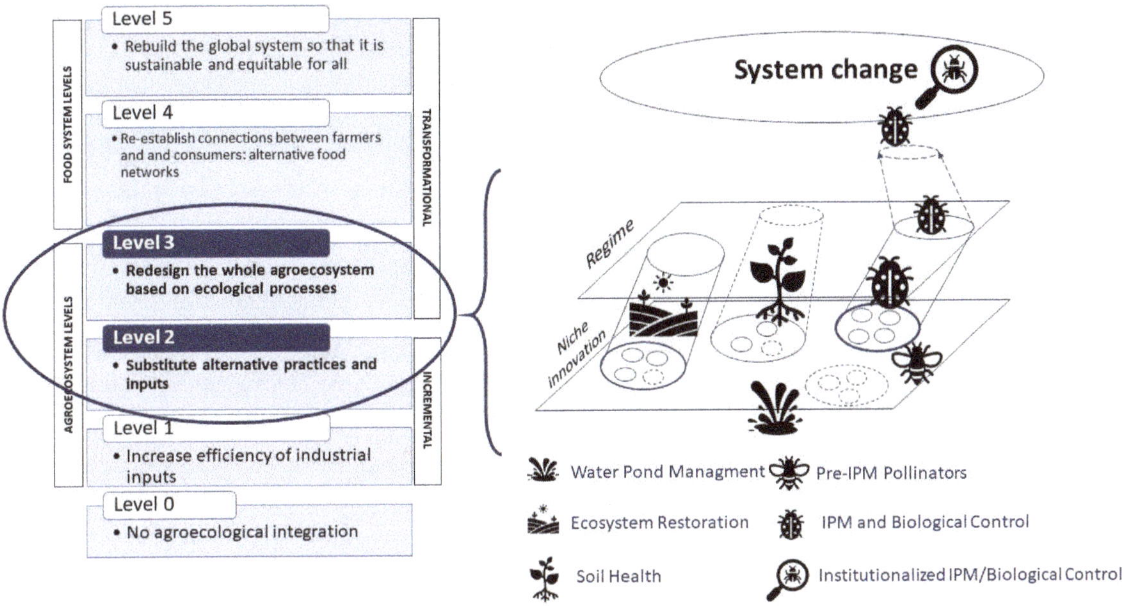

Figure 1. Focus of article and Almeria's location in the five levels of agroecology and four niche experiments in the multi-level perspective; authors' elaboration based on Gliessman, S. [53], Geels, F.W. [36], and Pereira, L.M., et al. [58].

This article is concerned with the implementation of level 2 (via niche innovations and experiments) and the transition to level 3, wherein growers transition away from conventional agricultural systems. The multi-level perspective, previously described above, will be utilised to set out the development necessary to effect the transformation from one level to another [36], considering niche activity, existing regime characteristics, and landscape, corresponding to: individual, collective, and organisational behaviour of farmers, their advisors, businesses, and other actors in discreet experiments, projects and activities; the dominant pattern of actors and structures in agri system; and broad exogenous factors; or prevailing meta narratives, respectively.

The inclusion of AKIS provides a context in which to describe in more detail the multi-actor approach, and more importantly, the local system in which the niche activities operate to bring farmers and their advisors closer to nature, so as to bring about sustainable transitions. Activation of resources and capabilities within AKIS, along with co-creation and collective action through niche activities, have begun to create shared visions of more sustainable (and realisable) paths [24,33,60]. More specifically, projects are able to "dynamically configure capabilities" for agricultural systems innovation [61].

In line with the above approach, this consideration of several niche initiatives within the Almeria AKIS related to the greenhouse sector in southeast Spain, is an example of how agricultural transitions can utilise niche activities and the dynamic characteristics of farming systems and specific contexts to "reset" narratives and development paths. We also refer to various regional, national, and European projects and initiatives that have been leveraged to accelerate such change, providing a "safe space" to experiment outside of a dominant market logic [61].

Finally, we will consider the outcomes of the four case studies in terms of their transition pathways from niche to influencing a new regime, relying on Geels and Kemp [62], Joffre et al. [59] and Pereira et al. [58].

3. Results and Discussion

3.1. Application to Almeria Agricultural Production System

Almeria represents a particular agricultural development model, one which in the last 50 years has undergone a profound technical, socio-economic, and agricultural system transition and transformation. Five decades ago, one would have found subsistence farming conditions. It was one of the poorest regions of Europe, suffering from inept policies of autarky and the isolation of Franco's dictatorship. Many of its inhabitants had to migrate to other parts of Europe and the Americas to survive. However, in the 1960s and 1970s, the installation of electrical infrastructure allowed more efficient exploitation of underground aquifers and the Franco government put into place development schemes, which lured farmers to the area. Rudimentary greenhouses were set up using cane sticks and old posts from a defunct table grape trade, with plastic strung over the top. Cooperative finance, based on knowledge of the German credit cooperatives, was quietly set up for the farmers by a small group of people who were searching for alternatives between fascism and communism [63,64], as well as markets so that farmers did not have to sell individually, at the low prices of the farm gate, to intermediaries. With the death of Franco in 1975, other institutional structures were put in place, such as an export association, an experimental agricultural farm initiated by the cooperative bank, and the establishment of farmer owned agricultural cooperatives, encouraged by local and state entities [65].

Throughout the remainder of the 1970s, technical, institutional, and socio-economic infrastructure was put into place. By the 1980s and 1990s, Almeria's production was increasingly influenced by European (and then global) trade, subject to increased competition and downward pressure on prices and increased calls for higher production [66]. Almeria agriculture has often been characterised as an agricultural industrial district or cluster [67].

Over the last five decades, Almeria has fostered an economic base of family farmers, which still survives to this day. The farmland is divided into small side-by-side parcels amongst 16,000 small-scale family farmers, each cultivating an average of approximately two hectares. These small growers are predominantly organised around packaging and marketing cooperatives and producer organisations, and the production activity is complemented by a significant agricultural auxiliary industry (plastics, seeds, fertilisers, natural enemies and pollinators, etc.) which is almost exclusively non-cooperative. The cooperative entities play an important role in the AKIS, as do public and private advisors, public and private research centres, universities, and companies.

The share of product marketed by cooperatives, as opposed to capitalistic companies, has actually increased [68]. Over 70% of sales are handled by marketing cooperatives, with an annual turnover of over EUR 2200 million and annual production of over 3.5 million tons (the total being somewhat higher if auctions are included). More than 90 farmer-owned cooperatives are currently operating in areas providing either specialised or general services. The agricultural cooperatives of Almeria represent 21% of all fruit and vegetable (F&V) cooperative turnover in Spain, and about 75% of Almeria's production is for export, accounting for 25% of the total Spanish F&V exports.

The farmers and their cooperatives have evolved locally, generating networks among the farms and interrelationships with the other stakeholders in the sector and territory.

There are cooperative associations (e.g., the Association of Fruit and Vegetable Producers Organisations of Almeria, COEXPHAL, and the Federation of Agrarian Cooperative Entities of Andalusia, CAA), as well as research institutions such as IFAPA, the Cajamar cooperative bank experimental station, and the University of Almeria. Collectively, they form an intertwined AKIS cluster.

In the last ten years, due to both market and social forces, various actors in the sector have become aware of the need for increased economic, social, and environmental sustainability [69]. Like much of European agriculture, the product has become commoditised and the margins increasingly small. Farmers and their cooperatives and producer organisations constantly seek new ways to optimise resources, differentiate products, and compete within the intensive agricultural system in which they find themselves. However, there is a growing sense that the optimisation of resources (Level 1), is not enough to ensure sustainability and that the system itself is in need of change.

Over the last 50 years, agriculture has been under scrutiny due to the impact it has on natural ecosystems and its repercussions on global warming. Several decades of "Green Revolution" principles have led to a high level of intensification of agricultural systems. Almeria's intensification began in the late 1970s as a response to poverty and food insecurity, and technologies that allowed easier exploitation of natural resources, gaining strength in the 1980s when markets liberalised, generated a sudden growth of the greenhouse area during the 1990s. Recent works highlight the involvement of such intensive agricultural models in the decline of biodiversity worldwide. For instance, insect biomass is falling, with an average of 2.5% annually, and every year, 1% of species are added to the list of declining ones. Among the main potential determining factors of this decline are habitat loss, agrochemical pollution, the introduction of invasive species, and climate change [70,71]. Habitat loss is the primary driver of insect declines in 49.7%. Chemical pollution, mainly due to synthetic fertilisers and pesticides, not only lowers insect numbers and other biodiversity, but also degrades their ecosystem functions [72]. This trend is worrying, as insects constitute essential items for trophic networks and secure the integrity and sound functioning of the world's terrestrial ecosystems—e.g., by providing USD 400 billion annually worth in 'natural biological pest control' [73].

The Green Revolution was founded on technological ability, based on scientific principles, with the underlying assumption that science was better than nature at providing better conditions for crops. While this assumption has been erroneous, for avoidance of doubt it does not signify a return to subsistence farming. The total area of protected cultivation is steadily increasing in the EU. In 2015 the estimated total area in the EU was about 175,000 ha, and the rate of increase was close to 4.5% between 2005 and 2013. Although in Almeria province, greenhouses constitute about 4% of the total surface [74], the development of intensive horticulture in the province has registered a huge increase in greenhouse surface area along last decades, reaching 32,368 ha in 2020 [75], where two agricultural regions (i.e., Campo de Dalías and Campo de Níjar) comprise up to 85–90% of that area. The greenhouses in Almeria produce over 3.7 million tons per year of F&V providing direct employment to more than 40,000 workers annually. More than 250 complementary or auxiliary businesses, both cooperative and investor-owned have been created with a turnover of more than EUR 1500 million. In a relatively short period (50 years), the people of this province went from suffering abject poverty to having a thriving, internationally focused economy [65]. However, this socio-economic development has had certain negative environmental impacts on a region with exceptional ecological value, and at times, this, together with the loss of natural capital, including the maintenance of ecosystem functions, has not been fully recognised by its population or policymakers.

To help to redress this situation, a critical rethinking of the current agro-production paradigm is necessary. During the early 1980s, agroecology emerged as a reaction opposing the current paradigm proposed by the Green Revolution. At that time, the focus of agroecology occurred at the farm level, where farmers were encouraged to increase productivity by substituting agrochemical inputs for sustainable, ecological principles provided by biodiversity. The concept of agroecology evolved during the 1990s to become the 'ecology of the entire food system' [76]. Currently, the agroecosystem view has expanded beyond the farm level to include all participants in the food system, by re-establishing the connection between farmers and consumers, while minimising the negative impact of all the actors between both groups. A further step was taken more recently by including a political economy focus to include all aspects of the food system and develop alternatives to the lock-ins that prevent a change in the food systems. The term agroecology is now defined as:

> "The integration of research, education, action and change that brings sustainability to all parts of the food system: ecological, economic, and social. It's transdisciplinary in that it values all forms of knowledge and experience in food system change. It's participatory in that it requires the involvement of all stakeholders from the farm to the table and everyone in between. And it is action-oriented because it confronts the economic and political power structures of the current industrial food system with alternative social structures and policy action. The approach is grounded in ecological thinking where a holistic, systems-level understanding of food system sustainability is required". [77]

In this study, we incorporate Gliessman's vision of agroecology (and its incremental levels) to characterise four areas (each one presented as a separate niche study). We explore how agroecology can transform conventional agricultural systems into systems that shield arthropods and other biodiversity, as well as consistently improve both production and environmental outcomes in which farmers are increasingly experimenting, relying on biodiversity-derived solutions through the gradual recognition and appreciation of nature, supported by their local AKIS. Section 3.2 sets out the current problems presented by the conventional agricultural systems, followed by each of the four niche transitions in Section 3.3.

3.2. Current Problems in Conventional Agricultural Systems

3.2.1. Pest Management

Due to the Mediterranean climate, the high concentration of greenhouses, and the overlap of different crop cycles, the area is extremely vulnerable to pests and diseases. Pest and disease management in greenhouse horticulture is more complicated than in most open field crops because the greenhouse microclimatic conditions and high plant density favour them [78]. In most horticultural crops, there is a prolonged period of harvest, so high pest levels cannot be tolerated by growers at any moment during the crop cycle. Furthermore, most crops are vulnerable to insect-transmitted viruses, potentially causing considerable crop losses [79]. These characteristics have influenced risk-aversion attitudes among most farmers towards pests and diseases, who, until the mid-2000s, based their management on chemical control with pesticides.

However, excessive reliance on chemical management of pests has led to an intensive use of pesticides and to pests developing resistance to insecticides [80]. Excessive pesticide treatments can produce multiple negative impacts on the health of farmers, workers, consumers, biodiversity, and ecosystems. In addition to these direct negative effects, there are potential indirect socio-economic consequences, such as the accumulation of pesticide residues on harvested vegetables. In the past, this has led to severe commercial and image problems, mostly in the form of rejection of Spanish horticultural products in other EU countries [81]. An alternative strategy to managing greenhouse pests with chemicals is biological control. Biological control is based on releasing natural enemies of pests (predators and parasitoids) in the greenhouse, to reduce their numbers and associated damage. The use of biological control in the greenhouse has proven to be a viable alternative to pesticide use from both an economic and environmental perspective [82]. Currently, thirty billion beneficial arthropods are released annually to deliver pest control protection in the greenhouse crops of Almeria. This practice enhances the quality of F&V consumed in the EU, while promoting pesticides reduction and environmental sustainability.

3.2.2. Habitat Management

Greenhouse horticulture is the biggest cause of habitat loss in Almeria. In particular, protected horticulture is clustered over the distribution area of a semi-arid habitat characterised by the endemic shrub *Maytenus senegalensis* subsp. *europaea* (also known as Artineras). This habitat is under serious threat because over 26,000 ha, representing ca. 43% of its distribution area is nowadays occupied by greenhouses. As a result, the landscape is highly fragmented and only a few poorly conserved patches of native vegetation remain in the area [15,83,84].

The role of biodiversity in securing crop protection has gained recognition, focusing on the regulation of undesirable organisms by their natural enemies (indigenous predators and parasitoids). The natural biological pest control enabled by biodiversity provides an ecosystem service of extremely high economic value [85]. However, enhancing pest control ecosystem services via the conservation and/or plantation of native vegetation is still a pending subject in the integrated pest management (IPM) programs in greenhouses in Almeria. Currently, as a result of intensive cooperative research and training activity, there are an increasing number of growers interested in planting hedgerows with native vegetation around their greenhouses. These initiatives, coupled with the ecological restoration of vegetation at the landscape level, could serve as green corridors for biodiversity and reduce the habitat fragmentation in the region.

3.2.3. Soil Management

In Almeria, sand mulch (or "arenado", as it is known locally) is the main type of soil management with more than 85% of the greenhouses using it [86–88]. "Arenado" consists of covering the surface of the crop field with an upper layer of silica sand, and a second layer of manure or organic matter underneath. Organic matter amendment application (locally called "retranqueo") is becoming less common over time, mainly due to the work involved in removing the sand layer and the increasing use of mineral fertilisers, which have replaced organic fertilisers in most cases. This fact, together with the common practice of monoculture in the greenhouses of Almeria, has led to a loss of soil health, which has resulted in an increased occurrence of soil-borne diseases and plant-parasitic nematodes in some cases, as well as soil fatigue [54,89]. For several years, the dominant strategy to avoid soil fatigue and reduce pathogen load in soils of Almeria greenhouses has been chemical soil disinfestation, mixing solarisation (i.e., covering the crop area with a thin transparent plastic sheet) with chemical fumigants, mainly metam sodium and dichloropropene, which negatively affect soil biodiversity. This practice is also carried out for preventive purposes with no evidence of pathogens or symptoms of soil fatigue. In addition, the agricultural waste biomass produced by horticulture in Almeria, estimated at 1,370,743 tons annually [90], has led to environmental problems [74,90]. One smart and sustainable practice is the reutilisation in situ of this valuable material as organic amendment, which can also be used for the control of soil pathogens in a more sustainable way compared to chemical soil fumigants [91–94]. Moreover, with this practice, growers contribute to cost savings [95] through the elimination and reduction of dependence on new (and in many cases harmful) inputs.

3.2.4. Management of Irrigation Ponds

The United Nations has forecast that freshwater resources will be reduced by 40% by 2030, generating a critical situation worldwide due to water crisis [96]. Strategies to optimise water management, particularly in agricultural systems, are necessary. In Almeria, most of the water resources for greenhouse agriculture come from underground water (80%) and the remaining 20% from seawater desalination plants, transfers, and other sources [97].

The lack of water resources, especially surface waters in this semi-arid region has generated the need to store water for its intensive cultivation system, which has resulted in the creation of a large number of water reservoirs and artificial irrigation ponds, which feed efficient drip irrigation. The expansion and intensification of irrigated agriculture has given rise to the construction of a high number of artificial irrigation ponds in Mediterranean regions. Over 8700 ponds have been inventoried in Almeria, most of which are used for irrigation and located in the coastal areas where greenhouses have spread [98].

Clogging of irrigation tube emitters is considered to be one of the most serious problems for drip irrigation systems and is usually more severe when treated wastewater is used [99]. This problem is closely related with water quality characteristics, particularly with the presence of suspended solids and algae [99,100]. In this sense, water and pond management by farmers is crucial to achieve good water transparency and quality characteristics. Maintaining submerged aquatic vegetation (SAV) (i.e., plants that grow completely under water) in water ponds can reduce clogging while improving water quality, without the need to use biocides, but to date a minority of farmers are aware of this technique [98].

3.3. Four Niche Experiments and Innovations to Transition from Conventional Agriculture to Agroecological Systems

In the transition to a more sustainable society through a reconnection to nature through agriculture (the subject of this special issue), the relationship that farmers have with nature, expressed through production approaches and practices, and their collaborative activities with related organisations is fundamental to changing the agroecosystem–nature relationship. To document these transitions, we explore below the components and process of four niche developments, which create the foundation for a transition to a more sustainable system.

3.3.1. Niche 1: Transitions in IPM and Biological Control

The first important step before the uptake of IPM in Almeria greenhouses was the introduction of bumblebees for pollination in tomato crops. Until bumblebees were available for natural pollination, tomato plants were pollinated manually [101] consuming 20 labour days per hectare per month. In contrast, natural pollination by bumblebees offered a more profitable option, with lower production costs, increased yields and improved fruit quality [102,103]. In Almeria, after a few years of commercial small-scale trials, bumblebees were massively introduced in tomato crops around 1995 [104,105]. Some initial tests with bumblebees were not successful because of the lack of technical knowledge. Furthermore, there was little experience concerning the transition towards successful biocontrol in greenhouses with a long history of chemical management. For these reasons, the results with biological control initially were unpredictable, and the success rate was low [106], often leading to contrasting opinions on biocontrol efficiency amongst the growers. However, the emergence of the auxiliary industry and technical support they brought, as well as a supportive AKIS, helped to enable change. The presence of natural pollinators obliged greenhouse tomato farmers in Spain to radically change their pest control system, since many of the formerly used pesticides were totally incompatible with the bumblebees [105,107]. Thus, for the first time, growers reduced the number of pesticide treatments, and substituted broad spectrum products for more selective and/or less persistent pesticides to respect the bumblebees [104]. In crops other than tomato, pesticide use was extremely high in greenhouses in the 1990s, exposing farmers and workers to toxicological risks. According to Parron et al. [108], 37% of farmers who applied pesticides, showed toxic signs and symptoms after spraying.

In this regard, working with pollinators helped farmers to experience the behaviour and associated benefits of bumblebees, possibly improving their attitude towards insects, reducing pesticide usage, and facilitating the transition towards an IPM system, as it similarly occurred in other countries (e.g., Beck et al. [109]). In other words, the introduction of pollinators facilitated the transition to the biological control of pests and diseases in greenhouses.

In Spanish greenhouses, biological pest control first became important in sweet pepper crops (*Capsicum annuum*), where the most important pest was thrips (*Frankliniella occidentalis*), which damages fruits and spreads economically important viruses [79,110,111].

Perceived as a major threat by farmers, their presence caused a significant increase in pesticide use, which produced complete resistance of thrips to most available active ingredients [80,112]. The massive use of insecticides caused intolerable levels of pesticide residues on the fruits, leading to major economic consequences for the horticultural sector. In 2007, the rapid alert on food safety was issued by European authorities after isofenphos-methyl residues were found in sweet pepper from Spain [81,113]. Immediately after this incident, the regional government reacted by creating a surveillance program to monitor illegal pesticide practices [114], and fines were imposed on 24 fruit export companies and 25 farmers in Almeria [115]. In response to market pressure, companies and growers looked for alternatives to chemical control, but the transition was neither easy nor automatic, since new tools were needed for pests and diseases management.

Ultimately, a number of factors resulted in the broad adoption of biological control programmes in sweet pepper, which is primarily based on the release of mass-reared predatory mites and bugs. This system had already been researched and fully working in the early 1990s in greenhouse crops in northern European countries [81]. However, this system needed thorough adaptations to enable massive implementation in the south of Europe. New biocontrol tools had to be developed against pests that were not present in the Northern countries. Since the first successes of IPM in northern countries, there was a clear interest from official research and extension institutes, such as the Universities of Granada and Almeria, and the regional government's institute for applied horticultural research (IFAPA) [116,117]. Research projects, such as BIOPROTECT, 'Biological Control in Sweet Pepper and Strawberries', were valuable for the optimisation of biocontrol, while other results showed the growth inhibiting effects of pesticides [118].

During this period, and until the mid-2000s, important advances were made: biocontrol companies made significant economic investments to guarantee the availability, quality and competitive selling prices from 2004 onwards, of key natural enemy species, as well as to better understand the effects of pesticide residues on the development of natural enemy populations [106,114]. Other AKIS actors and initiatives included government policy on economic subsidies and the coordination of technical seminars, where public research centres, cooperative organisations, and the private sector taught more than 750 field technicians over two years how to successfully establish biological control programs in greenhouses [114]. The result of all these factors produced an exponential increase in farmers' adoption of biological control in Almeria, from 8% in the 2006–2007 harvest to almost 85% of a total of 8000 ha of sweet pepper in 2008–2009 [106,114].

A few years afterwards, the application of the Directive 2009/128/EC in 2014 resulted in an additional reduction of the use of pesticides and increased use of biocontrol agents. Growers increasingly moved from chemical-based crop protection to novel biological pest control practices, reconnecting with the benefits of nature-based solutions. Research to test the efficiency of new strategies and/or new biological control agents was carried out by universities and biological control companies after the wide scale adoption took place around 2007 [113,114,119]. These studies, combined with the technical assistance carried out by technicians from biocontrol companies and field advisors contributed to reinforce biocontrol uptake by farmers and expansion into other crops.

More recently, H2020 projects, such as NEFERTITI (www.nefertiti-h2020.eu; accessed on 5 October 2021) and IPMWorks (www.ipmworks.net; accessed on 5 October 2021), have contributed to increase farmers' and technicians' understanding of the functionality of biological control techniques in most crops through demonstrative, participatory activities in commercial greenhouses. Both projects share a common view (also shared with the EU Green Deal) about the need to reduce the amount of pesticides used and to substitute chemical control for alternative, sustainable practices, such as biological control. Feedback received during field demonstrations suggest that a large majority of participant farmers understand and are keen on improving their IPM practices. It is becoming common among farmers in Almeria to be able to recognise a wide diversity of species of natural enemies, secondary pests and other insects, which shows increasing interest in biodiversity.

Throughout the activities carried out in the field, groups of farmers, technicians and other actors share their experiences and know-how on aspects related with biodiversity and biocontrol strategies, thus deepening their connection to nature through the appreciation of the ecosystem services (i.e., pest regulation) that beneficial arthropods provide. COEXPHAL, the association of cooperatives and producer organisations, have actively contributed to training and knowledge exchange among >1000 farmers, and the interprofessional, HortiEspaña has launched a collaborative and multi-actor campaign of biological control campaign (www.ilovebichos.com accessed on 5 October 2021). The joint effort between cooperatives, auxiliary industry, administration, field advisors, research centres and universities, and, most of all, an important number of proactive farmers, made possible the transition to IPM. Currently, in Almeria, there is a significant adoption of biological control in most greenhouse crops, with 50.5% of the crop surface using it [75].

3.3.2. Niche 2: Transitions in Ecological Restoration and Landscaping for Pest Control

Non-crops habitats could provide growers with ecosystem services like natural pest control and benefit biodiversity at a small scale [120]. However, in protected horticultural systems growers have been reluctant to incorporate margin habitats near their greenhouses because numerous species of weeds are known to harbour both pests and plant viruses transmitted by insects [121]. On the other hand, results have demonstrated that the main viruses affecting greenhouse crops in Almeria are not found on native perennial plants [122]. Thus, restoration habitat by design in Almeria landscapes seeks to replace these weeds by integrating perennial native vegetation with greenhouse landscape to disadvantage pests, and advantage the natural enemies that attack pests, while biodiversity is enhanced. In this sense, IFAPA together with the cooperative Cajamar Foundation co-led the knowledge, innovation, key findings, and outlined plans with local governments for future adoption measures of integrating native vegetation around greenhouses.

With the reduction of broad-spectrum insecticides since 2007, and application of the general principles of IPM in the EU (Directive 2009/128/EU), naturally occurring beneficial insects began to colonise the crops spontaneously, but also, other pests simultaneously emerged in protected horticulture. With the increasing recognition that pest management needs to be considered beyond the greenhouse boundary, growers sought information on the role of biodiversity as a source habitat for natural enemies and pests. Therefore, the following step to biodiversity implementation was to ascertain how pests and natural enemies used native vegetation in greenhouse surroundings. To fill this knowledge gap, the RECUPERA 2020 European funded project entitled "New technologies to increase the efficiency of biological pest control in greenhouse areas" sought to identify the key native shrubby insectary-plant species for revegetation programs among greenhouses. To do this, four plots arranged in a semi-natural patch were newly planted among greenhouses. The patch was designed to simulate plant species associations naturally [123,124]. Plant species chosen had to meet a range of criteria for enhancing pest control: (1) non-host for plant viruses; (2) provide refuge and food for natural enemies, e.g., phenology flowering, architecture, provide nectar and pollen, shelter, flower colour, flower morphology, etc.; (3) native to the region; (4) commercial availability; and (5) workable around greenhouse practices [125]. Finally, 28 key native plants were identified from 18 different botanical families, suitable for habitat restoration by design in Almeria horticultural production systems. More than 1000 growers, advisors, agricultural school students, researchers, and others, have visited these ecological infrastructures since they were planted in 2010 and expanded in 2015.

Research outcomes indicated that two of the most important greenhouse pests had consistently lower abundance through the year in native plants than in crops. These results confirmed that the studied plant species were not a major source for pests' reproduction [123]. In addition, models showed that certain predators were using these plants to prey on both pests [123]. Native plants also hosted other potential providers of pest regulating services [124]. Research also revealed the tight interdependence of diversity above ground vegetation and soil biodiversity [126] and that soil biotic communities might provide benefits to pest biological control, for instance, by improving indirect plant defences and enhancing recruitment of generalist natural enemies [127]. The RECUPERA project subsequently triggered the project "New Biological Control Strategies against Aphids in Greenhouses: Ecological Infrastructures and Disruption of the Ant-Aphid Mutualism", funded by the Spanish National Institute for Agricultural and Food Research and Technology (INIA), which aims to use stable isotopes as an insect marking technique to explore pest and natural enemy movement from native vegetation to horticultural crops.

Two technology transfer projects aimed to encourage growers to change their practices towards biodiversity implementation: (1) "Participatory innovation for protected agriculture (PP.TRA.TRA201600.9)" funded by the EU-FEDER program, and (2) the BIOPLAN project "Biodiversity and biological control against the effects of global warming in intensive agriculture in the Mediterranean coastal areas", were presented to the climate change call of the Spanish Biodiversity Foundation. The BIOPLAN project involved several pilot projects where growers were interested in setting up hedgerows with native vegetation on their greenhouses. Finally, a recent Spanish EIP-Agri Operating Group project GOIDEAS "Implementation of ecological developments for sustainable agriculture" (www.goideas.es; accessed on 5 October 2021) illustrated the benefits of vegetation management around greenhouses, which will encourage more growers to plant hedgerows in their farms. Scientific impact, training impact, and dissemination impact has been achieved by the development of two applications for smart phones for promoting applied (IFAPA GUIA) and natural pest control (PlantEN), and a webpage (https://www.goideas.es/; accessed on 5 October 2021) for promoting the plantation of hedgerows in Spanish Mediterranean agricultural systems. Technology transfer talks (>50 talks), training for students (>2500 h) and dissemination additionally supported the initiative. Finally, all of these changes carried significant implications for local-level governance: for instance, regulations that oblige the establishment of hedgerows in the surroundings of new creation greenhouses (B.O.P. of Almeria number 148 of 3 August 2017); and the existence of incentives by regional administration to subsidise the establishment of green infrastructures (BOJA number 69 of 11 April 2017).

As a result, the horticultural sector is increasingly interested in the biodiversity concept and associated benefits of native vegetation, with over 80 growers applying for design of ecological infrastructures. Once growers have expressed interest in these strategies, they require support to design appropriate functional hedgerows able to act as habitats for the natural enemies, but also adapted to their needs. To respond to this need, IFAPA and Cajamar Foundation developed in 2020 a new tool, DiseñEN, which is a free DSS, (www.dise~nen.es; accessed on 5 October 2021), which helps anyone interested, without any knowledge about plants or arthropods, to design a habitat for the natural enemies adapted to the characteristics of their farms. Interest is increasing, and in the 2021/22 season, over 100 km have been planted in total. For instance, the Association of Producer Organizations of Andalucía (APROA) offers a specific service of revegetation by design to their affiliated members. Increasing interest in revegetation by design has created new job opportunities for local nursery companies that specifically offer native plants for pest control. Currently, most biological control companies offer insectary plants and other services to promote biodiversity for biological control management in greenhouses. Similarly, a number of farmers' cooperatives, cooperative associations (COEXPHAL), and environmental associations (ANSE), are engaging growers about how to integrate biodiversity in and around their greenhouses. Finally, local agricultural schools include landscape management for biodiversity issues in their formation programmes.

Nevertheless, important work remains to be done related to the connectivity of natural vegetation patches through hedgerows. One important task is to give awareness about the available agri-environmental aids which let us increase revegetation by design at a higher scale. A survey carried out about 91 growers showed that they considered these aids useful (GO IDEAS, unpublished data). Moreover, almost 80% observed an increase in useful fauna present in the crops, and more than 70% of growers reduced the number of phytosanitary applications. The most notable result was the willingness of growers to carry out actions by themselves, aimed at achieving more sustainable horticulture even though they had not applied for any aid, for example, using biological control, or implementing measures to promote functional biodiversity in their crops. Many growers are starting to consider other biodiversity conservation and are starting to set up nesting boxes to encourage the presence of pollinators, birds, and bats, which indicates significant improvements in growers' perception of nature.

3.3.3. Niche 3: Transitions in Sustainable Soil Management

Soil health [128] takes into account the importance of a living soil. Healthy soils contain a high soil biodiversity and are more resilient to constraints, such as pests and diseases. Sufficient soil organic matter (SOM) content is the basic factor for this because it is the first level of the soil food web. Consequently, SOM and soil biota are considered essential to maintain and improve agricultural soil health and fertility [129,130]. Thus, sustainable greenhouse soils management should favour soil biodiversity and avoid the use of environmentally harmful substances and materials [131]. The norm among the majority of Almeria greenhouse farmers is to disinfect the soil to reduce pathogen load, but also to avoid soil fatigue. A total of 98% of farmers disinfected their greenhouse soils in the 2013/2014 season, once per year for 63% of the cases, and every two years for 28%. The largest proportion of farmers prefer using a combination of solarisation with broad spectrum chemical disinfectants (46.1%), [87,88,132], which negatively affect soil biodiversity. However, there has been a slight gradual transition in the techniques used to manage the soils, from a paradigm based on using chemical disinfectants to avoiding their use or substituting them for other, more sustainable methods. The impact of soil health loss, together with the EU restrictions on the use of fumigants, as well as the demands of European markets, demanding healthier and safer products, including organics, has over the last 20 years changed the perception of soil as being an inert substrate to an agroecological concept, where living organisms with different functions, are necessary to maintain the fertility and health of soils and crops. Local research and extension actions have assisted farmers and advisors to become aware of soil health. An additional role has been played by the auxiliary industry, promoting its own biological products.

Biofumigation and (bio)solarisation, as well as green manures and cover crops are considered among the best practices for promoting soil health, with a high efficacy on controlling soil-borne diseases [133,134]. In Almeria, an increase in practising soil solarisation (with no organic amendments) was observed between 2006 and 2013, with 30–45% of farmers applying it in their greenhouses in 2013 [87,88]. During the same year, the use of biosolarisation began to be referenced (i.e., 0.4% of farmers; [88]). In a national congress held in Almeria in 2004, local scientists reported promising results regarding the management of soil-borne diseases by means of biofumigation and biosolarisation. In 2005, only 180 ha of greenhouses were certified for organic agriculture in the province of Almeria [86]. In 2020, this grew to 3693 ha [135]. This significant increase of organic farming indicates a higher interest of greenhouse growers in biodiversity and sustainability, without losing sight of business opportunities.

Polyethylene plastic (black type) is the most common material used for biosolarisation, also used as the main material for mulching in greenhouses [87]. However, due to the difficulties and costs of plastic residues management, there are few farmers starting to use alternative products based on biodegradable or biocompostable films, paper-based or hay mulches [136,137]. However, these alternative products are more expensive for the farmer, thus limiting their extended use. The involvement of the auxiliary industry (sustainable inputs) reflects the demand of greenhouse growers willing to be more connected to nature, thus avoiding non-degradable materials.

Arenado (sand mulch) is the predominant soil type in the greenhouse sector in Almeria [138]. The frequency of organic matter repositioning (mainly sheep manure) is decreasing [87], and it is increasingly common for farmers to incorporate organic material only below the crop lines, mainly to reduce costs [88]. In addition, a lack of organic amendments use was reported for 28.5% of farmers surveyed in 2012/2013 season in contrast to 6.5% for season 1999/2000 [87], and a survey conducted in 2020, showed that only 22% of farmers incorporated manure or plants before planting season (Hortyfruta, unpublished data). In this regard, the benefits derived from the organic amendments [139,140], which has strongly contributed to maintain soil health and fertility in the greenhouses of the province for decades, could also be obtained from other sources of organic matter, such as plant debris obtained at the end of the cropping season. The practice of directly removing and chopping plant debris, but letting the debris on the surface of the sand, is now broadly used mainly by pepper growers. This contributes to cost savings through a more sustainable agricultural practice according to principles of circular economy [95]. Furthermore, the incorporation of the vegetal biomass into the soil provides fertilising elements and improves soil quality [141,142]. When combined with solarisation (i.e., soil biosolarisation) it reduces, or even eliminates, the extra inputs of fertilisers for the correct development of a greenhouse tomato crop [92,143,144], reducing nitrate leaching [94,143], mitigating the soil fatigue and monoculture effects through the restoration of the soil productive capacity [145], while also being economically beneficial for growers. It has also been proven to be an alternative method for the control of relevant soil pathogens in horticultural crops in the area, even when the material used was infested by the pathogen itself [93]. Indeed, studies focused on the impact of biosolarisation technique on soil microbiota conclude that, even when treatments have a detrimental effect on soil fungi and bacteria population, they tend to regrow along the crop cycle [146–148]. In this context, since 2017 there are public subsidies for farmers who use their own crop debris for organic amendments [149], and according to data from season 2013/2014, more than 11% of farmers reported self-management of plant debris [86], which included its use as organic amendment, as well as for composting. In 2020, new measures had been incorporated as part of the European operational funds for producers' organisations dealing directly with the measure reincorporation of plant debris into the soil. During 2021, a total of 68 ha of greenhouses from farmers affiliated to producer organisations have received this public subsidy (APROA, unpublished data).

In terms of biological control of diseases and nematodes, there is also noticeable change in recent years, with an increase of registered products based on biological microorganisms to control diseases, moving from zero biological antagonists registered in Spain in 2003 to 24 different microorganisms species authorised as fungicides or nematicides in 2021 [150]. In addition to biopesticides, microbiological fertilisers have also increasingly being used in greenhouses in recent years. Most of these products and organisms require specific management, different to the conventional chemical treatments, implying in many cases an advanced knowledge and awareness about life in soils by growers.

The AKIS actors and initiatives include the University of Almeria, IFAPA, and the Cajamar foundation, among others, who together with private research institutions, have been working in cooperation with farmers' associations, growers and advisors, interested on improving soil knowledge and soil agroecological management. In the last 20 years, a number of growers' associations and farm companies have funded research projects focused on soil management and soil microbiology. Other stakeholders have also funded research linked with soil microbiology. These projects always combined research and transfer, and normally concluded with seminars or workshops open to farmers and advisors. Together with other local specific dissemination actions in terms of agroecological soil management, there were over 2200 attendees in more than 30 activities held since 2014, indicating a rather successful outreach. It is worth noting the increase of workshops and seminars in the last two years, under the framework of the H2020 project 'Best4Soil' (https://www.best4soil.eu/; accessed on 5 October 2021), which is actively disseminating knowledge on soil health, and promoting, via a growing network of stakeholders, knowledge exchange about the best real-world practices by organising regional and local workshops in which growers, advisers, researchers, students, and educators interact and learn from each other. For the time being, the real impact of these workshops on ready-to-use soil management practices carried out in the greenhouse sector in Almeria is not clear, but their promotion may encourage growers to improve their soil health management, by better understanding the benefits of promoting soil biodiversity, ultimately supporting them to connect with nature. The feedback obtained shows a great interest for soil topics and for biological control of diseases, as well as a high receptivity of growers for incorporating innovations. However they are reluctant to incorporate practices that increase the workforce, even if the cost is lower than other practices.

3.3.4. Niche 4: Transitions in Ecological Management of Irrigation Ponds in Greenhouses of Almeria

Greenhouse farmers have traditionally used two main pond management techniques in Almeria: dredging and biocide treatment [151]. On the one side, dredging is done to avoid sediment accumulation and preserve all pond volume. On the other side, biocide treatments are used to avoid algae and aquatic vegetation development. These two main techniques have been carried out since the first irrigation ponds were built in the 1970s. There are two main types of ponds in the area, concrete made (oldest and shallowest) and polyethylene-lined (comparatively newer and deeper) [151].

In a study conducted in Almeria to better understand pond management practices among farmers, two thirds of the interviewed growers dredged their ponds. However, this practice was not effective at eradicating the presence of microalgae in the long term, as it was found in 38% of the ponds after a few weeks, even when biocides were applied [151]. The reduction or suppression of this practice would significantly reduce the economic cost of pond management, while increasing biodiversity (particularly SAV) in ponds [98]. As it occurred with dredging, two thirds of the interviewed farmers in Almeria used biocides [151]. According to farmers, they used this technique to prevent the appearance of biological activity in the water column (i.e., mainly algae and SAV, both perceived by this group as detrimental). Juan et al. [151] found that most farmers with SAV (almost 60%) applied biocides despite the fact that they noticed that water turbidity increased after their removal [152]. However, some farmers had a positive opinion about SAV (almost 26% of those who had aquatic vegetation), and conserved it by stopping the use of biocides. Based on their experience, they concluded that SAV presence had a positive influence on the water quality, showing significantly better water quality (clearer water) for drip irrigation. Indeed, results on chemical pond treatment have shown a worsening of water quality for drip irrigation systems since biocides increase the values of planktonic chlorophyll *a* and total suspended solids [152], key parameters to ensure a correct uniformity of irrigation and reduce the probability of clogging by algae and other organic matter. The use of biocides, such as copper sulphate, keep nutrients available to microalgae in the water column, so their use may favour episodes of algae bloom and death and microorganisms decomposition that lead to reduced water quality. Thus, research indicates that biocide treatments do not improve water quality, and sometimes can worsen it. In this sense, the presence of SAV (e.g., Chara spp.) can improve water clarity for drip irrigation systems by reducing the concentrations of planktonic chlorophyll *a* and total suspended solids [152]. In addition, many studies show the benefits of the presence of SAV versus phytoplankton, competing with it by reducing available nutrients and producing a clear water column free of microalgae, but also free of sediment, since the roots of aquatic plants help to fix the substrate [153]. Therefore, avoiding or reducing treatments with biocides would favour the presence of SAV in ponds, improving water quality for drip irrigation, while increasing the ecological value of water ponds for biodiversity conservation.

Although there is no data available on the use of pond covers, it appears that it is becoming more predominant among farmers in Almeria, partly because this technique is subsidised through the operative funds for those affiliated to producer organisations (APROA, unpublished data).

However, some farmers have shown positive opinions about SAV, because they realised (through practical experience) that without doing anything (i.e., covering, dredging or treating with biocides), they had optimal water quality in their ponds (and they saved money). Water ponds managed ecologically, by preserving SAV showed better water quality values compared to the other three techniques [152], while at the same time developing an authentic aquatic ecosystem with greater biodiversity. Currently, most farmers do not know the advantages offered by the maintenance of aquatic vegetation in irrigation ponds, quite possibly because of the scarce knowledge dissemination activities that have been carried out based on the studies previously done by various research organisations.

The University of Almeria studied different techniques of pond management in the province of Almeria and concluded that of all of them, the maintenance of the SAV contributes significantly to maintaining better water quality while greatly enhancing the environmental value of them [152]. In this way, SAV could provide, at the same time, a shelter for many other species of animals as macroinvertebrates, amphibians, fish, and birds [154]. If adequate dissemination of studies were done, it could help to change the farmers' attitudes with regard to pond management techniques and nature, favouring those that are environmentally sustainable, while avoiding the use of biocides.

The AKIS actors involved in this initiative were public universities in Almeria and Seville (the importance of pond management in aquatic ecology, especially the presence of SAV and phytoplankton and water quality for drip irrigation systems), laboratories of COEXPHAL, the association of producer organisations (study of relationship between fungal communities and pond management), and the EU project IPMWorks (monitor farmers who implement SAV in their ponds with the objective of improving water quality using aquatic vegetation). Dissemination was carried out through local and national farmers' technical journals. Most importantly was the involvement of over 100 farmers, all of whom based their pond management in their practical experience, without any previous knowledge. Techniques used in their ponds were learned from their parents or peers.

3.4. *Summary of Niche Initiatives, Transitions, and Farmers Relationship to Nature*

The transition towards a sustainable agricultural system is not a linear path, but a complex process consisting of distinct phases towards system transitions. Regime systemic change generally emerges after a long period of preparation and trial and practice, where niche experimental innovations form a new coalition at the local production or micro-level of the system. Learning processes and knowledge exchange amongst niche actors are fundamental to create synergies and navigate transitions. Innovations, particularly those tested and demonstrated, start to change the agroecosystem, although some innovations become stunted for lack of uptake. When a "window of opportunity" emerges [58], for example, a new technology, new legislation, or even a crisis, innovation can become more rapidly institutionalised at the regime level. The symbols below in Figure 2 show how niche innovations have progressed in intensive greenhouse agriculture in Almeria.

The four niches referred to in Figure 2 above have had somewhat distinct development paths.

3.4.1. Niche 1: IPM and Biological Control

IPM and biological control are already established and are now widespread across most greenhouse crops in SE Spain, with the release of natural enemies for key pests. A consolidated regime exists, heavily supported by auxiliary industry and other AKIS actors (Figure 3). IPM continues to develop as new biopesticides become available across the EU and the pressure to reduce the use of synthetic pesticides continues.

Throughout the years, growers have gained the knowledge and experience needed to manage pests with biological control. Growers now recognise several other positive outcomes: that there are no phytotoxic effects associated with biological control; there is no withholding period after their release; no pest resistance build-up occurs; and releasing of natural enemies is far more amenable to workers, and positive for their own well-being and health [82,118].

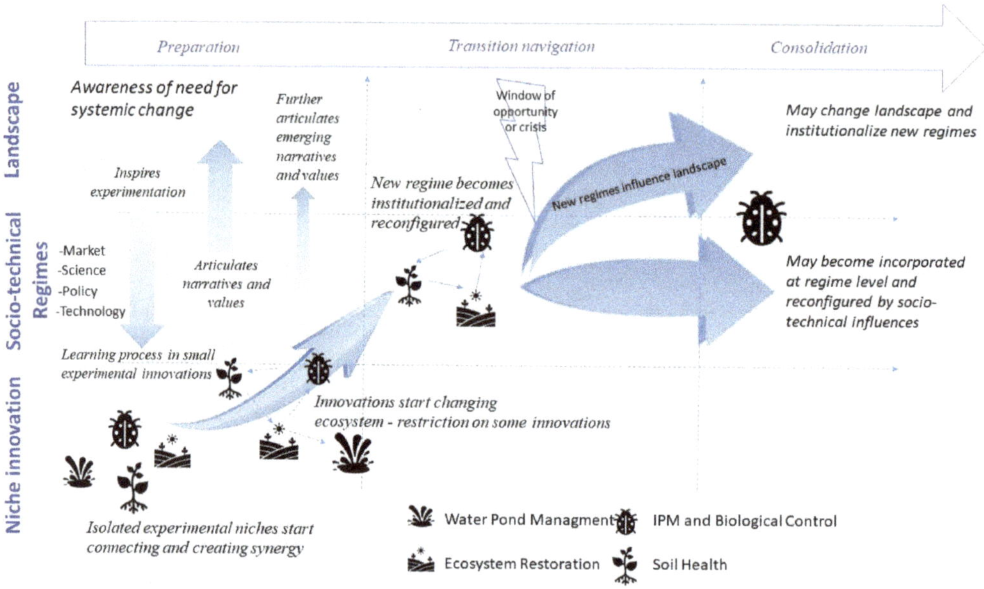

Figure 2. Authors' elaboration based on multi-level concepts offered by Geels, F.W. [36], and niche synergy offered by Pereira L.M. [58].

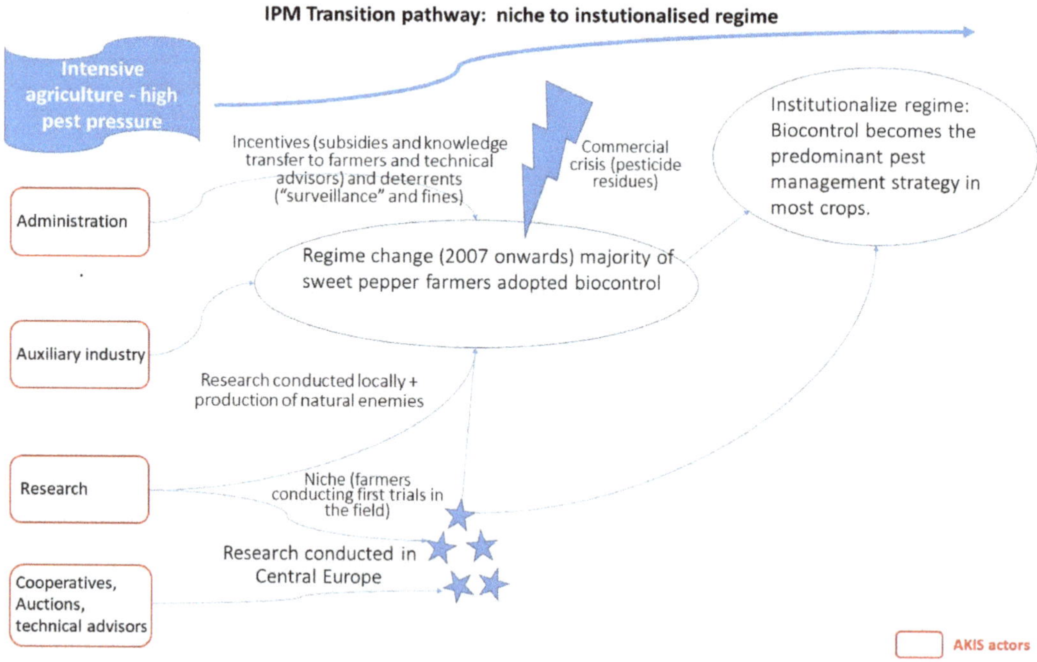

Figure 3. AKIS and IPM transition pathway (authors' elaboration).

The massive adoption of biological control with beneficial arthropods has proven to be an efficient method to address a sustainable pest management system. After starting with the introduction of commercially reared biological control agents, growers now recognise the fundamental role of beneficial species that appear spontaneously in the crops from the surroundings, and have come to at least a functional appreciation of nature. This has led to a whole new approach to pest management, with a leading role for conservation biological control and restoration of biodiversity. Thus, Niche 1 has set the stage for further development of Niche 2.

3.4.2. Niche 2: Ecological Restoration and Landscaping for Pest Control

Several examples of projects, experiments, demonstrations, and farmers having already planted hedgerows in multiple niche initiatives are evident (Figure 4), but there is no institutionalised regime as of yet, although local legislation requires new greenhouses to plant hedgerows. There is a progressive alignment of elements and a trend (via empirical evidence in the field and scientific/innovation projects) to consolidate the technical advances.

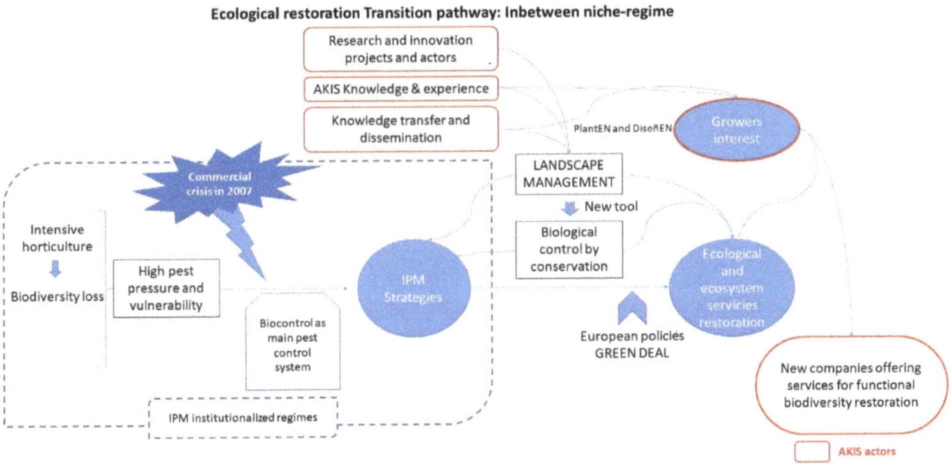

Figure 4. AKIS and ecological restoration transition pathways (authors' elaboration).

The growing criticism of intensive agricultural practices that lead to a deterioration of natural resources and a decrease of biodiversity has progressively led to pressure at the landscape level, where environmental constraints have been put on agricultural activities through more ecological European public policies. These landscape level policies have been geared towards protecting environmentally sensitive areas, improving groundwater quality, and more recently, developing organic farming and/or reducing pesticide use. However, these regulations are still very far from applying truly "ecological" agricultural practices. While the concept of integrated IPM has been accepted and incorporated in public policies and regulations in Europe, signifying not only a regime change, but also a consolidation at landscape levels, a holistic implementation of IPM has not yet been developed. For instance, though applied biological control programs have been successfully implemented in a wide range of horticultural crops in the greenhouse of Almeria, the loss of the natural habitats goes on, and the adoption of the strategies described herein for conservation biodiversity and revegetation by design is still slow and punctual rather than widespread.

Well-designed communication campaigns, applied (greenhouse-level) research, and supportive policies could ensure implementation. Small-scale IPM programs in greenhouses could include the active plantation of hedgerows with native plant species around greenhouses by growers and, at a landscape scale, the regional government should urgently invest in revegetating degraded land and reconnecting habitat remnants. These two actions together would generate interconnected green corridors, able to strengthen biocontrol services and other ecosystem services throughout the greenhouse landscapes.

These combined efforts would most likely improve farmers' appreciation of nature through an increased acknowledge of biodiversity and the inherent benefits derived from it.

3.4.3. Niche 3: Sustainable Soil Management

Soil health is a main concern of EU agricultural policies, but few programmes offer specifically the implementation of management practices linking farmers to Nature. However, several research, dissemination and networking initiatives on agricultural soil are currently ongoing in Mediterranean greenhouses because, compared to other greenhouse cropping systems in Northern EU countries, in which crops are mostly grown in artificial substrate or soilless systems, the vast majority of greenhouses in Almeria grow in soil. This fact clearly indicates growers' acknowledgement of the soil as a key factor linked to the horticultural activity. These soils have been managed for over 50 years following the principles of conventional (chemical) agriculture. The two predominant practices have been the continuous application of mineral soluble fertilisers and soil chemical fumigation as a preventive measure to prepare the soil for the new seasonal crop.

However, over the last years, these two practices have been confronted with sustainable alternatives that are gaining interest among the growers: the use of microbiological fertilisers and the implementation of biosolarisation and biofumigation, as opposed to soil fumigation. The increasing scientific and technical knowledge about biosolarisation, biofumigation and the importance of organic matter content in soils, on one side, together with the recognition of the microbiological fertilisers as specific inputs in the national regulations, are facilitating farmers and advisors to gain awareness of the biodiversity of soils and their associated benefits.

Thus, the transition of the model for soils, from a chemically-based model to an agroecological biologically-based model, occurs via two main paths: (1) a path of inputs substitution (level 2 of Figure 1), where mineral fertilisers are replaced by microbiological fertilisers. Simultaneously, as new biocides are being approved by the EU and promoted by auxiliary industry, it might trigger a gradual decrease of current fumigants in the short-to-mid term; and (2) a holistic approach path, improving soil health by means of increasing organic matter to enrich soil fertility (mineral and biological functions), managing soil-borne diseases and parasitic nematodes using biosolarisation, biofumigation, or organic amendments (level 3 of Figure 1). The coexistence of these two paths (inputs substitution and holistic) with the conventional (chemical-based) path is a realistic scenario for the next several years (Figure 5).

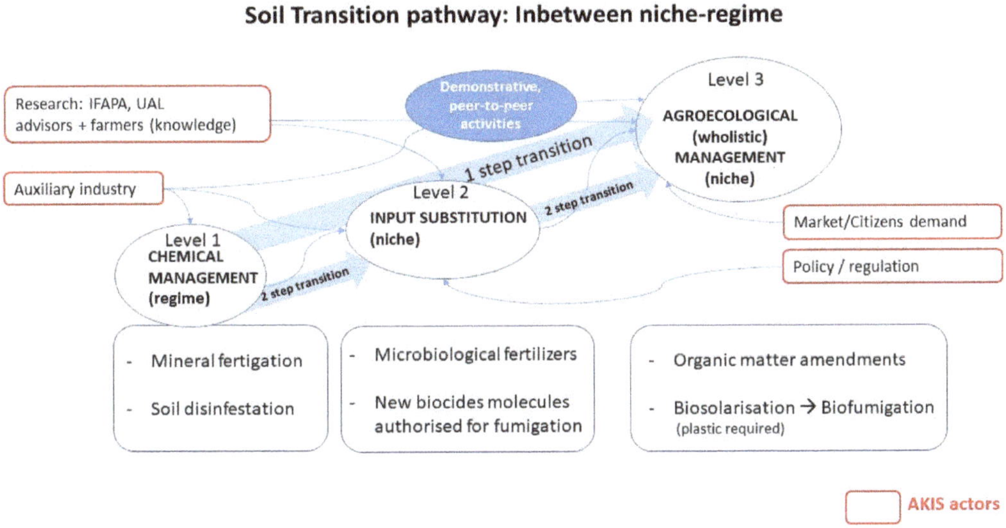

Figure 5. Soil transition pathway (authors' elaboration).

Growers' acceptance of any of these two paths will be primarily based on its profitability, as well as the regulatory (legal) limitations and subsidies. Technical constraints will likely be faced for the two alternative paths to progress, because of the high diversity of greenhouses and soil types in Almeria. Therefore, alternative paths must guarantee their success for different crops, crop cycles, water and soil types, etc. To enable most farmers to choose the agroecological transition, the big challenge involves recovering the biodiversity and fertility of natural soils, while maintaining economic profitability.

3.4.4. Niche 4: Ecological Management of Irrigation Ponds

Out of the four niches, the ecological pond management is the least developed (Figure 6). Knowledge is available (e.g., scientific papers and even technical reports published), but there seems to be a lack of scaling up, so it remains as an isolated niche experiment. The current pond management practices in Almeria's intensive horticultural systems are misguided in their target to eliminate biological water activity, and have clearly failed at improving the water quality for drip irrigation. The predominant conventional (chemical) practices are not compatible with biodiversity conservation. Pond dredging and biocide treatment are two of the most widespread management practices among farmers with the aim of avoiding biological activity, mainly SAV and algae development [151]. These management practices are not effective in the medium- or long-term as they do not improve nor maintain the water quality for drip irrigation systems. It would be advisable to reduce or eliminate these practices, which could achieve better water quality values of these ponds and increased potential for biological conservation [152]. Scientific studies have shown the importance of irrigation ponds as complementary habitats to natural wetlands, especially in semi-arid regions [155]. Therefore, the naturalisation of greenhouse water ponds (with spontaneous presence of SAV) could have an important environmental role at local, but also regional scale, while providing their important agronomical function by clearing the water for drip irrigation.

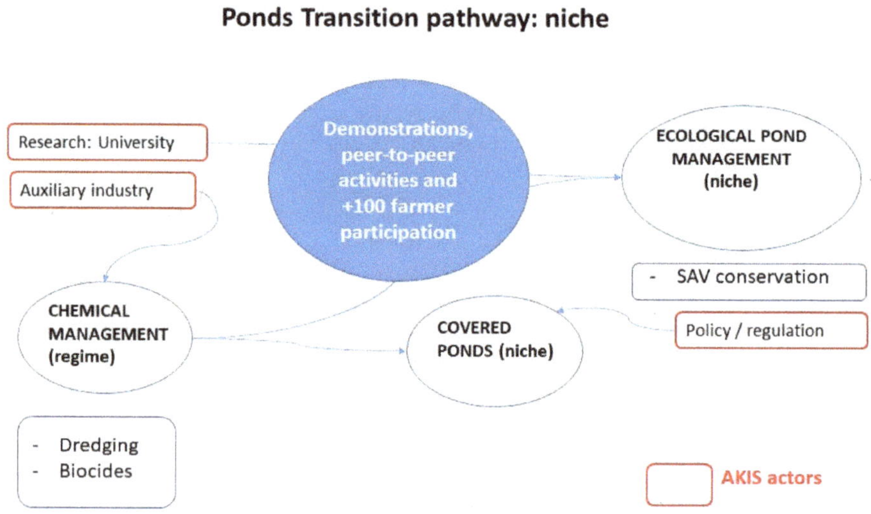

Figure 6. PONDs transition pathway (authors' elaboration).

However, there are some farmers who have already realised the benefits of maintaining aquatic biodiversity. They are aware that in their ponds the water quality values were better than those with biocide treatments or dredging. In addition, these farmers spent significant less money because they neither needed chemical treatments nor additional labour. Based on these experiences, it seems necessary to disseminate and demonstrate scientific results to farmers and other stakeholders, which could benefit both the greenhouse sector and the environment, and in doing so, reconnect farmers to nature.

4. Conclusions

During the last decade, sustainable strategies have been extensively studied and experimented on at the niche level, with the support of AKIS actors. These strategies and techniques are gradually being adopted by farmers and their cooperatives and produce organizations. Based on the wide-scale adoption of biocontrol, there has been a significant increase of farmers establishing auxiliary flora between their crops (e.g., flower strips and banker plants) and ecological infrastructures around their greenhouses (e.g., hedgerows, biodiversity islands, and even nest-boxes for bird and microbat species). These efforts, led by individual farmers, collectively foster the presence of spontaneously appearing arthropods and other animals that play a major role reinforcing biological pest control and other ecosystem services, while helping to preserve the natural habitats in the surrounding landscape. With far less pesticides currently being used, several species of arthropods can now be commonly observed inside and around greenhouses [156].

Moreover, soil health and water pond management have been key areas used to help farmers recognise the importance of biodiversity. Instead of using biocides, techniques, such as biosolarisation and biofumigation in combination with organic matter addition, help to preserve soil micro- and macrobiota, gradually restoring soil fertility. The conservation of SAV in ponds offer another profitable example in which biodiversity supports processes directly related with crop production and environmental quality. Altogether, the development of these four niches have also contributed to the rise of organic farming interest in greenhouse horticulture, which now occupies approximately 12% of the total production area in Almeria. Ultimately, the adoption of sustainable strategies and certifications (e.g., Demeter, Naturland, Bio Suisse) by farmers also favours the presence of autochthonous flora and fauna around greenhouses, transforming these anthropised landscapes and the people living and working on them more connected with nature.

From all four niche cases presented, we can conclude that the "connection to nature" of the majority of farmers in Almeria comes from an appreciation of the functionality (i.e., ecosystem services) that different biological elements (e.g., soil organisms, invertebrates, terrestrial and aquatic plants) provide for farm activity.

On-farm demonstrations have been the most influential tools used to support knowledge transfer to (and between) farmers, based (particularly) on the experiences drawn from three of the studied niches: biological control (e.g., IPMWorks, NEFERTITI), ecological restoration and soil health (e.g., experimental farm installations and demos, workshops and farm days, Best4Soil and NEFERTITI). In the majority of cases, farmers clearly understand the positive impacts that conserving and promoting biodiversity brings to their crops. In this regard, both research and innovation projects are useful to produce positive changes and to highlight the benefits of agricultural practices becoming more embedded in nature.

Based on the four niche case studies, and returning to Gliessman's five levels of agroecology, we consider that the greenhouse sector in Almeria is well advanced and continues experimentation at level 1 (increasing efficiencies of inputs), and is currently in level 2 (substitution of alternative practices and inputs) with the beginnings of a strong base for transition to level 3 (redesign of the whole agroecosystem based on ecological processes), aided by the reconnection with nature which is implicated in the transition from level 2 to level 3. In order to reach further levels, it is necessary to re-establish the communication between growers and consumers (level 4). Initiatives, such as Cute Solar (www.cutesolar.eu; accessed on 5 October 2021) or CO-FRESH (www.co-fresh.eu; accessed on 5 October 2021), are helping consumers and other stakeholders across Europe to visualise the process of how crops are grown in greenhouses by farmers, thus serving as a first step to enable a closer communication and ultimately a reconnection between both groups.

The multiple-level perspective has shed light on how transformations take place from one level to another. The analysis of local, multi-actor niche activity and experimentation has allowed us to consider the individual, collective and organizational behaviour of farmers, business, and related research and market institutions as well as policy actors. We have traced how niches or discreet experiments and projects have gone from isolated activities to dominate regime level patterns within the agri system. In some instances, we also demonstrated how the landscape or prevailing meta-narrative of HNC can be influenced, with farmers recognising ecosystem services as an integral part of their farming activity. The activation of resources and capabilities within the AKIS, along with co-creation, have been key factors in creating realisable and feasible paths and enabling sustainable transitions.

With respect to the roles of different types of AKIS actors, we can observe that research centres have provided a solid base in all four cases, but only IPM and biocontrol (and currently, to a lesser degree, ecosystem restoration and soil health) have received enough attention through dissemination/transfer activities to farmers to result in regime change and landscape influence. The auxiliary industry has also played an important role in IPM (but indirectly in ecological restoration) and soil health ("biostimulants"), but less so in ecosystem restoration and pond management niches. In this regard, water pond management is clearly the least developed area, but opportunities exist to broaden the adoption of this technique through the involvement of more AKIS actors and demonstrations. Public administrations have played an important role in promoting biocontrol, ecosystem restoration and soil health, but in the case of pond management, policies, such as subsidies for pond covers, have actually jeopardised the adoption of ecological pond management.

5. Limitations and Further Research

Further research is required to understand farmers' more profound appreciation of nature, not just the benefits, but the deep appreciation of living entities and the surrounding ecosystems in which their economic activities take place. Although these benefits (i.e., ecosystem services) are important, the farmers' knowledge and beliefs (i.e., cultural services) are determinants of whether they protect, ignore, neglect, or even destroy biodiversity [157]. In this regard, social scientists could work, documenting farmers' recognition of these cultural services. This research, together with the agroecological research carried out in greenhouses in the region, could be combined to create specific environmental education programs, to encourage farmers' appreciation of nature, and support biodiversity conservation.

We have illustrated that farmers' reconnection with nature has been mediated by on-farm practices. The main driver for adoption has generally either demonstrated benefits (farm demonstration or peer-to-peer learning) or external crises and "windows of opportunity" from landscape level influences. This implies that further research is needed on economic benefits and costs for adopting more sustainable practices that place farming within a recognition of nature, so as to encourage adoption by farmers. However, cost–benefit analysis is often a result of scale, and if there is not widespread adoption, then a lack of scale remains an issue.

In particular, with respect to IPM, although the use of biological control agents is common among farmers, there is still some reliance on the use of pesticides, particularly against some pests for which no natural enemies are currently commercially available. Further research is necessary to identify predators and/or parasitoids of these pest species and revise and optimize the already existing IPM protocols with additional compatible tools (e.g., biological products). Moreover, with respect to ecological restoration of Almeria's greenhouse area, it is necessary to quantify natural pest control, which is found in semi natural habitats and to determine which species or functional traits are responsible to exert this biological control. This knowledge would optimize revegetation designs and their functionality for pest control. More importantly it would encourage a paradigm shift in the agricultural sector and society regarding the importance of conserving semi natural habitats and biodiversity in the agricultural landscape. Indicators of healthy, living soils are scarce but they are a key to facilitate the reconnection with nature by means of soil management. The visualisation of the soil as a natural system is not so obvious. Traditional methods, i.e., the incorporation of animal manure for building the 'arenado' system, persist in the mind of most of the growers in Almeria, who appreciate and identify the characteristics of a good soil, by means of the structure and content of organic matter. Overall, there is willingness to change, but further research would support this transition, particularly surveys identifying the connection of farmers with nature, by means of their knowledge and beliefs towards soil. The main limitation to continue the spread of knowledge among farmers on ecological pond management techniques involves the lack of dissemination of results by public or private entities, particularly on-farm demonstrations to and between farmers. AKIS actors can move the transition forward to a more sustainable model in the most practical sense, by demonstrations in situ, showing agronomic, environmental, and economic advantages, motivating more farmers to take the step towards an agroecological transition.

Finally, even though the four niches are interconnected, the strategies to improve sustainability are managed separately. Further research on designing and organising coupled innovations, including the reconnection of farmers with nature as part of the process, is needed [158].

Author Contributions: Conceptualisation, C.G., S.M. and E.C.-M.; methodology, C.G., S.M. and E.C.-M.; investigation, C.G., S.M., E.C.-M., M.d.C.-G., M.G., J.I.M.-G., E.R., J.v.d.B. and M.J.; writing—original draft preparation, C.G., S.M., E.C.-M., M.d.C.-G., M.G., J.I.M.-G., E.R., J.v.d.B. and M.J.; writing—review and editing, all authors; visualisation, C.G., S.M., E.C.-M., M.d.C.-G., M.G., J.I.M.-G., E.R., J.v.d.B. and M.J.; supervision, C.G. and E.C.-M.; project administration, C.G. and E.C.-M.; funding acquisition, C.G., E.C-M., M.d.C.-G., M.G., J.I.M.-G., E.R. and J.v.d.B. All authors have read and agreed to the published version of the manuscript.

Funding: This research was funded by the following projects: Networking European Farms to Enhance Cross Fertilisation and Innovation Uptake through Demonstration (NEFERTITI) H2020, 772705; CO-creating sustainable and competitive FRuits and vEgetableS' value cHains in Europe (CO-FRESH) H2020, 101000852; IPMWorks (an EU-wide farm network demonstrating and promoting cost-effective IPM strategies) H2020, 101000339; Boosting 4 BEST practices for SOIL health in Europe (BEST4SOIL)

H2020 817696 Networking and dissemination of knowledge on soil health; Implementación de Desarrollos Ecológicos para una Agricultura Sostenible- Implementation of ecological developments for a sustainable agriculture (GOIDEAS) Spanish Rural Development Programme (2014–2020) 07434, co-funded European Union 20190020007434; Instituto Nacional de Investigación y Tecnología Agraria y Alimentaria RTA2015-00012-C02-01 INIA.

Data Availability Statement: Data sharing not applicable. No new data were created or analyzed in this study.

Conflicts of Interest: The authors declare no conflict of interest.

References

1. Melchior, I.C.; Newig, J. Governing transitions towards sustainable agriculture—Taking stock of an emerging field of research. *Sustainability* **2021**, *13*, 528. [CrossRef]
2. Crippa, M.; Solazzo, E.; Guizzardi, D.; Monforti-Ferrario, F.; Tubiello, F.N.; Leip, A. Food systems are responsible for a third of global anthropogenic GHG emissions. *Nat. Food* **2021**, *2*, 198–209. [CrossRef]
3. Rockström, J.; Edenhofer, O.; Gaertner, J.; DeClerck, F. Planet-proofing the global food system. *Nat. Food* **2020**, *1*, 3–5. [CrossRef]
4. Barrett, H.; Rose, D.C. Perceptions of the Fourth Agricultural Revolution: What's In, What's Out, and What Consequences are Anticipated? *Sociol. Rural.* **2020**. [CrossRef]
5. Hebinck, A.; Klerkx, L.; Elzen, B.; Kok, K.P.W.; König, B.; Schiller, K.; Tschersich, J.; van Mierlo, B.; von Wirth, T. Beyond food for thought—Directing sustainability transitions research to address fundamental change in agri-food systems. *Environ. Innov. Soc. Transit.* **2021**, *41*, 81–85. [CrossRef]
6. Klerkx, L.; Begemann, S. Supporting food systems transformation: The what, why, who, where and how of mission-oriented agricultural innovation systems. *Agric. Syst.* **2020**, *184*, 102901. [CrossRef] [PubMed]
7. Zurek, M.; Hebinck, A.; Selomane, O. Looking across diverse food system futures: Implications for climate change and the environment. *Q Open* **2021**, *1*. [CrossRef]
8. UNEP. *Making Peace with Nature: A Scientific Blueprint to Tackle the Climate, Biodiversity and Pollution Emergencies*; UNEP: Nairobi, Kenia, 2021.
9. Sala, O.E.; Chapin, F.S., III; Armesto, J.J.; Berlow, E.; Bloomfield, J.; Dirzo, R.; Huber-Sanwald, E.; Huenneke, L.F.; Jackson, R.B.; Kinzig, A.; et al. Global Biodiversity Scenarios for the Year 2100. *Science* **2000**, *287*, 1770–1774. [CrossRef] [PubMed]
10. Díaz, S.; Settele, J.; Brondízio, E.S.; Ngo, H.T.; Agard, J.; Arneth, A.; Balvanera, P.; Brauman, K.A.; Butchart, S.H.M.; Chan, K.M.A.; et al. Pervasive human-driven decline of life on Earth points to the need for transformative change. *Science* **2019**, *366*. [CrossRef] [PubMed]
11. Desneux, N.; Decourtye, A.; Delpuech, J.-M. The Sublethal Effects of Pesticides on Beneficial Arthropods. *Annu. Rev. Entomol.* **2007**, *52*, 81–106. [CrossRef] [PubMed]
12. Alavanja, M.C.R. Introduction: Pesticides Use and Exposure, Extensive Worldwide. *Rev. Environ. Health* **2009**, *24*, 303–309. [CrossRef]
13. Gomiero, T.; Pimentel, D.; Paoletti, M.G. Environmental Impact of Different Agricultural Management Practices: Conventional vs. Organic Agriculture. *CRC Crit. Rev. Plant Sci.* **2011**, *30*, 95–124. [CrossRef]
14. Vermunt, D.A.; Negro, S.O.; Van Laerhoven, F.S.J.; Verweij, P.A.; Hekkert, M.P. Sustainability transitions in the agri-food sector: How ecology affects transition dynamics. *Environ. Innov. Soc. Transit.* **2020**, *36*, 236–249. [CrossRef]
15. Mendoza-Fernández, A.J.; Martínez-Hernández, F.; Pérez-García, F.J.; Garrido-Becerra, J.A.; Benito, B.M.; Salmerón-Sánchez, E.; Guirado, J.; Merlo, M.E.; Mota, J.F. Extreme habitat loss in a Mediterranean habitat: Maytenus senegalensis subsp. europaea. *Plant Biosyst.—Int. J. Deal. Asp. Plant Biol.* **2015**, *149*, 503–511. [CrossRef]
16. Guterres, A. Secretary-General's Address at Columbia University: "The State of the Planet". Available online: https://www.un.org/sg/en/content/sg/speeches/2020-12-02/address-columbia-university-the-state-of-the-planet (accessed on 3 December 2021).
17. Eurostat. *Agriculture, Forestry and Fishery Statistics*, 2020th ed.; Cook, E., Ed.; EU: Brussels, Belgium, 2020; ISBN 978-92-76-21522-6.
18. Deconinck, K. *Concentration and Market Power in the Food Chain*; OECD Publishing: Paris, France, 2021; Volume 151.
19. Dasgupta, P. *The Economics of Biodiversity: The Dasgupta Review*; HM Treasury: London, UK, 2021; ISBN 978-1-911680-29-1.
20. Dupeux, B. Beyond net zero emission in agriculture EEB pathway for a net-zero agriculture and agriculture-related land emission. *Eur. Environ. Bur.* **2021**. Available online: chrome-extension://bocbaocobfecmglnmeaeppambideimao/pdf/viewer.html?file=https%3A%2F%2Feeb.org%2Fwp-content%2Fuploads%2F2021%2F07%2FEEB-pathways-for-a-net-zero-agriculture.pdf (accessed on 5 October 2021).
21. Poore, J.; Nemecek, T. Reducing food's environmental impacts through producers and consumers. *Science* **2018**, *360*, 987–992. [CrossRef] [PubMed]
22. Benton, T.; Bieg, C.; Harwatt, H.; Pudassaini, R.; Wellesley, L. *Food System Impacts on Biodiversity Loss Three levers for Food*; Chatham House: London, UK, 2021; ISBN 9783030023188.
23. FAO. *Guiding the Transition to Sustainable Food and Agricultural Systems the 10 Elements of Agroecology*; FAO: Rome, Italy, 2018.

24. Knierim, A.; Labarthe, P.; Laurent, C.; Prager, K.; Kania, J.; Madureira, L.; Ndah, T.H. Pluralism of agricultural advisory service providers—Facts and insights from Europe. *J. Rural Stud.* **2017**, *55*, 45–58. [CrossRef]
25. Pigford, A.A.E.; Hickey, G.M.; Klerkx, L. Beyond agricultural innovation systems? Exploring an agricultural innovation ecosystems approach for niche design and development in sustainability transitions. *Agric. Syst.* **2018**, *164*, 116–121. [CrossRef]
26. Köhler, J.; Geels, F.W.; Kern, F.; Markard, J.; Onsongo, E.; Wieczorek, A.; Alkemade, F.; Avelino, F.; Bergek, A.; Boons, F.; et al. An agenda for sustainability transitions research: State of the art and future directions. *Environ. Innov. Soc. Transit.* **2019**, *31*, 1–32. [CrossRef]
27. Geels, F.W. Socio-technical transitions to sustainability: A review of criticisms and elaborations of the Multi-Level Perspective. *Curr. Opin. Environ. Sustain.* **2019**, *39*, 187–201. [CrossRef]
28. Markard, J.; Raven, R.; Truffer, B. Sustainability transitions: An emerging field of research and its prospects. *Res. Policy* **2012**, *41*, 955–967. [CrossRef]
29. Aubert, B.A.; Schroeder, A.; Grimaudo, J. IT as enabler of sustainable farming: An empirical analysis of farmers' adoption decision of precision agriculture technology. *Decis. Support Syst.* **2012**, *54*, 510–520. [CrossRef]
30. El Bilali, H. Innovation-Sustainability Nexus in Agriculture Transition: Case of Agroecology. *Open Agric.* **2019**, *4*, 1–16. [CrossRef]
31. van Mierlo, B.; Beers, P.J. Understanding and governing learning in sustainability transitions: A review. *Environ. Innov. Soc. Transit.* **2020**, *34*, 255–269. [CrossRef]
32. Elzen, B.; Barbier, M.; Cerf, M.; Grin, J. Stimulating transitions towards sustainable farming systems. In *Farming Systems Research into the 21st Century: The New Dynamic*; Springer: Dordrecht, The Netherlands, 2012; pp. 431–455. [CrossRef]
33. Ingram, J. A food systems approach to researching food security and its interactions with global environmental change. *Food Secur.* **2011**, *3*, 417–431. [CrossRef]
34. Spaargaren, G.; Oosterveer, P.; Loeber, A. Sustainability Transitions in Food Consumption, Retail and Production. 2011. Available online: https://www.taylorfrancis.com/chapters/edit/10.4324/9780203135921-7/sustainability-transitions-food-consumption-retail-production-gert-spaargaren-peter-oosterveer-anne-loeber (accessed on 5 October 2021).
35. STRN. *Sustainability Transitions: Policy and Practices*; Publications Office of the European Union: Luxembourg, 2019; ISBN 978-92-9480-086-2.
36. Geels, F.W. Technological transitions as evolutionary reconfiguration processes: A multi-level perspective and a case-study. *Res. Policy* **2002**, *31*, 1257–1274. [CrossRef]
37. Smith, A.; Voß, J.-P.; Grin, J. Innovation studies and sustainability transitions: The allure of the multi-level perspective and its challenges. *Res. Policy* **2010**, *39*, 435–448. [CrossRef]
38. Rip, A.; Kemp, R. *Technological Change*; Rayner, S., Malone, E.L., Eds.; Battelle Press: Columbus, OH, USA, 1998; Volume II.
39. Geels, F.W. The multi-level perspective on sustainability transitions: Responses to seven criticisms. *Environ. Innov. Soc. Transit.* **2011**, *1*, 24–40. [CrossRef]
40. Geels, F.W. Processes and patterns in transitions and system innovations: Refining the co-evolutionary multi-level perspective. *Technol. Forecast. Soc. Chang.* **2005**, *72*, 681–696. [CrossRef]
41. El Bilali, H. Research on agro-food sustainability transitions: Where are food security and nutrition? *Food Secur.* **2019**, *11*, 559–577. [CrossRef]
42. Wigboldus, S.; Klerkx, L.; Leeuwis, C.; Schut, M.; Muilerman, S.; Jochemsen, H. Systemic perspectives on scaling agricultural innovations: A review. *Agron. Sustain. Dev.* **2016**, *36*, 46. [CrossRef]
43. El Bilali, H. Transition Heuristic Frameworks in Research on Agro-Food Sustainability Transitions. *Environ. Dev. Sustain.* **2020**, *22*, 1693–1728. [CrossRef]
44. Gaitán-Cremaschi, D.; Klerkx, L.; Duncan, J.; Trienekens, J.H.; Huenchuleo, C.; Dogliotti, S.; Contesse, M.E.; Benitez-Altuna, F.J.; Rossing, W.A.H. Sustainability transition pathways through ecological intensification: An assessment of vegetable food systems in Chile. *Int. J. Agric. Sustain.* **2020**, *18*, 131–150. [CrossRef]
45. Weber, H.; Poeggel, K.; Eakin, H.; Fischer, D.; Lang, D.J.; Von Wehrden, H.; Wiek, A. What are the ingredients for food systems change towards sustainability?—Insights from the literature. *Environ. Res. Lett.* **2020**, *15*. [CrossRef]
46. Soga, M.; Gaston, K.J. Extinction of experience: The loss of human-nature interactions. *Front. Ecol. Environ.* **2016**, *14*, 94–101. [CrossRef]
47. Bratman, G.N.; Daily, G.C.; Levy, B.J.; Gross, J.J. The benefits of nature experience: Improved affect and cognition. *Landsc. Urban Plan.* **2015**, *138*, 41–50. [CrossRef]
48. Gosling, E.; Williams, K.J.H. Connectedness to nature, place attachment and conservation behaviour: Testing connectedness theory among farmers. *J. Environ. Psychol.* **2010**, *30*, 298–304. [CrossRef]
49. Geng, L.; Xu, J.; Ye, L.; Zhou, W.; Zhou, K. Connections with Nature and Environmental Behaviors. *PLoS ONE* **2015**, *10*, e0127247. [CrossRef]
50. Schultz, W.P. The structure of environmental concern: Concern for self, other people, and the biosphere. *J. Environ. Psychol.* **2001**, *21*, 327–339. [CrossRef]
51. European Commission. Farm to Fork Strategy: For a Fair, Healthy and Environmentally-Friendly Food System. Available online: https://ec.europa.eu/food/horizontal-topics/farm-fork-strategy_en (accessed on 30 September 2021).
52. FAO; UNEP. The FAO-UNEP Sustainable Food Systems Programme. Available online: http://www.fao.org/fileadmin/templates/ags/docs/SFCP/Flyer_SP_01.pdf (accessed on 21 October 2021).

53. Gliessman, S. Transforming food systems with agroecology. *Agroecol. Sustain. Food Syst.* **2016**, *40*, 187–189. [CrossRef]
54. Guerrero, M.M.; Guirao, P.; Martinez-Lluch, M.C.; Tello, J.C.; Lacasa, A. Soil fatigue and its specificity towards pepper plants in greenhouses. *Span. J. Agric. Res.* **2014**, *12*, 644–652. [CrossRef]
55. Roditakis, E.; Vasakis, E.; García-Vidal, L.; del Rosario Martínez-Aguirre, M.; Rison, J.L.; Haxaire-Lutun, M.O.; Nauen, R.; Tsagkarakou, A.; Bielza, P. A four-year survey on insecticide resistance and likelihood of chemical control failure for tomato leaf miner Tuta absoluta in the European/Asian region. *J. Pest Sci.* **2018**, *91*, 421–435. [CrossRef]
56. AKIS. *Preparing for Future AKIS in Europe*; AKIS: Brussels, Belgium, 2019.
57. United Nations. 17 Sustainable Development Goals (SDGs). Available online: https://sdgs.un.org/goals (accessed on 1 September 2021).
58. Pereira, L.M.; Bennett, E.; Biggs, R.; Peterson, G.; McPhearson, T.; Norström, A.; Olsson, P.; Preiser, R.; Raudsepp-Hearne, C.; Vervoort, J. Seeds of the Future in the Present: Exploring Pathways for Navigating Towards "Good" Anthropocenes. In *Urban Planet: Knowledge towards Sustainable Cities*; Elmqvist, T., Bai, X., Frantzeskaki, N., Griffith, C., Maddox, D., McPhearson, T., Parnell, S., Romero-Lankao, P., Simon, D., Watkins, M., Eds.; Cambridge University Press: Cambridge, UK, 2018; pp. 327–350. ISBN 9781316647554.
59. Joffre, O.M.; Klerkx, L.; Khoa, T.N.D. Aquaculture innovation system analysis of transition to sustainable intensification in shrimp farming. *Agron. Sustain. Dev.* **2018**, *38*. [CrossRef]
60. Klerkx, L.; van Mierlo, B.; Leeuwis, C. Evolution of systems approaches to agricultural innovation: Concepts, analysis and interventions. In *Farming Systems Research into the 21st Century: The New Dynamic*; Springer: Dordrecht, The Netherlands, 2012; pp. 457–483. [CrossRef]
61. Turner, J.A.; Klerkx, L.; White, T.; Nelson, T.; Everett-Hincks, J.; Mackay, A.; Botha, N. Unpacking systemic innovation capacity as strategic ambidexterity: How projects dynamically configure capabilities for agricultural innovation. *Land Use Policy* **2017**, *68*, 503–523. [CrossRef]
62. Geels, F.W.; Kemp, R. Dynamics in socio-technical systems: Typology of change processes and contrasting case studies. *Technol. Soc.* **2007**, *29*, 441–455. [CrossRef]
63. Giagnocavo, C.; Aguilera, D.U. The case for proactive cooperative banks and local development: Innovation, growth, and community building in Almeria, Spain. In *Financial Cooperatives and Local Development*; Goglio, S., Alexopoulos, Y., Eds.; Routledge: New York, NY, USA, 2013; pp. 93–110.
64. Giagnocavo, C.; Fernández-Revuelta Pérez, L.; Uclés Aguilera, D. Modern agriculture, sustainable innovation and cooperative banks: The development of Almería (1963–2010). In Proceedings of the Financial Cooperative Approaches to Local Development through Sustainable Innovation, Trento, Italy, 10–11 June 2010; pp. 10–11.
65. Giagnocavo, C. The Development of the Cooperative Movement and Civil Society in Almeria, Spain: Something from Nothing? *Sustainability* **2020**, *12*, 9820. [CrossRef]
66. Giagnocavo, C.; Galdeano-Gómez, E.; Pérez-Mesa, J. Cooperative Longevity and Sustainable Development in a Family Farming System. *Sustainability* **2018**, *10*, 2198. [CrossRef]
67. Angel Aznar-Sanchez, J.; Sanchez-Picon, A. Innovation and District around a "Miracle": Configuration of the local productive system in intensive agriculture in Almeria. *Rev. Hist. Ind.* **2010**, *42*, 157–193.
68. COEXPHAL. *COEXPHAL Annual Report 2020–21*; COEXPHAL: Almería, Spain, 2021.
69. Castro, A.J.; López-Rodríguez, M.D.; Giagnocavo, C.; Gimenez, M.; Céspedes, L.; La Calle, A.; Gallardo, M.; Pumares, P.; Cabello, J.; Rodríguez, E.; et al. Six Collective Challenges for Sustainability of Almería Greenhouse Horticulture. *Int. J. Environ. Res. Public Health* **2019**, *16*, 4097. [CrossRef]
70. Dudley, N.; Alexander, S. Agriculture and biodiversity: A review. *Biodiversity* **2017**, *18*, 45–49. [CrossRef]
71. Sánchez-Bayo, F.; Wyckhuys, K.A.G. Worldwide decline of the entomofauna: A review of its drivers. *Biol. Conserv.* **2019**, *232*, 8–27. [CrossRef]
72. Lundgren, J.G.; Fausti, S.W. Trading biodiversity for pest problems. *Sci. Adv.* **2015**, *1*, e1500558. [CrossRef]
73. Costanza, R.; D'Arge, R.; de Groot, R.; Farber, S.; Grasso, M.; Hannon, B.; Limburg, K.; Naeem, S.; O'Neill, R.V.; Paruelo, J.; et al. The value of the world's ecosystem services and natural capital. *Ecol. Econ.* **1998**, *25*, 3–15. [CrossRef]
74. Egea, F.J.; Torrente, R.G.; Aguilar, A. An efficient agro-industrial complex in Almería (Spain): Towards an integrated and sustainable bioeconomy model. *N. Biotechnol.* **2018**, *40*, 103–112. [CrossRef]
75. Cajamar. Análisis de la campaña hortofrutícola de Almería. Campaña 2019/2020. Available online: https://www.cajamar.es/es/agroalimentario/innovacion/formacion/actividades-de-transferencia/ano-2020/webinar-analisis-de-la-campana-hortofruticola-campana-20192020/ (accessed on 5 October 2021).
76. Francis, C.; Lieblein, G.; Gliessman, S.; Breland, T.A.; Creamer, N.; Harwood, R.; Salomonsson, L.; Helenius, J.; Rickerl, D.; Salvador, R.; et al. Agroecology: The Ecology of Food Systems. *J. Sustain. Agric.* **2003**, *22*, 99–118. [CrossRef]
77. Gliessman, S.R. *Agroecology: The Ecology of Sustainable Food Systems*, 2nd ed.; CRC Press: Boca Raton, FL, USA, 2007.
78. Paulitz, T.C.; Richard, R.B. Biological control in greenhouse systems. *Annu. Rev. Phytopathol.* **2001**, *39*, 101–133. [CrossRef] [PubMed]
79. Messelink, G.J.; Labbé, R.; Marchand, G.; Tavella, L. Sweet Peppers. In *Integrated Pest and Disease Management in Greenhouse Crops*; Springer International Publishing: Cham, Switzerland, 2020; Volume 22, pp. 513–535.

80. Espinosa, P.J.; Bielza, P.; Contreras, J.; Lacasa, A. Insecticide resistance in field populations of Frankliniella occidentalis (Pergande) in Murcia (south-east Spain). *Pest Manag. Sci.* **2002**, *58*, 967–971. [CrossRef] [PubMed]
81. Glass, R. Biological Control in the Greenhouses of Almería and Challenges for a Sustainable Intensive Production. *Outlooks Pest Manag.* **2012**, *23*, 276–279. [CrossRef]
82. van Lenteren, J.C. Success in Biological Control of Arthropods by Augmentation of Natural Enemies Inundative release Seasonal inoculative release. In *Biological Control: Measures of Success*; Springer: Dordrecht, The Netherlands, 2000; pp. 77–103.
83. Mota, J.F.; Peñas, J.; Castro, H.; Cabello, J.; Guirado, J.S. Agricultural development vs biodiversity conservation: The Mediterranean semiarid vegetation in El Ejido (Almeria, southeastern Spain). *Biodivers. Conserv.* **1996**, *5*, 1597–1617. [CrossRef]
84. Mendoza-Fernández, A.J.; Martínez-Hernández, F.; Salmerón-Sánchez, E.; Pérez-García, F.J.; Teruel, B.; Merlo, M.E.; Mota, J.F. The relict ecosystem of maytenus senegalensis subsp. europaea in an agricultural landscape: Past, present and future scenarios. *Land* **2021**, *10*, 1. [CrossRef]
85. Losey, J.; Vaughan, M. The Economic Value of Ecological Services Provided by Insects. *Bioscience* **2006**, *56*, 311–323. [CrossRef]
86. Junta De Andalucia Consejeria de Agricultura, Ganaderia, Pesca y Desarrollo Rural. Agencia de Gestión Agraria y Pesquera de Andalucía. Available online: https://www.cap.junta-andalucia.es/agriculturaypesca/observatorio/servlet/FrontController?action=RecordContent&table=11031&element=1586184& (accessed on 3 December 2021).
87. Garcia-Garcia, M.C.; Céspedes, A.J.; Lorenzo, P. *El Sistema de Producción Hortícola Protegido de la Provincia de Almería*; Consejeria de Agricultura, Ganaderia, Pesca y Desarrollo Rural de Andalucía: Sevilla, Spain, 2016; 180p.
88. Valera, D.L.; Belmonte, L.J.; Molina, F.D.; Lopez, A. *Greenhouse Agriculture in Almería: A Comprehensive Techno-Economic Analysis*; Cajamar Caja Rural: Almería, Spain, 2016; 408p.
89. Bouhot, D. La fatigue des sols. Position du probléme et principe du diagnostic. In *La fatigue des sols. Diagnostic de la Fertilité dans le Systémes Culturaux*; Institut National de la Recherche Agronomique (INRA): Paris, France, 1983; pp. 9–21.
90. Duque-Acevedo, M.; Belmonte-Ureña, L.J.; Plaza-Úbeda, J.A.; Camacho-Ferre, F. The Management of Agricultural Waste Biomass in the Framework of Circular Economy and Bioeconomy: An Opportunity for Greenhouse Agriculture in Southeast Spain. *Agronomy* **2020**, *10*, 489. [CrossRef]
91. Castillo-Díaz, F.J.; Marín-Guirao, J.I.; Belmonte-Ureña, L.J.; Tello-Marquina, J.C. Effect of Repeated Plant Debris Reutilization as Organic Amendment on Greenhouse Soil Fertility. *Int. J. Environ. Res. Public Health* **2021**, *18*, 11544. [CrossRef]
92. García-Raya, P.; Ruiz-Olmos, C.; Marín-Guirao, J.; Asensio-Grima, C.; Tello-Marquina, J.; de Cara-García, M. Greenhouse Soil Biosolarization with Tomato Plant Debris as a Unique Fertilizer for Tomato Crops. *Int. J. Environ. Res. Public Health* **2019**, *16*, 279. [CrossRef]
93. Gómez-Tenorio, M.A.; Lupión-Rodríguez, B.; Boix-Ruiz, A.; Ruiz-Olmos, C.; Marín-Guirao, J.I.; Tello-Marquina, J.C.; Camacho-Ferre, F.; de Cara-García, M. Meloidogyne -infested tomato crop residues are a suitable material for biodisinfestation to manage Meloidogyne sp. in greenhouses in Almería (south-east Spain). *Acta Hortic.* **2018**, *1207*, 217–222. [CrossRef]
94. Marín-Guirao, J.I.; Tello, J.C.; Díaz, M.; Boix, A.; Ruiz, C.A.; Camacho, F. Effect of greenhouse soil bio-disinfection on soil nitrate content and tomato fruit yield and quality. *Soil Res.* **2016**, *54*, 200–206. [CrossRef]
95. Torres, J.M. Uso agronómico de restos de cosecha en los invernaderos enarenados de la cuenca mediterranea. *Fund. Cajamar* **2016**.
96. Connor, R. *The United Nations World Water Development Report 2015: Water for a Sustainable World*; UNESCO Publishing: Paris, France, 2015; Volume 1.
97. Caparrós-Martínez, J.; Rueda-Lópe, N.; Milán-García, J.; de Pablo Valenciano, J. Public policies for sustainability and water security: The case of Almeria (Spain). *Glob. Ecol. Conserv.* **2020**, *23*, e01037. [CrossRef]
98. Casas, J.; Toja, J.; Bonachela, S.; Fuentes, F.; Gallego, I.; Juan, M.; León, D.; Peñalver, P.; Pérez, C.; Sánchez, P. Artificial ponds in a Mediterranean region (Andalusia, southern Spain): Agricultural and environmental issues. *Water Environ. J.* **2011**, *25*, 308–317. [CrossRef]
99. Ayers, R.S.; Westcot, D.W. *Water Quality for Agriculture*; FAO Irrigation: Rome, Italy, 1985.
100. Nakayama, F.S.; Bucks, D.A. Water quality in drip/trickle irrigation: A review. *Irrig. Sci.* **1991**, *12*. [CrossRef]
101. Vergara, C.H.; Fonseca-Buendía, P. Pollination of Greenhouse Tomatoes by the Mexican bumblebee *Bombus ephippiatus* (Hymenoptera: Apidae). *J. Pollinat. Ecol.* **2012**, *7*, 27–30. [CrossRef]
102. van Ravestijn, W.; van der Sande, J. Use of bumblebees for the pollination of glasshouse tomatoes. *Acta Hortic.* **1991**, *288*, 204–212. [CrossRef]
103. Velthuis, H.H.W.; van Doorn, A. A century of advances in bumblebee domestication and the economic and environmental aspects of its commercialization for pollination. *Apidologie* **2006**, *37*, 421–451. [CrossRef]
104. Navarro Castillo, J.A. Introducción de abejorros como polinizadores en cultivos protegidos de Almería. *Boletín Inst. Estud. Almer.* **1994**, 101–118.
105. Van der Blom, J. La introducción artificial de la fauna auxiliar en cultivos agrícolas. *Boletín Sanid. Veg. Plagas* **2002**, *28*, 107–118.
106. van der Blom, J.; Robledo, A.; Torres, S.; Sánchez, J.A. Consequences of the wide scale implementation of biological control in greenhouse horticulture in Almeria, Spain. *Biol. Control* **2009**, *11*, 8.
107. Marletto, F.; Patetta, A.; Manino, A. Laboratory assessment of pesticide toxicity to bumblebees. *Bull. Insectol.* **2003**, *56*, 155–158.
108. Parrón, T.; Hernández, A.; Pla, A.; Villanueva, E. Clinical and biochemical changes in greenhouse sprayers chronically exposed to pesticides. *Hum. Exp. Toxicol.* **1996**, *15*, 957–963. [CrossRef]
109. Beck, N.G.; Martin, N.A.; Workman, P.J. IPM for greenhouse crops in New Zealand: Grower acceptance. *WPRS Bull.* **1993**, *16*, 1.

110. Rodríguez, M.D.; Belda, J.E. *Trips en los Cultivos Hortícolas Protegidos. Consejeria de Agricultura y Pesca*; Junta de Andalucía: Sevilla, Spain, 1989; ISBN 978-84-505-8381-6.
111. Cabello, T.; Pascual, F.; Abad, M.M. Capturas de *Frankliniella occidentalis* (Pergande) (Thys.: Thripidae) en trampas de distintos colores en cultivos en invernaderos. *Boletín Sanid. Veg. Plagas* **1990**, *17*, 265–270.
112. Espinosa, P.J.; Bielza, P.; Contreras, J.; Lacasa, A. Field and laboratory selection of *Frankliniella occidentalis* (Pergande) for resistance to insecticides. *Pest Manag. Sci.* **2002**, *58*, 920–927. [CrossRef] [PubMed]
113. Calvo, F.J.; Bolckmans, K.; Belda, J.E. Biological control-based IPM in sweet pepper greenhouses using *Amblyseius swirskii* (Acari: Phytoseiidae). *Biocontrol Sci. Technol.* **2012**, *22*, 1398–1416. [CrossRef]
114. Vila, E.; Cabello, T. Biosystems engineering applied to greenhouse pest control. In *Biosystems Engineering: Biofactories for Food Production in the Century XXI 2014*; pp. 99–128. Available online: https://www.researchgate.net/publication/292356356_Biosystems_Engineering_Biofactories_for_Food_Production_in_the_Century_XXI (accessed on 5 October 2021).
115. Rincón-Amat, S. *Possible Changes in Order to Solve the Pesticides Problem*; Swedish University of Agricultural Sciences: Uppsala, Sweden, 2007.
116. Rodríguez, M.D.; Moreno, R.; Rodríguez, M.P.; Mirasol, E.; Lastres, J.M.; Téllez, M.M. IPM Tomate. In *Programa de Manejo Integrado en el Cultivo del Tomate Bajo Plástico en Almeria*; Junta de Andalucía: Sevilla, Spain, 1993; ISBN 84-87564-91-7.
117. Moreno Vázquez, R.E. *Sanidad Vegetal en la Horticultura Protegida*; Junta de Andalucía: Sevilla, Spain, 1994; p. 440. Available online: https://www.researchgate.net/profile/Tomas-Cabello/publication/256445529_Noctuidos_Lep_Noctuidae_plagas_en_cultivos_horticolas_en_invernaderos/links/54e3886c0cf2b2314f5d4492/Noctuidos-Lep-Noctuidae-plagas-en-cultivos-horticolas-en-invernaderos.pdf (accessed on 17 January 2022).
118. Giménez–Moolhuyzen, M.; van der Blom, J.; Lorenzo–Mínguez, P.; Cabello, T.; Crisol–Martínez, E. Photosynthesis inhibiting effects of pesticides on sweet pepper leaves. *Insects* **2020**, *11*, 69. [CrossRef]
119. Cabello, T.; Gámez, M.; Varga, Z.; Garay, J.; Carreño, R.; Gallego, J.R.; Fernández, F.J. Selection of Trichogramma spp. (Hym.: Trichogrammatidae) for the biological control of Tuta absoluta (Lep.: Gelechiidae) in greenhouses by an entomo-ecological simulation model. *Integr. Control Prot. Crop. Mediterr. Clim. IOBC-WPRS Bull.* **2012**, *80*, 171–176.
120. Gurr, G.M.; Wratten, S.D.; Luna, J.M. Multi-function agricultural biodiversity: Pest management and other benefits. *Basic Appl. Ecol.* **2003**, *4*, 107–116. [CrossRef]
121. Gilbertson, R.L.; Batuman, O.; Webster, C.G.; Adkins, S. Role of the Insect Supervectors Bemisia tabaci and Frankliniella occidentalis in the Emergence and Global Spread of Plant Viruses. *Annu. Rev. Virol.* **2015**, *2*, 67–93. [CrossRef] [PubMed]
122. Rodríguez, E.; van der Blom, J.; González, M.; Sánchez, E.; Janssen, D.; Ruiz, L.; Elorrieta, M.-A. Plant viruses and native vegetation in Mediterranean greenhouse areas. *Sci. Hortic.* **2014**, *165*, 171–174. [CrossRef]
123. Rodríguez, E.; González, M.; Paredes, D.; Campos, M.; Benítez, E. Selecting native perennial plants for ecological intensification in Mediterranean greenhouse horticulture. *Bull. Entomol. Res.* **2018**, *108*, 694–704. [CrossRef]
124. Cotes, B.; González, M.; Benítez, E.; De Mas, E.; Clemente-Orta, G.; Campos, M.; Rodríguez, E. Spider Communities and Biological Control in Native Habitats Surrounding Greenhouses. *Insects* **2018**, *9*, 33. [CrossRef]
125. Rodríguez, E.; Schwarzer, V.; van der Blom, J.; Cabello, T.; González, M. The selection of native insectary plants for landscaping in greenhouse areas of SE Spain. *IOBC/WPRS Bull.* **2012**, *75*, 73–76.
126. Benítez, E.; Moreno, B.; Paredes, D.; González, M.; Campos, M.; Rodríguez, E. Ecological Infrastructures among greenhouses: Revegetation and soil quality. *Ecosistemas* **2019**, *28*, 54–62. [CrossRef]
127. Benítez, E.; Paredes, D.; Rodríguez, E.; Aldana, D.; González, M.; Nogales, R.; Campos, M.; Moreno, B. Bottom-up effects on herbivore-induced plant defences: A case study based on compositional patterns of rhizosphere microbial communities. *Sci. Rep.* **2017**, *7*, 6251. [CrossRef] [PubMed]
128. FAO. *Addressing the Impacts of COVID-19 in Food Crises. April–December 2020: FAO's Component of the Global COVID-19 Humanitarian Response Plan*; FAO: Rome, Italy, 2020.
129. Marín-Guirao, J.I.; De Cara-García, M.; Crisol-Martínez, E.; Gómez-Tenorio, M.Á.; García-Raya, P.; Tello-Marquina, J.C. Association of plant development to organic matter and fungal presence in soils of horticultural crops. *Ann. Appl. Biol.* **2019**, *174*, 339–348. [CrossRef]
130. Pankhurst, C.E.; Lynch, J.M. 12 The role of soil microbiology in sustainable intensive agriculture. *Adv. Plant Pathol.* **1995**, *11*, 229–247.
131. Norris, C.E.; Congreves, K.A. Alternative Management Practices Improve Soil Health Indices in Intensive Vegetable Cropping Systems: A Review. *Front. Environ. Sci.* **2018**, *6*. [CrossRef]
132. Gómez, A.J.C.L.; García, M.C.G.; Pérez-Parra, J.J.; Cuadrado, I.M. *Caracterización de la Explotación Hortícola Protegida de Almería*; Fundación para la Investigación Agraria en la Provincia de Almería (FIAPA); Fundación Cajamar: Almería, Spain, 2009; ISBN 84-88246-32-5.
133. de Cara, M. Mini-Paper: Anaerobic Soil Disinfestation and other Techniques of Non Chemical Soil Disinfestation Techniques. Available online: http://ec.europa.eu/eip/agriculture/sites/default/files/5_eip_sbd_mp_asd_final.pdf (accessed on 3 December 2021).
134. EIP-AGRI Focus Group on Soil-Borne Diseases: Final Report. Available online: https://ec.europa.eu/eip/agriculture/sites/default/files/eip-agri_focus_group_on_ipm_practices_for_soil-borne_diseases_final_report_2015.pdf (accessed on 8 December 2021).

135. Junta De Andalucia Producción Ecológica en Andalucía Balance 2020. Available online: https://www.juntadeandalucia.es/export/drupaljda/DECO21_Balance_Estadistico_Produccion_Ecologica_2020_v1.pdf (accessed on 13 December 2021).
136. Larkin, R.P. Effects of Selected Soil Amendments and Mulch Type on Soil Properties and Productivity in Organic Vegetable Production. *Agronomy* **2020**, *10*, 795. [CrossRef]
137. Marín-Guirao, J.I.; Hernández-Medina, J.A.; Martín-Expósito, E.; García-García, M.C.; de Cara-García, M. Efecto de tres tipos de acolchado en el suelo y el cultivo de tomate ecológico (Solanum lycopersicum L.) en invernadero. In Proceedings of the XXVIII Jornadas Tecnicas S.E.A.E., Online, 28–29 October 2020; ISBN 978-84-949844-2-6.
138. Bretones, F. El enarenado. In: Técnicas de producción en cultivos protegidos (Coord.: F. Camacho Ferre). *Ed. Caja Rural Intermediterránea Cajamar Inst. Estud. Cajamar.* **2003**, *I*, 109–118.
139. Diacono, M.; Montemurro, F. Long-term effects of organic amendments on soil fertility. A review. *Agron. Sustain. Dev.* **2010**, *30*, 401–422. [CrossRef]
140. Gamliel, A.; van Bruggen, A.H.C. Maintaining soil health for crop production in organic greenhouses. *Sci. Hortic.* **2016**, *208*, 120–130. [CrossRef]
141. Contreras, J.I.; Baeza, R.; Segura, M. Cuantificación de los nutrientes aportados al suelo por la incorporación de los restos de los cultivos hortícolas de invernadero. In Proceedings of the XI Congreso SEAE, Vitoria-Gasteiz, Spain, 1–4 October 2014.
142. Salinas, J.; Meca, D.; del Moral, F. Short-Term Effects of Changing Soil Management Practices on Soil Quality Indicators and Crop Yields in Greenhouses. *Agronomy* **2020**, *10*, 582. [CrossRef]
143. Gómez-Tenorio, M.A.; Ruiz-Olmos, C.A.; Marín-Guirao, J.I.; Martín, F.; Camacho-Ferre, F.; Tello-Marquina, J.C. La biodesinfección de un suelo arenado y relación con el lixiviado de nitratos. In *Actas de Horticultura*; Sociedad Española de Ciencias Hortícolas: Córdoba, Spain, 2015; pp. 16–21. ISBN 978-84-606-8547-0.
144. Martín-Expósito, E.; Marín-Guirao, J.I.; Hernández-Medina, J.A.; García-García, M.C.; de Cara-García, M. ¿Es suficiente un único abonado de fondo con estiércol en un cultivo de tomate ecológico bajo invernadero? In Proceedings of the VIII Congreso Internacional Agroecología, Vigo, Spain, 1–3 July 2020; pp. 685–686, ISBN 978-84-09-21743-4.
145. Marín-Guirao, J.I.; Tello-Marquina, J.C. Microbiota edáfica y fatiga de suelo en invernaderos de la provincia de Granada. In *I Jornadas de Transferencia Hortofrutícola de CIAMBITAL*; Camacho-Ferre, F., Valera-Martínez, D.L., Belmonte-Ureña, L., Herrero-Sánchez, C., Reca-Cardeña, J., Marín-Membrive, P., del Pino-Gracia, A., Casa-Fernández, M., Eds.; Universidad de Almeria, CIAMBITAL: Almeria, Spain, 2017; pp. 17–36. ISBN 978-84-16389-98-8.
146. Marín-Guirao, J.I.; de Cara, M.; Tello-Marquina, J.C. Effect of soil biodisinfection on soil fungal communities associated to horticultural crops. *Ecosistemas* **2019**, *28*, 63–72. [CrossRef]
147. Marín-Guirao, J.; Gómez-Tenorio, M.A.; Castillo -Díaz, F.J.; Magdaleno-González, J.; Tello-Marquina, J.C. Incidencia de la biodesinfección de suelos de invernadero usando restos vegetales del cultivo de pimiento sobre la microbiota edáfica (hongos y bacterias). In Proceedings of the Congreso Internacional de Sistemas Agrarios Tradicionales, V Encuentro InterVegas, La Laguna, Tenerife, Spain, 24–26 October 2019; pp. 23–26.
148. Marín-Guirao, J.I.; Martín-Expósito, E.; Hernández-Medina, J.A.; García-García, M.C.; de Cara-García, M. Incidencia de la biosolarización en suelo desnudo y arenado sobre la microbiota edáfica. In Proceedings of the VIII Congreso Internacional Agroecología, Online, 2–3 July 2020; Simón, X., Pérez-Neira, D., Copena, D., Eds.; pp. 683–684.
149. BOJA, Orden del 6 de abril. *Boletín Of. Junta Andal.* **2017**, *69*, 96–118. Available online: https://www.juntadeandalucia.es/boja/2017/66/index.html (accessed on 17 January 2022).
150. MAPA. Lista Comunitaria de Sustancias Activas Aprobadas, Excluidas y en Evaluación Comunitaria, Sustancias de Bajo Riesgo, Sustancias Candidatas a la Sustitución y Lista de Sustancias Básicas. Available online: https://www.mapa.gob.es/agricultura/pags/fitos/registro/fichas/pdf/Lista_Sustancias_activas_aceptadas_excluidas.pdf (accessed on 3 December 2021).
151. Juan, M.; Casas, J.; Bonachela, S.; Fuentes-Rodríguez, F.; Gallego, I.; Elorrieta, M.A. Construction characteristics and management practices of in-farm irrigation ponds in intensive agricultural systems—Agronomic and Environmental Implications. *Irrig. Drain.* **2012**, *61*, 657–665. [CrossRef]
152. Bonachela, S.; Juan, M.; Casas, J.J.; Fuentes-Rodríguez, F.; Gallego, I.; Elorrieta, M.A. Pond management and water quality for drip irrigation in Mediterranean intensive horticultural systems. *Irrig. Sci.* **2013**, *31*, 769–780. [CrossRef]
153. Scheffer, M. The story of some shallow lakes. In *Ecology of Shallow Lakes*; Population and Community Biology Series; Springer: Dordrecht, The Netherlands, 2004; Volume 22.
154. Casas, J.; Sánchez-Oliver, S.J.; Sanz, A.; Furné, M.; Trenzado, C.; Juan, M.; Paracuellos, M.; Suárez, M.D.; Fuentes, F.; Gallego, I.; et al. The paradox of the conservation of an endangered fish species in a Mediterranean region under agricultural intensification. *Biol. Conserv.* **2011**, *144*, 253–262. [CrossRef]
155. Casas, J.J.; Toja, J.; Peñalver, P.; Juan, M.; León, D.; Fuentes-Rodríguez, F.; Gallego, I.; Fenoy, E.; Pérez-Martínez, C.; Sánchez, P.; et al. Farm Ponds as Potential Complementary Habitats to Natural Wetlands in a Mediterranean Region. *Wetlands* **2012**, *32*, 161–174. [CrossRef]
156. Robledo, A.; Pekas, A. *Guía de Campo de los Artrópodos de los Invernaderos Hortícolas*; Phytoma-España: Valencia, Spain, 2020.
157. Horgan, F.G.; Mundaca, E.A.; Crisol-Martínez, E. Emerging Patterns in Cultural Ecosystem Services as Incentives and Obstacles for Raptor Conservation. *Birds* **2021**, *2*, 185–206. [CrossRef]
158. Meynard, J.-M.; Jeuffroy, M.-H.; Le Bail, M.; Lefèvre, A.; Magrini, M.-B.; Michon, C. Designing coupled innovations for the sustainability transition of agrifood systems. *Agric. Syst.* **2017**, *157*, 330–339. [CrossRef]

agriculture

MDPI

Article

The Role of Consumer-Citizens and Connectedness to Nature in the Sustainable Transition to Agroecological Food Systems: The Mediation of Innovative Business Models and a Multi-Level Perspective

Sepide Mehrabi [1], Juan Carlos Perez-Mesa [2,3] and Cynthia Giagnocavo [2,4,*]

1 Department of Economics and Business, University of Almería, La Cañada, 04120 Almería, Spain; sepidemehrabi@ual.es
2 Cátedra COEXPHAL-UAL Horticulture, Cooperative Studies and Sustainable Development, Department of Economics and Business, Universidad de Almería, La Cañada de San Urbano, 04120 Almería, Spain; juancarl@ual.es
3 Centro de Investigación Mediterráneo de Economía y Desarrollo Sostenible (CIMEDES), Edificio B, Despacho 1.030, Universidad de Almería, La Cañada de San Urbano, 04120 Almería, Spain
4 Centro de Investigación en Agrosistemas Intensivos Mediterráneos y Biotecnología Agroalimentaria (CIAMIBITAL), Edificio de Servicios Técnicos 2.13.0, Universidad de Almería, La Cañada de San Urbano, 04120 Almería, Spain
* Correspondence: cgiagnocavo@ual.es

Citation: Mehrabi, S.; Perez-Mesa, J.C.; Giagnocavo, C. The Role of Consumer-Citizens and Connectedness to Nature in the Sustainable Transition to Agroecological Food Systems: The Mediation of Innovative Business Models and a Multi-Level Perspective. *Agriculture* **2022**, *12*, 203. https://doi.org/10.3390/agriculture12020203

Academic Editors: José Luis Vicente-Vicente, Cristina Quintas-Soriano and María D. López-Rodríguez

Received: 11 January 2022
Accepted: 28 January 2022
Published: 31 January 2022

Abstract: Conventional agricultural systems have contributed to social, economic and environmental problems and are the main threat to global sustainability. In response, theoretical frameworks to describe the transition to sustainable food systems have been proposed, emphasizing the necessity to shift from farm-level solutions to a focus on interactions within the entire value chain, from production to consumption. Despite the emphasis on the importance and potential of consumers to contribute to sustainable agri-food transitions, approaches to their role have remained within the traditional, linear supply chain framework. Marketing approaches view consumers as passive actors, limited to voting with their wallets, which has deepened the disconnection between consumers, producers and nature, resulting in a triple fracture. We analyze the role of the consumer in agri-food systems, contrasting marketing approaches with other consumers/citizens concepts and locate them within sustainability transition frameworks and a multi-level perspective. We discuss the re-establishment of the connection between farmers and consumers and human–nature connectedness and explore this connection mediated through innovative business models, which act as niche innovations with the capacity to influence regimes and landscapes within the multi-level perspective. The role of consumers/citizens in the co-creation of innovative business models is also addressed.

Keywords: agroecology; sustainable agricultural transition; consumers/citizens; sustainable consumption; innovative business models; alternative agri-food networks (AAFNs); human–nature connectedness (HNC); multi-level perspective (MLP); community-supported agriculture (CSA); cooperatives

1. Introduction

Current conventional agri-food systems are among the major threats to global sustainability [1–4]. In addition, the industrial approach to food and farming has helped unsustainable agri-food systems to evolve and become firmly established [5]. Conventional agri-food systems seek to produce large amounts of standardized foods to achieve economies of scale, where production volume and yield outputs are indicators of productivity [6]. This fact results in negative environmental impacts, such as climate change, environmental degradation, stressed resources and biodiversity loss [7–10], as well as socio-economic problems, such as "demographic change, urbanization, growing inequality, unequal access to resources, unhealthy eating habits and poverty" [11,12]. Despite the massive volume of food production in the global markets for an ever-increasing population (expected to rise to nearly 10 billion by 2050 [13]), unequal access to nutritious food has led to an increase in the number of undernourished people in the world, rising to 811 million in 2020 [14,15]. These socio-economic challenges suggest that agri-food value chains should embrace more sustainable objectives and measures.

Repositioning current agri-food systems from being the largest drivers of global environmental change [16] to becoming an agent of global sustainability transition requires a major shift from farm-level solutions to a focus on the entire value chain [17–19]. This shift concerns not only the production and processing stages but, more importantly, human–nature connectedness (HNC), or what Berti [20] refers to as the "triple fracture" in the agri-food value chain, namely, a disconnection between elements of nature, consumers and producers in the agri-food systems.

Concurrently, the important role of consumers and citizens in sustainability transition is gaining recognition, as evidenced by the increasing number of publications about this issue across sectors [21–23]. Furthermore, the agri-food sector is not an exception. In 2018, the United Nations Food and Agriculture (FAO) and the United Nations Environmental Programme (UNEP) jointly framed sustainable agriculture as "a consumer-driven, holistic concept that refers to the integrated implementation of sustainable patterns of food consumption and production", emphasizing that consumers around the world can be a powerful force for change towards more sustainable and equitable agri-food systems [24] (p. 2). In addition, on 20 May 2020, the European Union launched the Farm-to-Fork (F2F) strategy [25], comprehensively addressing food sustainability from primary production to the final consumers [26]. The F2F journey is not a simple connection between farmers and consumers, since it involves a vast range of different actors, stakeholders and agents. Moreover, it acknowledges the role of individuals as both consumers and citizens with the agency that allows them to build a coalition rather than act solely as consumers. For this reason, we refer to "consumer/citizen" herein. In addition, as a form of food democracy initiative, it provides space for the interconnection of individual and collective consumers/citizens [19,26,27].

However, despite the recognition of the high importance of consumers, citizens, users and, more broadly, civil society, in the agri-food sustainability transition, studies offer a very fragmented perspective when defining their various roles in this transition [21]. Several studies and global schemes have focused on the proximity of the consumers/citizens and producers and their potential to facilitate the sustainable transition of agri-food systems. This has been coupled with an increasing focus on transparency [28,29], traceability [30,31], a wide range of sustainable and "green" certifications and other initiatives. Consequently, there is an increased emphasis on consumer marketing initiatives and sophisticated techniques in order to stimulate "pro-environmental" consumer behavior.

This predominantly liberal market, demand-driven approach assumes that well-informed consumers will make the correct choices based on transparent information, aided by appropriate technologies and innovations such as blockchain technology [32] (p. 179), thus creating a market for sustainable agricultural products resulting from communications with consumers through marketing measures [33,34]. Yet, the physical and knowledge-based disconnection between consumers/citizens and agricultural production, or what Widener et al. [35] refer to as the absence of "food system literacy", which might stem from the commoditization of agricultural products in global supply chains [36,37], makes the meaningful role of the consumer citizen even less tenable.

This article explores the research question of how the re-establishment of HNC in the form of consumer/citizen connection with sustainable growers and nature may further enable the active role of consumers in sustainability transitions. We explore innovative business models that mediate between consumers/citizens and growers and which allow consumers/citizens to re-connect with agriculture and the natural environment. We also consider how these niche innovative business models and alternative networks become institutionalized to connect with broader social changes in order to share knowledge and meaning among actors and to facilitate the active participation of consumers in the co-creation of more sustainable food systems. The widely referenced framework of Gliessman's 5 levels of agroecological transformation [38] and FAO's 10 elements of agroecology for the sustainable transition of agri-food systems [39] are used as a starting point to anchor our research question. Within this framework, we focus on level four, the re-connection among consumers/citizens, growers and alternative food networks, which is seen to be transformative and which precedes level five, which refers to the rebuilding of the food system itself.

Against such backdrop, this article is a theoretical analysis of such level four and the contribution of the concept of HNC and re-connectedness to nature through grower–consumer relationships, weaving together and triangulating findings from literature reviews from diverse disciplines implicated in our research question. In Section 2, we describe the methodology implemented for reviewing multidisciplinary literature. Section 3 sets out the results and discussion. Section 3.1 outlines the perception of consumers in agri-food supply chain systems (including those that are considered "sustainable") and critically discusses the limitations of current marketing approaches to consumers. In addition, we distinguish between the individual and collective roles of consumers. In Section 3.2, we introduce sustainability transitions and the framework of Gliessman's five levels of agroecology and corresponding FAO elements, locating our research focus in level four. In Section 3.3, we describe the concept of and theories on HNC and how the active role of consumers in sustainability transitions may be strengthened by such connection, that is, given a medium through which to do so. Section 3.4 outlines how consumers/citizens may be connected to growers, thus, nature, through innovative business models. The multi-level perspective (MLP) is used to demonstrate how innovative business models, niche experiments and networks may become institutionalized so as to influence agri-food systems [40–42]. Business models discussed in this article include alternative and/or innovative food networks, social enterprise and cooperative businesses, collective producer groups that share knowledge and land, collective-food-buying groups, community-supported agriculture, collective-food-buying groups, short supply chains, etc. The business models in Section 3.4 are also categorized with respect to the degree of consumers/citizens' engagement with nature (HNC) and FAO elements of agroecology.

2. Materials and Methods

In considering the role of the consumer/citizen and their relationship to nature as a bridge to sustainable transitions, we used a framework for analysis, combining Gliessman's five levels of agroecology [38] and FAO's 10+ elements of sustainable food systems [39], as well as sustainable transition literature applied to agriculture [9,14,31,41,43–47]. Gliessman's five levels of agroecology (see below in Section 3.2) comprise incremental, transformational and system-level changes. The first two levels focus on increasing the efficiency of inputs and substituting alternative practices at the farm level. The third level focuses on a redesign of agroecosystems through diversity, resilience and creating synergies. Levels four and five focus on food system change through the re-connection of consumers/citizens to farmers and the rebuilding of the global food system, respectively. Herein, our focus is on the re-connection of consumers/citizens to farmers (level four). Although the re-connection of consumers/citizens with farmers and producers does not necessarily lead to a re-connection to nature (HNC), for example, in highly industrialized farming, it does provide an increased opportunity for consumers/citizens to engage with those sustainable producers and agroecological systems that do exist as well as actively participating in the co-creation of alternative value chains [38,39].

Starting from this framework, we carried out a literature review to analyze the fundamental aspects of the three points of analysis relevant to such framework, i.e., (i) sustainability transitions in agri-food systems; (ii) the role of consumers/citizens in sustainable transitions of agri-food and connection to nature (HNC); and (iii) innovations in agri-food systems, which mediate between consumers/citizens and growers/nature. This review formed a basis from which to triangulate these three distinct areas, illustrating the different approaches to sustainable transitions and the innovative business models that mediate the re-connection of agri-food consumers with sustainable growers and, by extension, nature. The MLP (outlined in Section 3.4) was then utilized to illustrate how incidences of niche innovative and alternative business models which make possible and enable consumer/citizen HNC may be institutionalized to contribute to the transition to more sustainable agriculture and, ultimately, influence system change.

The literature review was based on the Scopus and Web of Science databases. Selected keywords were based on the research question and the multidisciplinary nature of the subject. The initial search was run based on title, abstract and keywords in the time span of the last 20 years. According to the aim of this article, which is to combine insights from multiple disciplines, a different set of keywords was used for each of the above points of analysis.

Based on keywords in query 1 (Table 1), the first point of analysis regarding the definition of sustainability transitions in agri-food systems was addressed. A total number of 1195 results were found. Due to the high number of results, filters were implemented to limit the results to "review articles" from 2018 to 2021. Out of the review articles, 15 references were selected based on relevance and on avoiding repetition.

Table 1. Description of research queries on Scopus and WOS.

Database	Scopus	Web of Science
Research query 1:	(TITLE-ABS-KEY (agri* OR agro*) AND TITLE-ABS KEY (sustainab* W/3 (transition* OR transformat*)))	TS = ("agri*-food" OR "agro*-food") AND TS=((sustainab* transition) or (sustainab* transformation))
Results	874	365
Refined (Review only)	63	46
Total selected without duplicates	15 references	
Research query 2:	(TITLE-ABS-KEY ("Connect* people" W/2 nature) OR TITLE-ABS-KEY (connect* W/2 nature) OR TITLE-ABS-KEY ("Human nature connectedness") AND TITLE-ABS-KEY (consumer OR citizen) AND TITLE-ABS-KEY (sustainab*))	TS = (((Connect NEAR/0 people) NEAR/3 (Nature)) OR (nature NEAR/3 connect*) OR ("Human nature connectedness")) AND TS = (consumer or citizen) AND TS = (sustainab*)
Results	36	40
Total selected without duplicates	10 references	
Research query 3:	(TITLE-ABS-KEY ((sustainab* AND innovat* AND agri* AND chain*)) AND KEY ((consumer* OR citizen*)))	TS = (sustainab* AND innovat* AND "agri-food" AND chain*) AND TS = (consumer or citizen)
Results	27	40
Total selected without duplicates	6 references	

Second, to address the impact of consumers/citizens connectedness with nature in agri-food sustainable transitions, keywords in query 2 (Table 1) were implemented. A total number of 76 articles were found. After review and duplication reductions, 10 core articles were selected.

Based on keywords in query 3 (Table 1), sustainable innovations in agri-food systems focused on consumers and citizens were addressed. In total, 67 references were found and reviewed and 6 references were selected as highly relevant.

Finally, given the complexity in choosing keywords that captured the subject matter, the backward and forward snowball method [48,49] was implemented and 35 articles were selected by this method. The Google Scholar database was also consulted to fill in any gaps, particularly with respect to business models found in gray literature.

3. Results and Discussion

3.1. Approaches towards the Role of Consumers in Agri-Food Systems

Despite the recognition of the role of consumers in transforming current food systems, the widely used linear approach in the agri-food supply chain, from inputs to consumers (Figure 1), does not adequately capture the inter-relationships between all actors and multi-stakeholders, particularly that of consumers and farmers. Hence, many initiatives have remained within the framework of the traditional food system in which consumers have been relegated to the end of the value chain as passive individuals confined to using their purchasing power, or what Hatanaka refers to as "voting with a wallet" [50], to influence upstream practices and sustainability practices [51].

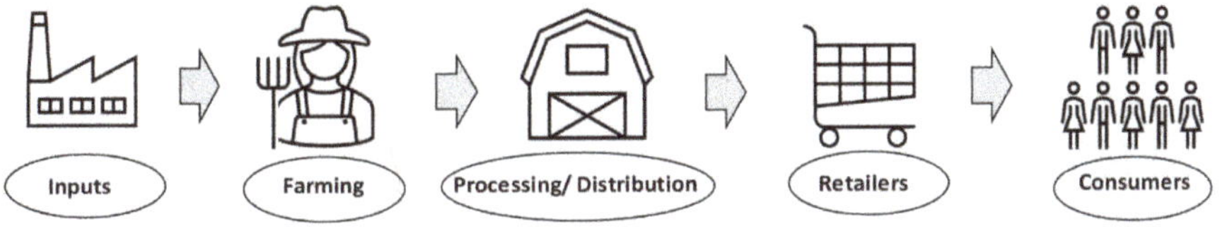

Figure 1. Consumers in the conventional agri-food supply chain (authors' elaboration).

In the agri-food supply chain, communication with consumers is generally achieved through marketing (including those supply chains which aspire to be more sustainable). A process of "consumer segmentation" is generally carried out in marketing studies, with a focus on increasing sales [52–54]. To promote environmentally responsible consumer behavior, the marketing literature has provided a conceptualization of environmentally sustainable consumer behavior, offering different dimensions for it, including "consumer acceptance", "consumer perception", "consumer attitude", "in-purchasing behaviour" and "willingness to pay", to find the best way to stimulate consumers' purchasing decision. These marketing studies concentrate on consumer segmentation in order to target each segment based on its characteristics and explore driving factors and methods to motivate more sustainable behavior and change purchasing habits through different marketing strategies [55–57]. For example, consumers are assessed in terms of perception of sustainability attributes to shape different clusters. The assessment can be based on their perceived value about the procedure of production assigned to the "fair trade" cluster, the local origin of the product assigned to the "local" cluster and based on readiness to pay for sustainable products assigned to "price-sensitive" clusters. Thus, these clusters provide useful information for corporations to implement marketing strategies directed to target consumers [55]. Moreover, exploring patterns and data obtained from consumer segmentation allows one to predict and analyze what will happen or is likely to happen, forecasting consumer demand or behavior [56]. In addition, forecasting consumers´ acceptance of innovative technologies ensures the successful implementation of new marketing strategies [53].

Although these innovative studies emphasize consumers in the agri-food systems, analyzing "consumer preferences" and "in-purchasing behaviour" to stimulate purchasing certain "green" or sustainable products, this approach tends to have limited power to result in meaningful transitions or allow consumers to engage in collective social activities with sustainability aims actively. This view portrays consumers as individual passive actors who can be manipulated and treated as simply economic actors whose participation in sustainability transition is limited solely to purchasing decisions, however well intentioned they may be [50].

Consumers/citizens, on the other hand, are becoming more concerned about the impact of agricultural activities, production and distribution of food on the environment, human wellbeing and social and economic implications [57,58]. In spite of the fact that food in developed countries has never been safer [59,60], "consumer perception on the safety of the food supply, the control infrastructure, and existing and new process technologies is often not positive" and has led to sensitized consumers who are wary of their food supply [54]. The increase in awareness and sensitivity has resulted in consumers demanding food and ingredients free of synthetic fertilizers or pesticides, negligibly processed, easily accessible and affordable and with minimum environmental impacts [61]. Some studies have shown that, when it comes to product selection by consumers, price and taste indicators no longer necessarily outweigh environmental and health considerations [62] and consumers' strong "green" preferences increase their willingness to pay a premium price for such products [63]. In fact, consumers feel better when they purchase products from brands with an environmentally and socially responsible image [64]. Hence, corporate social responsibility (CSR) has emerged in response to consumers' needs for intangible attributes of food products [58]. Nevertheless, according to Boccia and Sarnacchiaro [65,66], CSR initiatives´ effects on consumers are low due to the lack of awareness. Moreover, some studies have shown that consumers/citizens' preoccupations with environmental protection do not necessarily drive agri-food purchasing motivations [67]. Even consumers who are committed to specific sustainable and ethical ideals may prioritize personal interests and needs (such as price or taste) over sustainability ideals when it comes to the time of purchasing [68], raising the question about the long-term commitment of consumers in their consumption practices.

In response to the traditional approach of the role of consumers in the agri-food systems, several concepts and initiatives have been developed considering a more active role for both individual and collective consumers/citizens in the agri-food value chain. For instance, sustainable consumption has emerged in marketing scholarship as a pressing matter [67,69], followed by other initiatives, such as ethical consumption [70], responsible consumer behavior [71], reflexive consumption [72], green procurement/consumption [73] and green certification [74]. However, these initiatives have also narrowed the focus only on the consumption part of the agri-food value chain and the proposed solutions have focused on merely consumers rather than a systemic alternative that encompasses a broader context involving other stages of the value or supply chain.

Additionally, it appears there is a gap in marketing studies, as very few have focused on the social dimensions of sustainability [75]. Most studies focus on individual consumer behavior, such as pro-environmental behavior in everyday personal life, individual environmental knowledge, individuals´ green product attachment and green value, personal anticipated pride and guilt, perceived effectiveness and individual aspects of connectedness to nature [76]. However, according to Verhees and Verbong [21], individual and collective consumers' roles in adapting to sustainable innovations are different due to the dynamics and behavior mechanisms. They offer a model concerning sustainable and innovative business models based on a dichotomy between individual and collective behaviors. In this model, there is a spectrum that goes from one extreme to the other, including the passive role of consumers (consumers as buyers), more active involvement of consumers in co-production and consumer-led innovations on the other side. Based on this spectrum, the collective role can range from collaborative consumption, such as "collective purchasing power" business models, where groups of people cooperate for mutual benefit, to a more active collective role, beyond simply buying, such as active participation of consumers/citizens in farming activities. The other extreme of consumers' collective role is "self-organized citizen groups" who are initiators, designers and maintainers of innovative sustainable business models in their locality (see Section 3.4). Verhees and Verbong [21] note that collective consumers use group power to create large-scale social movements. In addition, engaging in collective altruistic behavior provides opportunities for socializing, building the network and acquiring common goals and meaning among people, which is a powerful force for social movements. Thus, in order to address current environmental, social and economic challenges, there is a need to go beyond just individual responsible consumption from the marketplace to structural and large-scale societal change [77,78].

Therefore, the concept of food citizenship, referring to the collective role of consumers/citizens, attempts to proportionally distribute the burden of sustainable responsibilities across the entire agri-food value chain. Food citizenship means the responsible act of consumers/citizens and producers who actively participate in the configuration of food systems in a myriad of ways. This includes co-producing sustainable food through engaging in the governance of the food system [50,79]. Food citizenship can encompass consumers/citizens and all other actors in the entire food system [80]. The emphasis of food citizenship is on the active engagement in decision making over the kinds of food produced and production procedure [81–84]. Another aspect of food citizenship refers to "civic engagement", meaning the individual sacrifice for collective goals [85] and transparency in food production and consumption practices, processes and relations. Food citizenship is associated with localization and the idea of short supply chains and localism enables more personalized relations direct and participatory forms of governance, which are often viewed as conducive to people acting as citizens [86].

Above, we discuss the characterization of and importance given to consumers/citizens as participants in the agri-food value chain and emphasize the necessity of changing the traditional linear approach towards consumers/citizens in the agri-food value chain. In the next step, below, we explore how system change may occur and what the role of consumers/citizens in facilitating the transition to a sustainable agri-food system is.

3.2. Sustainability Transition Framework and the Role of Consumers

Many scholars have argued that the current agri-food system should be changed through sustainable transition practices or agroecological transformation [16,46]. Weber et al. [87] used the phrase "deep change" as an umbrella term for both transition and transformation in the sustainable food system. According to Markard et al. [88], "sustainable transition" is defined as "long-term, multidimensional and fundamental transformation processes through which established socio-technical systems shift to more sustainable modes of production and consumption". The arguments presented indicate that the way food is produced should be changed, but the manner in which food is consumed is also of importance. Gaitán-Cremaschi et al. [89] maintain that a "sustainability transition" is needed to transform existing food regimes into alternative regimes.

Out of this diverse literature, different frameworks and approaches have emerged in order to analyze sustainability transition. For example, in 2017, the United Nations defined 17 goals for sustainable development (SDGs) as a roadmap to effect changes before 2030 and offered broad goals for transforming the world [90]. FAO's common vision for a sustainable food and agriculture framework, which is called FAO 10+ elements of agroecology [39], has identified a scientific pathway to achieve SDGs. In addition, the high-level panel of experts (HLPE) published a report in July 2019 [45], offering 13 agroecological principles for sustainable agriculture and food systems that enhance food security and nutrition based on SDGs. Moreover, Gliessman [91] set out five stages of transformation to agroecology (Figure 2) that also portray a pathway and different levels of transition to sustainable agriculture. According to Wezel et al. [92], FAO 10+ element and HLPE13 agroecological principles are in alignment with Gliessman's five levels of agroecology. Agroecology is understood as a science, a set of practices and a social movement [46,85,86] and is considered as an approach that can address multiple crises in the food system while addressing environmental, economic and social aspects of sustainability transitions [93].

In Gliessman's five levels of agroecology (Figure 2), the first two levels, considered "agroecosystems", involve incremental change, whereby farming systems convert from industrial agriculture systems to more ecological systems through the increase in input efficiencies and substituting alternative practices and inputs. Level three, also at the agroecosystem level, is considered transformational, given that it redesigns the whole agroecosystem based on ecological processes. However, levels four and five go beyond the farm to broader food systems and the societies in which they are embedded, emphasizing the "system change" for transformational alteration.

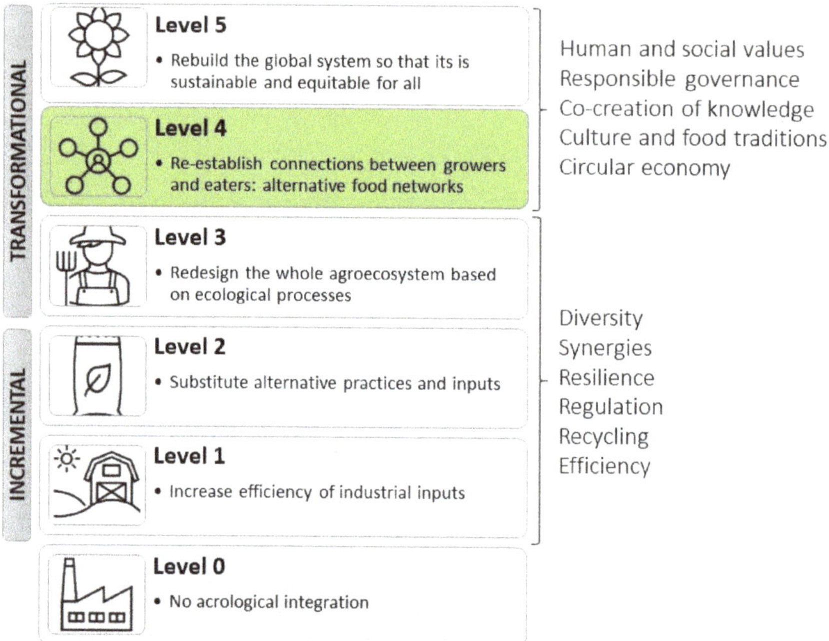

Figure 2. Authors' elaboration based on Gliessman´s five levels of agroecology and FAO 10+ elements alignment (Level 4 indicates the main focus of the article).

What is notable for this article is the reference, in level four, to re-establishing the connections between growers and consumers and the development of alternative food networks. The re-connection between consumers/citizens and farmers and, by extension nature, is part of the process of agroecological transition.

The re-connection between consumers/citizens and farmers does not necessarily mean that consumers/citizens are re-connected to nature. This depends very much on the connection to the nature of the farmers or growers. However, if growers have a connection to nature, it would be reasonable to assume that consumers/citizens would also be able to access such connections through their relationship with such growers. The latter topic of the connection of farmers with nature is outside the scope of this article, but the relationships of consumers/citizens and farmers usually involve some type of business model, food networks, or systems, through which their relationship is mediated, which is discussed below in Section 3.4.

3.3. *Connectedness to Nature*

The current modern lifestyle and the lack of proximity with nature have been identified as deepening the disconnection between humans and nature. Rapid urban growth, which utilizes natural areas and industrial agricultural intensification, can explain the ever-weakening HNC and, consequently, severe environmental and social problems [44,94]. There has been an increase in the number of research articles that are in favor of strengthening consumers/citizens connectedness to nature [95], in addition to articles discussing various scaling methods, such as the new environmental paradigm scale to research the relationship between environmental concern and sustainable behavior [96,97].

The separation of consumers/citizens from nature could explain the deterioration of the environment, since a decrease in individual emotional connection, simultaneous with a decline in the opportunity to experience nature directly, discourages positive attitude and emotions towards the environment and creates a cycle of disaffection [76,94].

Indeed, consumers/citizens often have little idea of the food source, initial steps of the production methods and the possible direct impact of food purchasing decisions on the environment [44,98]. This state of affairs raises the question of how consumers/citizens may have an active role in sustainable transition [67]. This disconnection is in line with the observations of Aldo Leopold, the 18th American philosopher that said that "We can only be ethical in relation to something we can see, feel, understand, love, or otherwise have faith in" [43] (p. 26). Leopold's thinking formed a basis for the objectives that the environmental conservation community and other similar environmental, "Deep Ecology" and ecological movements have long emphasized, that is, engaging people more closely with nature so as to increase their care and concern for the natural environment [94].

To this end, studies suggest different interpretations of the HNC concept. Wesley [99] described it as "the extent to which an individual includes nature within his/her cognitive representation of self", while Geng et al. [100] used belongingness for describing individuals' feelings about the connection with nature from both an emotional and a cognitive perspective. Frantzeskaki et al. [101] portrayed the concept as people's affective and experimental connection with nature. In general, there are two main views about people and nature, (i) an anthropocentric one, which deems nature as a source of materials, services or commodities; and (ii) an eco-centric view of nature that considers nature as valuable in itself, including considering nature as a stakeholder with its own rights [102]. For enabling sustainability transition of the agri-food value chain, scholars have highlighted that the scale of change is beyond what can be achieved via anthropocentric views in the incremental level of agroecology (e.g., water or pesticide management) and eco-centric conceptions of nature entail a transformational change in the relations of humans with the natural environment [103]. Consequently, to change from an anthropocentric to an eco-centric view of nature, the cognitive, emotional and philosophical dimensions of consumers/citizens' experience with nature provide us with conceptual lenses necessary for creating a sustainable transition pathway.

According to Zylstra et al. [95], individuals' experience with nature can range from merely possessing "information about nature" and having "experience in nature" to being "committed towards nature". On the other hand, Dickinson [104] recommended that, in order to promote change in social actors involved in a territory, "place attachment" has a major impact on the identity and sense of place. Place attachment has four main dimensions, place identity, place dependence, place social bonding and place nature bonding [105]. Ramkissoon et al. [106] emphasized the pluralistic nature of pro-environmental behavior in the above-mentioned four dimensions of place attachment, suggesting that the meaning of environmental actions and pro-environmental behavior is constructed through social interactions in different settings.

Consumers/citizens' place attachment might intensify the co-creation of knowledge and dialogue among different actors through nature-based environmental education, thus increasing engagement in pro-environmental behavior and circular economy [76,107]. In addition, research has demonstrated the significance of HNC with consumers/citizens engagement in agrarian landscapes and stewardship practices [108,109]. According to Auer et al. [37], agricultural landscapes are important in human wellbeing, impacting social capital. More specifically, Pérez-Ramírez et al. [110] found that the human values associated with agricultural landscapes linked with farming activities might explain a stronger sense of place, thus a sense of responsibility towards it.

As a result, HNC could provide an inherent motivation for developing ecological behavior and efforts that might last throughout people's lives. Previous research has demonstrated that solid connectedness to the natural environment in consumers results in pro-environmental behavior such as a willingness to preserve the natural environment and active engagement in environmentally sustainable consumption behavior [97,111–113] and positively impacts individual and social wellbeing [110,114]. In fact, connectedness to nature has proven to have a similar or an even more important role in stimulating consumers/citizens' environmentally friendly behavior rather than socio-demographic segmentation [99]. In other words, the more people are connected to nature, the more they are expected to behave sustainably [111] and the less they are prone to harm it, since this damage would be considered as damage to themselves [76].

Thus, consumers/citizens' connection with nature would increase their sensitivity to the natural environment and might enhance their engagement as initiators, designers and maintainers of sustainable innovations.

3.4. Connecting Consumers/Citizens to Growers and Nature through Innovative Business Models

The need for fundamental changes in the way farmers and consumers/citizens interact with nature to achieve sustainable transformation is receiving more attention [115]. Moreover, the recent global crises have raised the question of the impact of existing corporate business models on the sustainability of the agri-food system and what initiatives may be necessary [116]. This entails realizing the organization of commercial structures in a way where not only are consumers/citizens more connected to nature, but business models and food networks are intrinsically transformative to make the transition to sustainability possible. In addition, it is important to explore how new initiatives and alternative business models may be developed and gain power in order to make transformative changes in the food system.

To understand the development mechanism in the emerging sustainable business models, a framework that conceptualizes socio-technical transitions as an interaction of social, environmental, political and economic changes is needed. The multi-level perspective (MLP) is the first and foremost approach adopted in recent sustainability transition studies [41,42]. The MLP conceptualizes transitions as an interconnection among three levels of relative stability, that is, niches that contribute at the "micro" level, regimes that contribute at the "meso" level and landscapes that contribute at the "macro" level [117] (Figure 3).

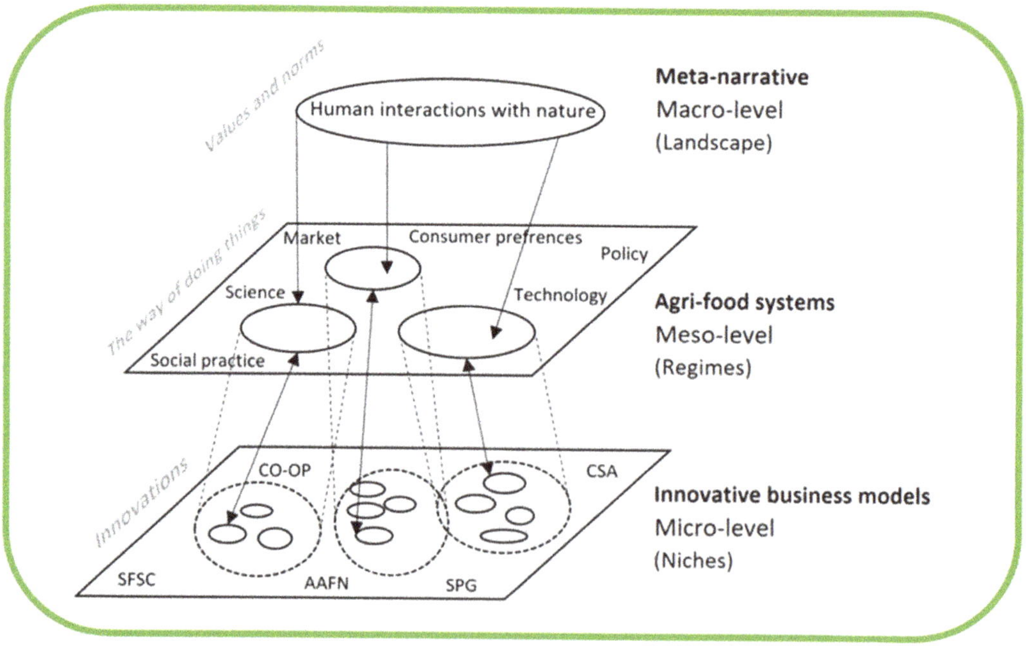

Figure 3. Authors' elaboration of the multi-level concept based on Geels, 2002 [118] (p. 1261), Pereira, et al., 2018 [40] (p. 330), and Garcia, et al., 2020 [119] (p. 420) illustrating level 4 of agroecological transition.

According to Darnhofer et al. [120], "niches" have been defined as changes that "new technologies and practices, new configurations of actor groups, new beliefs and values, new networks, new policies" might bring about [121]. The strength and maturity of the niche is necessary to reach a greater number of people and, consequently, provide the conditions to scale up and out to the meso level contributing to the agri-food transition. "Regime", in agri-food systems, refers to the intensive, conventional, industrial agro-food sector and its associated rules and practices, food safety laws, existing business networks, logistics transport and infrastructure [122]. Moreover, "landscape" refers to exogenous major social, cultural, worldwide values and norms that are difficult to influence (Figure 3). The fundamental change of the dominant regime (agri-food, for instance) is explained inter alia by pressure from niche innovations (for example, innovative business models) on the one hand and by pressure from the landscape level (meta-narrative or values and norms) on the other hand (Figure 3), which represents the slowly changing regime context [123]. This resistance at the regime level, which is also known as system lock-ins or path dependency, could be related to regime elements such as policies, practices, technologies, knowledge, or social values that stabilize each other, making change a challenging task [123].

As outlined above in Section 3.1, consumer marketing efforts have often relied on standard business models but with different motivations for consumers to purchase a particular product. However, innovative business models offer the opportunity to change the role of the consumer in the standard, linear supply chain. According to Khanagha, innovative business models may be referred to as "incremental changes in individual components of business models, the extension of the existing business model, introduction of parallel business models, right through to disruption of the business model, which may potentially entail replacing the existing model with a fundamentally different one" [124] (p. 324).

HNC innovative business models attempt to alter the modality of current human interactions with nature in agri-food systems. By shaping alternative agri-food networks (AAFNs), they prioritize local markets, support local economies and try to enable a "circular economy" by developing a virtuous cycle. In addition, they have the potential to rebalance traditional and modern food habits, promote healthy consumption and support cultural identity to enable a "cultural and food economy". Moreover, these business models might include participatory processes in which farmers´ practical knowledge blends in indigenous knowledge to promote formal and non-formal education, resulting in the co-creation and sharing of knowledge. Furthermore, by protecting and improving social wellbeing, they build autonomy and adaptive capacities to empower consumers/citizens and communities to have shared "social values" and "responsible governance". Renting et al. [84], critical of the limitations of AAFNs, advocated for those innovations that represent a shift in the role of consumers and producers as "civic food networks", which include an enhanced role of civil society from the perspective of governance.

For the scope of this article, the selected business models below are limited to those that, by their enabling of HNC and proximity with farmers, enable consumers/citizens to have an active role in the sustainable transformation; we also highlight the FAO 10+ elements where applicable.

Self-organized citizen groups (as introduced in Section 3.1) are social innovations that capture self-management and self-mobilization in supplying agri-food products. "Collective food buying groups", such as "solidarity purchasing groups" (SPG), are examples of business models which are formed for the aim of supplying food based on the mutual values and needs of the local collective citizens [70]. Solidarity purchasing groups are defined as groups of households that establish mutual coordination for the purpose of purchasing food directly from sustainable producers, who are selected in accordance with ethical and solidarity principles such as environment and social values [125] (p. 232). These collective groups have a significant role in changing both the dietary habits of consumers/citizens and farmers' production systems. They support small local producers associated with place attachment and enhancement from conventional farming to more organic and low-input systems. They might offer food baskets to consumers/citizens through networking with organizations, finding producers and distributing to consumers/citizens in addition to creating multiple communication channels for them to meet informally, communicate and share information, learn about food systems and be involved in the governance of the organization [126].

Furthermore, other consumer-based business models go further than purchasing schemes and offer a different legal form with various levels of formality. "Community-supported agriculture (CSA)" and consumer cooperatives are examples of democratic business models that also invert the traditionally perceived flows of the agri-food value chain. For example, the consumer cooperative is one well known business model that allows consumer participation in ownership and governance [127]. As well, there are associations for the support of peasant agriculture, such as Association de Maintien de l'Agriculture Paysanne (AMAPs), which is a French CSA organization that, by the partnership between urban citizens and farmers, advocates against large-scale traditional food supply chains [128]. CSA acts as a business strategy in which consumers/citizens are members of food production procedures and share associated costs and risks. CSA is defined as "a direct partnership between a group of consumers and producers, whereby the risks, responsibilities and rewards of farming activities are distributed through long-term agreements" [27]. By avoiding intermediaries, consumers and farmers communicate directly and, in the end, not only consumers/citizens gain a portion of the food production but also CSA activities educate them about sustainable agriculture. As a result, producers may receive higher incomes due to consumers/citizens' participation in harvesting, consumers' willingness to pay a premium price and fewer intermediaries [129]. These innovative business models that strengthen people's knowledge of the multiple links between food and nature, planting, harvesting and preparation, may serve to increase HNC and place attachment [130].

Additionally, consumer cooperation contributes to mutual understanding between producers and consumer/citizens, a sense of partnership and a sense of ownership—thus food citizenship—by promoting more sustainable agri-food systems in which consumers become co-responsible, in financial and organizational terms, for the production of food, participating in farming activities when needed [50]. This organizational innovative business model enables consumers to be a part of farming activities, increasing their proximity with farmers, enhancing consumers´ place attachment and social bonding in their geographical living area, thus becoming institutionalized to connect with broader social changes [131].

In addition, there are other types of business models that make possible a closer approximation of consumers and producers. For example, short food supply chains (SFSC) bring consumers and farmers closer together either geographically or by reducing intermediaries from farm to fork. "Farmers markets", "on-farm selling" (to individual consumers, not the commercial sector) and "pick-your-own" schemes are examples of a local SFSC that allows consumers/citizens to create a social bond with local farmers, obtain information about local food and its origin directly from farmers and, in return, farmers might receive consumer feedback [57,132]. "Box schemes" and "prepaid baskets" are other direct-to-consumer e-commerce business models that connect local farmers and producers directly to consumers. Aside from the variety of products in these boxes, consumers/citizens may receive information about seasonal products and traditional recipes adapted to the modern lifestyle that create more value for consumers/citizens and de-commoditize agri-food products, an important step in HNC. As Marsden et al. [133] point out, in these supply chains, the emphasis is on the type of relationship between the producer and the consumers and on the potential of this relationship in constructing value and meaning, rather than solely selling product. Although e-commerce business models and direct selling from a producer web page and/or mobile application do not necessarily connect consumers/citizens with nature, the information provided about the natural aspects of agri-products, such as "ugly food", and production conditions, such as water use, pest management, biodiversity and open grazing for animals, creates awareness and knowledge for consumers/citizens. Additionally, e-commerce significantly reduces intermediaries between consumers/citizens and farmers, which, particularly, in rural areas, might solve poverty to a great extent and bring vitality to these areas. Despite some challenges such as lack of well-developed logistics, human resources' talent and internet-based infrastructure in rural areas, these business models may reduce waste, improve farmers' income and increase productivity [134].

In addition, another form of innovative business model is related to participatory harvesting schemes, such as the "self harvested gardens", that might be implemented in community gardens [135] or private lands and require the active involvement of consumers/citizens in the harvesting steps of food production. In addition, crowd farming refers to financial sponsorship (e.g., adoption) of a tree, vines, etc., the delegation of the harvest to farmers and the receipt of a portion of products. In this process, consumers are able to observe growers and feel like a part of the harvesting procedure whilst providing needed financing [136]. These innovative business models foster environmental and socio-cultural sustainability in agricultural landscapes through HNC [112] and provide the opportunity for consumers to experience nature and have an active role in producing their food. Internet services and platforms might play a crucial role in this type of business model, for example, to connect consumers/citizens with landowners interested in sharing part of their land and the creation of peer food networks targeting surplus production of non-commercial farmers to reach interested consumers/citizens and avoid food waste [33].

Based on consumer/citizens' demand for transparency, traceability, information and knowledge about the way agri-food products are grown, the kind of labor involved, the relationship to nature, or even the public research investments which are implied in production, the need for a new integrated approach has led to the emergence of third-party labels or certifications [23]. However, in response to criticism about the passive role of consumers who do not have control over criteria or indicators that make a product eligible to receive a certification [137], "participatory guarantee systems (PGS)" have been introduced. PGS means a group of consumers/citizens as a "represented consumer organization" who indicate criteria and standards to certify producers. In business models based on PGSs, participation is an essential, created value for consumers/citizens, as well as farmers and other stakeholders, in the definition, implementation and verification of standards and rules. Thus, a consumers' represented group (who can be other farmers and agronomists) collectively takes responsibility to peer review the production process, ensuring the integrity of products verified by the PGS. A PGS leads to equal sharing of power and responsibilities, the formation of trust and a common vision and a permanent learning process through the engagement of all stakeholders [138,139].

In Figure 4, below, the business models referred to above are categorized based on the degree of consumers' engagement and their individual or collective role. As discussed at the beginning of this section, small niche innovative business models generally do not have the economic power to transform the agri-food value chain. However, the strength and maturity of niche business models provide an opportunity for them to scale up and make a change at the regime level. The niche business models' ability to reach a greater number of people might provide the conditions for innovation to become institutionalized and add pressure on the mainstream [128]. In addition, these business models promote citizens' active participation, leading to increased awareness and construction of common goals and meanings amongst diverse social groups, which is a prerequisite in overcoming the social lock-in mechanism. Furthermore, Bennett et al. [140] point out the importance of communication in creating synergies and forming networks and alliances between niche business models in prevailing lock-in mechanisms and scaling up to the meso level (see Figure 3). The creation of new coalitions among local and small niches might lead to the decentralized power of stabilized actors and alterations in regulations, standards and policies [40] (p. 330).

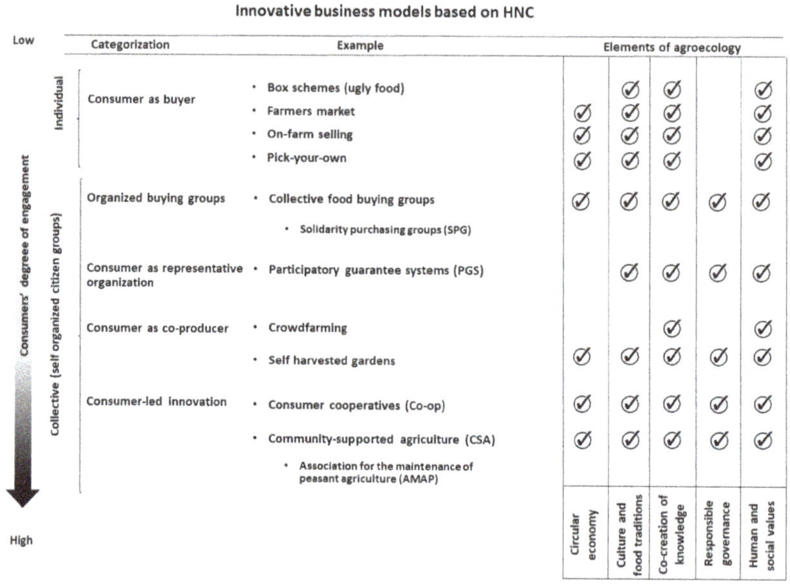

Figure 4. Innovative business models categorization (authors' elaboration).

4. Conclusions

The importance of consumers' role in sustainability transition of the agri-food value chain has gained more attention in recent years, as evidenced by the F2F strategy launched by the EU in 2020 [25], which addresses the connection between farmers and consumers, in addition to FAO and UNEP's announcement in 2017 [24], emphasizing the consumer-driven approach and integrated implementation of sustainable food consumption. Despite recognizing the role of consumers/citizens in transforming current agri-food systems, studies have remained dominated by the traditional linear supply chain framework whereby the consumers´ role is merely limited to "voting with their wallets". This linear approach from inputs to consumers does not sufficiently capture interrelationships between actors and multiple stakeholders. In addition, marketing studies generally focus on the individual consumer through consumer segmentation, perception, acceptance, behavioral attitudes and willingness to pay to promote sustainable products and increase the adaptation of innovative sustainable solutions. This approach portrays consumers as individual, passive end users who can be manipulated by green marketing initiatives and treats them as simply economic actors whilst disregarding the collective power of consumers/citizens and their potentially more active role in sustainability transitions.

In order to understand this potential for a more active consumer/citizen role, here, we focus on sustainability transition frameworks such as Gliessman's five levels of agroecology in combination with FAO 10+ elements and HLPE13 agroecological principles, which all point to the necessity for a closer relationship between consumers/citizens and growers. We focused on level four and on alternative business models and consumer-based innovations that mediate the relationship between consumers/citizens and farmers and, ultimately, the environment and nature. Although the re-connection of consumers/citizens with farmers and growers does not necessarily lead to HNC (e.g., in highly industrialized farming), if growers have a connection to nature [141], it would be reasonable to assume that consumers/citizens would also be able to access such connections through their relationship with growers.

Alternative business models reshape consumer–farmers relationships beyond merely "commodity and economic" transactions. The farmers' role significantly shifts from seller of primary or raw material to the food industry and large retailers to a source of first-hand information about food provision and the natural environment, creating diversified values for consumers/citizens and changing the commoditization approach to agri-food products. The consumers/citizens' role, on the other hand, goes beyond that of the end user and passive buyer to one of proactive co-creator of alternative value chains through the mobilization of social learning, increased awareness and knowledge, participation in governance, and creation of common goals and meaning among different actors in the value chain, which all results in a better understanding of food system dynamics and higher HNC.

We propose a categorization for alternative business models that enable consumers/citizens to have an active involvement in the sustainable transformation of agri-food systems by proximity with nature and farmers. This categorization is based on the degree of consumer engagement, both as individual and collective citizens. Business models that create value for individual consumers, including box schemes, farmers' markets, on-farm selling and pick-your-own, are considered to provide fewer degrees of engagement in comparison with innovative business models that enable the collective role of consumers/citizens. Collaborative business models in our categorization include the following: first, collective food buying groups, such as solidarity purchasing groups that provide the opportunity for consumers/citizens to support smart local food producers; second, participatory guarantee systems in which a group of consumers (who can be other farmers and agronomists) acts as a represented consumer organization, to indicate criteria and standards for certifying producers and peer-reviewing the production process; third, participatory harvesting business models that provide the opportunity for consumers/citizens to be closer with farmers and nature as co-producers in initiatives such as crowdfarming and self-harvested gardens; lastly, consumer cooperatives and community-supported agriculture, such as AMAP, are business models that self-organized citizen groups initiate, design and maintain.

These increasingly visible and innovative business models, with different legal forms and formality levels, as illustrated in the MLP in Figure 3, may provide both a breeding ground for local community innovations to mature and reach a wider community, so as to countervail the dominance of industrial agri-food regimes in the context of the current human–nature interaction meta-narrative.

5. Future Research

Based on Gliessman's five levels of agroecology, this article concentrates on level 4 of agroecology (re-connection between consumers and growers and the creation of alternative food networks). This framework assumes that consumers' relationship with nature is implicit in a closer connection between "growers" and "eaters", with farmers being considered a mediator between consumers and nature. We point out that not all farmers and farming methods mediate connectedness to nature, as in the case of highly industrialized farming systems and, in this article, we explore alternative business models where consumers/citizens may have a closer connectedness to nature. This is an area for further study in specific contexts, particularly with respect to the categorization of the degree of consumers/citizens involvement in various business models referred to herein. Related to this, is the interaction between collective consumer activity and collective producer activity, e.g., consumer cooperatives relationship to producer cooperatives.

Furthermore, determining the impact of innovative business models on HNC in different economic, social and environmental contexts depends on each area's socio-economic aspects, technology development and cultural dimensions, which are outside the scope of this article and could be studied in a comparative analysis in future research. Moreover, the role of policymakers and policy tools in shaping the consumer adoption of innovative business models which make possible HNC and their effects on accelerating sustainable transition would be an interesting area for future research, although, due to the complexity of food systems and consumer behavior, the devising of appropriate solutions and policies would need to be informed by more transdisciplinary research.

Author Contributions: Conceptualization, C.G. and S.M.; methodology, all authors; investigation, S.M. and C.G.; writing—original draft preparation, S.M.; writing—review and editing, all authors; visualization, all authors; supervision, C.G.; project administration, C.G. and J.C.P.-M.; funding acquisition, C.G. and J.C.P.-M. All authors have read and agreed to the published version of the manuscript.

Funding: This research study relied on funding from the following projects: (1) Co-Fresh (CO-creating sustainable and competitive fruits and vegetables' value chains in Europe) under Horizon 2020-EU.3.2.1.3, with grant ID number 101000852. (2) SmartRed project UAL2020-SEJ-D2026 (UAL/CTEICU/FEDER). (3) European Commission LIFEWatch2019-10-UGR-01_WP-6.2. (4) Spanish Ministry of Science, Innovation and Universities, AGROFOODSCAPES PID2020-117198RB-100.

Conflicts of Interest: The authors declare no conflict of interest.

Abbreviations

HNC	Human–nature connectedness
AAFNs	Alternative agri-food networks
CSR	Corporate social responsibility
SDGs	Sustainable Development Goals
FAO	United Nations Food and Agriculture Organization
UNEP	United Nations Environmental Programme
HLPE	High-Level Panel of Experts
F2F	Farm to Fork strategy
NEP	New environmental paradigm
MLP	Multi-level perspective
SPG	Solidarity purchasing groups
CSA	Community-supported agriculture
SFSCs	Short food supply chains
PGS	Participatory guarantee systems
CO-OP	Cooperation
AMAPs	associations for the support of peasant agriculture (Association de Maintien de l'Agriculture Paysanne)

References

1. Brunori, G.; Barjolle, D.; Dockes, A.-C.; Helmle, S.; Ingram, J.; Klerkx, L.; Moschitz, H.; Nemes, G.; Tisenkopfs, T. CAP Reform and Innovation: The Role of Learning and Innovation Networks. *EuroChoices* **2013**, *12*, 27–33. [CrossRef]
2. Woodcock, B.A.; Firbank, L.G.; Attwood, S.; Eory, V.; Gadanakis, Y.; Lynch, J.M.; Sonnino, R.; Takahashi, T. Grand challenges in Sustainable intensification and ecosystem Services. *Front. Sustain. Food Syst.* **2018**, *2*, 28. [CrossRef]
3. Bennett, E.M.; Balvanera, P.; Folke, C. Toward a more resilient agriculture. *Solutions* **2014**, *5*, 65–75.
4. Vanloqueren, G.; Baret, P.V. How agricultural research systems shape a technological regime that develops genetic engineering but locks out agroecological innovations. *Res. Policy* **2009**, *38*, 971–983. [CrossRef]
5. Ramankutty, N.; Mehrabi, Z.; Waha, K.; Jarvis, L.; Kremen, C.; Herrero, M.; Rieseberg, L.H. Trends in Global Agricultural Land Use: Implications for Environmental Health and Food Security. *Annu. Rev. Plant Biol.* **2018**, *69*, 789–815. [CrossRef]
6. Benton, T.G.; Bailey, R. The paradox of productivity: Agricultural productivity promotes food system inefficiency. *Glob. Sustain.* **2019**, *2*, e6. [CrossRef]
7. Carrington, D. Nearly All Global Farm Subsidies Harm People and Planet—UN. Available online: https://www.theguardian.com/environment/2021/sep/14/global-farm-subsidies-damage-people-planet-un-climate-crisis-nature-inequality (accessed on 12 September 2021).
8. Bongaarts, J. IPBES, 2019. Summary for policymakers of the global assessment report on biodiversity and ecosystem services of the Intergovernmental Science-Policy Platform on Biodiversity and Ecosystem Services. *Popul. Dev. Rev.* **2019**, *45*, 680–681. [CrossRef]
9. HLPE. *Food Security and Nutrition: Building a Global Narrative towards 2030*; Food and Agriculture Organization of the United Nations: Rome, Italy, 2020.
10. Rockström, J.; Edenhofer, O.; Gaertner, J.; DeClerck, F. Planet-proofing the global food system. *Nat. Food* **2020**, *1*, 3–5. [CrossRef]
11. Dury, S.; Bendjebbar, P.; Hainzelin, E.; Giordano, T.; Bricas, N.E. *Food Systems At Risk. New Trends and Challenges*; Food and Agriculture Organization: Rome, Italy, 2019; ISBN 9782876147515.
12. Tittonell, P.; Piñeiro, G.; Garibaldi, L.A.; Dogliotti, S.; Olff, H.; Jobbagy, E.G. Agroecology in Large Scale Farming—A Research Agenda. *Front. Sustain. Food Syst.* **2020**, *4*, 214. [CrossRef]
13. United Nations. *World Population Prospects*; United Nations Publications: New York, NY, USA, 2019.

14. FAO. The State of Food Security and Nutrition in the World 2021. Available online: http://www.fao.org/state-of-food-security-nutrition (accessed on 5 October 2021).
15. KC, K.B.; Dias, G.M.; Veeramani, A.; Swanton, C.J.; Fraser, D.; Steinke, D.; Lee, E.; Wittman, H.; Farber, J.M.; Dunfield, K.; et al. When too much isn't enough: Does current food production meet global nutritional needs? *PLoS ONE* **2018**, *13*, e0205683. [CrossRef]
16. Rockström, J.; Williams, J.; Daily, G.; Noble, A.; Matthews, N.; Gordon, L.; Wetterstrand, H.; DeClerck, F.; Shah, M.; Steduto, P.; et al. Sustainable intensification of agriculture for human prosperity and global sustainability. *Ambio* **2017**, *46*, 4–17. [CrossRef] [PubMed]
17. Ingram, J. A food systems approach to researching food security and its interactions with global environmental change. *Food Secur.* **2011**, *3*, 417–431. [CrossRef]
18. Bilali, H. El Research on agro-food sustainability transitions: Where are food security and nutrition? *Food Secur.* **2019**, *11*, 559–577. [CrossRef]
19. Riccaboni, A.; Neri, E.; Trovarelli, F.; Pulselli, R.M. Sustainability-oriented research and innovation in 'farm to fork' value chains. *Curr. Opin. Food Sci.* **2021**, *42*, 102–112. [CrossRef]
20. Berti, G. Sustainable Agri-Food Economies: Re-Territorialising Farming Practices, Markets, Supply Chains, and Policies. *Agriculture* **2020**, *10*, 64. [CrossRef]
21. Verhees, B.; Verbong, G. *Users, Consumers, Citizens: A Systematic Review of their Roles in Sustainability Transitions*; Technische Universiteit Eindhoven: Eindhoven, The Netherlands, 2015; Volume 201504.
22. Tolkamp, J.; Huijben, J.C.C.M.; Mourik, R.M.; Verbong, G.P.J.; Bouwknegt, R. User-centred sustainable business model design: The case of energy efficiency services in The Netherlands. *J. Clean. Prod.* **2018**, *182*, 755–764. [CrossRef]
23. Macrae, R.; Szabo, M.; Anderson, K.; Louden, F.; Trillo, S. Empowering the Citizen-Consumer: Re-Regulating Consumer Information to Support the Transition to Sustainable and Health Promoting Food Systems in Canada. *Sustainability* **2012**, *4*, 2146–2175. [CrossRef]
24. FAO. UNEP The FAO-UNEP Sustainable Food Systems Programme. Available online: http://www.fao.org/fileadmin/templates/ags/docs/SFCP/Flyer_SP_01.pdf (accessed on 21 October 2021).
25. European Commission. Farm to Fork Strategy: For a Fair, Healthy and Environmentally-Friendly Food System. Available online: https://ec.europa.eu/food/horizontal-topics/farm-fork-strategy_en (accessed on 30 September 2021).
26. Schebesta, H.; Candel, J.J.L. Game-changing potential of the EU's Farm to Fork Strategy. *Nat. Food* **2020**, *1*, 586–588. [CrossRef]
27. Mowlds, S. The EU's farm to fork strategy: Missing links for transformation. *Acta Innov.* **2020**, *2*, 5–7. [CrossRef]
28. George, R.V.; Harsh, H.O.; Ray, P.; Babu, A.K. Food quality traceability prototype for restaurants using blockchain and food quality data index. *J. Clean. Prod.* **2019**, *240*, 118021. [CrossRef]
29. Van Rijswijk, W.; Frewer, L.J. Consumer needs and requirements for food and ingredient traceability information. *Int. J. Consum. Stud.* **2012**, *36*, 282–290. [CrossRef]
30. Olsen, P.; Borit, M. The components of a food traceability system. *Trends Food Sci. Technol.* **2018**, *77*, 143–149. [CrossRef]
31. Kittipanya-Ngam, P.; Tan, K.H. Production Planning & Control The Management of Operations A framework for food supply chain digitalization: Lessons from Thailand A framework for food supply chain digitalization: Lessons from Thailand. *Prod. Plan. Control* **2019**, *31*, 158–172. [CrossRef]
32. Valentini, R.; Sievenpiper, J.L.; Antonelli, M.; Dembska, K. *Achieving the Sustainable Development Goals Through Sustainable Food Systems*; Valentini, R., Sievenpiper, J.L., Antonelli, M., Dembska, K., Eds.; Springer International Publishing: Cham, Switzerland, 2019; ISBN 978-3-030-23968-8.
33. El Bilali, H.; Allahyari, M.S. Transition towards sustainability in agriculture and food systems: Role of information and communication technologies. *Inf. Process. Agric.* **2018**, *5*, 456–464. [CrossRef]
34. Mishra, N.; Singh, A. Use of twitter data for waste minimisation in beef supply chain. *Ann. Oper. Res.* **2018**, *270*, 337–359. [CrossRef]
35. Widener, P.; Karides, M. Food system literacy: Empowering citizens and consumers beyond farm-to-fork pathways. *Food Cult. Soc.* **2014**, *17*, 665–687. [CrossRef]
36. Cucagna, M. *Value Creation in the Agri-Food Value Chain*; University of Illinois at Urbana-Champaign: Champaign, IL, USA, 2014.
37. Auer, A.; Maceira, N.; Nahuelhual, L. Agriculturisation and trade-offs between commodity production and cultural ecosystem services: A case study in Balcarce County. *J. Rural Stud.* **2017**, *53*, 88–101. [CrossRef]
38. Gliessman, S. Transforming food systems with agroecology. *Agroecol. Sustain. Food Syst.* **2016**, *40*, 187–189. [CrossRef]
39. FAO. *Guiding the Transition to Sustainable Food and Agricultural Systems the 10 Elements of Agroecology*; Food and Agriculture Organization of the United Nations: Rome, Italy, 2018.
40. Pereira, L.M.; Bennett, E.; Biggs, R.O.; Peterson, G.; McPhearson, T.; Norström, A.; Olsson, P.; Preiser, R.; Raudsepp-Hearne, C.; Vervoort, J. Seeds of the Future in the Present: Exploring Pathways for Navigating Towards "Good" Anthropocenes. In *Urban Planet: Knowledge towards Sustainable Cities*; Elmqvist, T., Bai, X., Frantzeskaki, N., Griffith, C., Maddox, D., McPhearson, T., Parnell, S., Romero-Lankao, P., Simon, D., Watkins, M., Eds.; Cambridge University Press: Cambridge, UK, 2018; pp. 327–350. ISBN 9781316647554.
41. El Bilali, H. Research on agro-food sustainability transitions: A systematic review of research themes and an analysis of research gaps. *J. Clean. Prod.* **2019**, *221*, 353–364. [CrossRef]

42. El Bilali, H. *Transition Heuristic Frameworks in Research on Agro-Food Sustainability Transitions*; Springer: Dordrecht, The Netherlands, 2020; Volume 22, ISBN 0123456789.
43. Leopold, A. *A Sand County Almanac*; Oxford University Press: New York, NY, USA, 1949; ISBN 9780197500262.
44. Fröhlich, G.; Sellmann, D.; Bogner, F.X. The influence of situational emotions on the intention for sustainable consumer behaviour in a student-centred intervention. *Environ. Educ. Res.* **2013**, *19*, 747–764. [CrossRef]
45. HLPE. *Agroecological and Other Innovative Approaches for Sustainable Agriculture and Food Systems that Enhance Food Security and Nutrition*; Food and Agriculture Organization of the United Nations: Rome, Italy, 2019.
46. Borsellino, V.; Schimmenti, H.E.B.E. Agri-Food Markets towards Sustainable Patterns. *Sustainability* **2020**, *12*, 2193. [CrossRef]
47. Gliessman, S. Agroecology and the transition to sustainability in West African food systems. *Agroecol. Sustain. Food Syst.* **2021**, *45*, 157–158. [CrossRef]
48. Van Wee, B.; Banister, D. How to Write a Literature Review Paper? *Transp. Rev.* **2016**, *36*, 278–288. [CrossRef]
49. Jalali, S.; Wohlin, C. Systematic literature studies. In Proceedings of the ACM-IEEE International Symposium on Empirical Software Engineering and Measurement—ESEM '12, Lund, Sweden, 19–20 September 2012; ACM Press: New York, NY, USA, 2012; p. 29.
50. Hatanaka, M. Beyond consuming ethically? Food citizens, governance, and sustainability. *J. Rural Stud.* **2020**, *77*, 55–62. [CrossRef]
51. De Burgh-Woodman, H.; King, D. Sustainability and the human/nature connection: A critical discourse analysis of being "symbolically" sustainable. *Consum. Mark. Cult.* **2013**, *16*, 145–168. [CrossRef]
52. Yang, L.; Dong, S. Sustainable Product Strategy in Apparel Industry with Consumer Behavior Consideration. *Sustainability* **2017**, *9*, 920. [CrossRef]
53. Kamrath, C.; Wesana, J.; Bröring, S.; Steur, H. What Do We Know About Chain Actors' Evaluation of New Food Technologies? A Systematic Review of Consumer and Farmer Studies. *Compr. Rev. Food Sci. Food Saf.* **2019**, *18*, 798–816. [CrossRef]
54. Wall, P.G.; Chen, J. Moving from risk communication to food information communication and consumer engagement. *NPJ Sci. Food* **2018**, *2*, 21. [CrossRef]
55. Bollani, L.; Bonadonna, A.; Peira, G. The Millennials' Concept of Sustainability in the Food Sector. *Sustainability* **2019**, *11*, 2984. [CrossRef]
56. Lezoche, M.; Hernandez, J.E.; del Mar Eva Alemany Díaz, M.; Panetto, H.; Kacprzyk, J. Agri-food 4.0: A survey of the supply chains and technologies for the future agriculture. *Comput. Ind.* **2020**, *117*, 103187. [CrossRef]
57. Barska, A.; Wojciechowska-Solis, J. E-Consumers and Local Food Products: A Perspective for Developing Online Shopping for Local Goods in Poland. *Sustainability* **2020**, *12*, 4958. [CrossRef]
58. Sgroi, F.; Donia, E.; Franco, M.; Mineo, A.M. Marketing strategy, social responsibility, and value chain in the agri-food system. *HortScience* **2020**, *55*, 208–215. [CrossRef]
59. Verbeke, W. Functional foods: Consumer willingness to compromise on taste for health? *Food Qual. Prefer.* **2006**, *17*, 126–131. [CrossRef]
60. Trienekens, J.; Zuurbier, P. Quality and safety standards in the food industry, developments and challenges. *Int. J. Prod. Econ.* **2008**, *113*, 107–122. [CrossRef]
61. Falguera, V.; Aliguer, N.; Falguera, M. An integrated approach to current trends in food consumption: Moving toward functional and organic products? *Food Control* **2012**, *26*, 274–281. [CrossRef]
62. Blanco-Gutiérrez, I.; Varela-Ortega, C.; Manners, R. Evaluating Animal-Based Foods and Plant-Based Alternatives Using Multi-Criteria and SWOT Analyses. *Int. J. Environ. Res. Public Health* **2020**, *17*, 7969. [CrossRef] [PubMed]
63. He, J.; Lei, Y.; Fu, X. Do Consumer's Green Preference and the Reference Price Effect Improve Green Innovation? A Theoretical Model Using the Food Supply Chain as a Case. *Int. J. Environ. Res. Public Health* **2019**, *16*, 5007. [CrossRef]
64. Toussaint, M.; Cabanelas, P.; González-Alvarado, T.E. What about the consumer choice? The influence of social sustainability on consumer's purchasing behavior in the Food Value Chain. *Eur. Res. Manag. Bus. Econ.* **2021**, *27*, 100134. [CrossRef]
65. Sarnacchiaro, P.; Boccia, F. Some remarks on measurement models in the structural equation model: An application for socially responsible food consumption. *J. Appl. Stat.* **2018**, *45*, 1193–1208. [CrossRef]
66. Boccia, F.; Sarnacchiaro, P. The Impact of Corporate Social Responsibility on Consumer Preference: A Structural Equation Analysis. *Corp. Soc. Responsib. Environ. Manag.* **2018**, *25*, 151–163. [CrossRef]
67. Robu, M.; Robu, A.D.; Chiran, A.; Costuleanu, C.L.; Leonte, E. Environmental concern factors and consumers' purchase decision on the local agri-food market. *Environ. Eng. Manag. J.* **2021**, *20*, 405–418. [CrossRef]
68. Vermeir, I.; Verbeke, W. Sustainable Food Consumption: Exploring the Consumer "Attitude—Behavioral Intention" Gap. *J. Agric. Environ. Ethics* **2006**, *19*, 169–194. [CrossRef]
69. Fuchs, D.; Giulio, A.D.; Glaab, K.; Lorek, S.; Maniates, M.; Princen, T.; Røpke, I. Power: The missing element in sustainable consumption and absolute reductions research and action. *J. Clean. Prod.* **2016**, *132*, 298–307. [CrossRef]
70. Sciarelli, M.; Tani, M.; Prisco, A.; Caputo, F. Fostering ethical consumption in food sector: Insights from the Italian Solidarity Purchasing Groups. *Br. Food J.* **2021**, *123*, 3100–3115. [CrossRef]
71. Yoshikawa, N.; Fujiwara, N.; Nagata, J. Scenario analysis of greenhouse gases reduction by changing consumer's shopping behavior. *Energy Procedia* **2014**, *61*, 1532–1535. [CrossRef]
72. Spaargaren, G.; Oosterveer, P. Citizen-consumers as agents of change in globalizing modernity: The case of sustainable consumption. *Sustainability* **2010**, *2*, 1887–1908. [CrossRef]

73. Guerreiro, J.; Pacheco, M. How Green Trust, Consumer Brand Engagement and Green Word-of-Mouth Mediate Purchasing Intentions. *Sustainability* **2021**, *13*, 7877. [CrossRef]
74. Higgins, V.; Dibden, J.; Cocklin, C. Building alternative agri-food networks: Certification, embeddedness and agri-environmental governance. *J. Rural Stud.* **2008**, *24*, 15–27. [CrossRef]
75. Cecchini, L.; Torquati, B.; Chiorri, M. Sustainable agri-food products: A review of consumer preference studies through experimental economics. *Agric. Econ.* **2018**, *64*, 554–565. [CrossRef]
76. Ibáñez-Rueda, N.; Guillén-Royo, M.; Guardiola, J. Pro-Environmental Behavior, Connectedness to Nature, and Wellbeing Dimensions among Granada Students. *Sustainability* **2020**, *12*, 9171. [CrossRef]
77. Grasseni, C.; Forno, F.; Signori, S. *Beyond Alternative Food Networks: Italy's Solidarity Purchase Groups*; Bloomsbury Publishing: Geneva, Switzerland, 2013.
78. Crompton, T.; Kasser, T. *Meeting Environmental Challenges: The Role of Human Identity*; WWF-UK: Godalming, UK, 2009; Volume 29, ISBN 9781900322645.
79. Lozano-Cabedo, C.; Gómez-Benito, C. A Theoretical Model of Food Citizenship for the Analysis of Social Praxis. *J. Agric. Environ. Ethics* **2017**, *30*, 1–22. [CrossRef]
80. Hassanein, N. Practicing food democracy: A pragmatic politics of transformation. *J. Rural Stud.* **2003**, *19*, 77–86. [CrossRef]
81. Hassanein, N. Locating food democracy: Theoretical and practical ingredients. *J. Hunger Environ. Nutr.* **2008**, *3*, 286–308. [CrossRef]
82. Lang, T. Food control or food democracy? Re-engaging nutrition with society and the environment. *Public Health Nutr.* **2005**, *8*, 730–737. [CrossRef] [PubMed]
83. McMichael, P. Historicizing food sovereignty. *J. Peasant Stud.* **2014**, *41*, 933–957. [CrossRef]
84. Renting, H.; Schermer, M.; Rossi, A. Building Food Democracy: Exploring Civic Food Networks and Newly Emerging Forms of Food Citizenship. *Int. J. Sociol. Agric. Food* **2012**, *19*, 289–307. [CrossRef]
85. Poulsen, M.N. Cultivating citizenship, equity, and social inclusion? Putting civic agriculture into practice through urban farming. *Agric. Human Values* **2017**, *34*, 135–148. [CrossRef]
86. Sieveking, A. Food policy councils as loci for practising food democracy? Insights from the case of Oldenburg, Germany. *Polit. Gov.* **2019**, *7*, 48–58. [CrossRef]
87. Weber, H.; Poeggel, K.; Eakin, H.; Fischer, D.; Lang, D.J.; Von Wehrden, H.; Wiek, A. What are the ingredients for food systems change towards sustainability?—Insights from the literature. *Environ. Res. Lett.* **2020**, *15*, 113001. [CrossRef]
88. Markard, J.; Raven, R.; Truffer, B. Sustainability transitions: An emerging field of research and its prospects. *Res. Policy* **2012**, *41*, 955–967. [CrossRef]
89. Gaitán-Cremaschi, D.; Klerkx, L.; Duncan, J.; Trienekens, J.H.; Huenchuleo, C.; Dogliotti, S.; Contesse, M.E.; Rossing, W.A.H. Characterizing diversity of food systems in view of sustainability transitions. A review. *Agron. Sustain. Dev.* **2019**, *39*, 1. [CrossRef]
90. United Nations. 17 Sustainable Development Goals (SDGs). Available online: https://sdgs.un.org/goals (accessed on 1 September 2021).
91. Gliessman, S.R. Transforming food and agriculture systems with agroecology. *Agric. Human Values* **2020**, *37*, 547–548. [CrossRef] [PubMed]
92. Wezel, A.; Herren, B.G.; Kerr, R.B.; Barrios, E.; Gonçalves, A.L.R.; Sinclair, F. Agroecological principles and elements and their implications for transitioning to sustainable food systems. A review. *Agron. Sustain. Dev.* **2020**, *40*, 40. [CrossRef]
93. Castellanos-Navarrete, A.; Jansen, K. Is Oil Palm Expansion a Challenge to Agroecology? Smallholders Practising Industrial Farming in Mexico. *J. Agrar. Chang.* **2018**, *18*, 132–155. [CrossRef]
94. Soga, M.; Gaston, K.J. Extinction of experience: The loss of human-nature interactions. *Front. Ecol. Environ.* **2016**, *14*, 94–101. [CrossRef]
95. Zylstra, M.J.; Knight, A.T.; Esler, K.J.; Le Grange, L.L.L. Connectedness as a Core Conservation Concern: An Interdisciplinary Review of Theory and a Call for Practice. *Springer Sci. Rev.* **2014**, *2*, 119–143. [CrossRef]
96. Bratman, G.N.; Daily, G.C.; Levy, B.J.; Gross, J.J. The benefits of nature experience: Improved affect and cognition. *Landsc. Urban Plan.* **2015**, *138*, 41–50. [CrossRef]
97. Mayer, F.S.; Frantz, C.M. The connectedness to nature scale: A measure of individuals' feeling in community with nature. *J. Environ. Psychol.* **2004**, *24*, 503–515. [CrossRef]
98. Wunderlich, S.; Smoller, M. Consumer awareness and knowledge about food sources and possible environmental impact. *Int. J. Environ. Impacts Manag. Mitig. Recover.* **2019**, *2*, 85–96. [CrossRef]
99. Schultz, P.W. The structure of environmental concern: Concern for self, other people, and the biosphere. *J. Environ. Psychol.* **2001**, *21*, 327–339. [CrossRef]
100. Geng, L.; Xu, J.; Ye, L.; Zhou, W.; Zhou, K. Connections with Nature and Environmental Behaviors. *PLoS ONE* **2015**, *10*, e0127247. [CrossRef]
101. Frantzeskaki, N.; van Steenbergen, F.; Stedman, R.C. Sense of place and experimentation in urban sustainability transitions: The Resilience Lab in Carnisse, Rotterdam, The Netherlands. *Sustain. Sci.* **2018**, *13*, 1045–1059. [CrossRef] [PubMed]
102. Boyd, D.R. *The Rights of Nature: A Legal Revolution that Could Save the World*; ECW Press: Toronto, ON, Canada, 2017; p. 272, ISBN 1-77041-239-5/978-1-77041-239-2.

103. Fischer, J.; Manning, A.D.; Steffen, W.; Rose, D.B.; Daniell, K.; Felton, A.; Garnett, S.; Gilna, B.; Heinsohn, R.; Lindenmayer, D.B.; et al. Mind the sustainability gap. *Trends Ecol. Evol.* **2007**, *22*, 10–13. [CrossRef] [PubMed]
104. Dickinson, E. The Misdiagnosis: Rethinking "nature-deficit Disorder". *Environ. Commun.* **2013**, *7*, 315–335. [CrossRef]
105. Budruk, M.; Wilhelm, S.A. Journal of Outdoor Recreation and Tourism Place attachment and recreation experience preference: A further exploration of the relationship. *J. Outdoor Recreat. Tour.* **2013**, *1–2*, 51–61. [CrossRef]
106. Ramkissoon, H.; Smith, L.D.G.; Weiler, B. Relationships between place attachment, place satisfaction and pro-environmental behaviour in an Australian national park. *J. Sustain. Tour.* **2013**, *21*, 434–457. [CrossRef]
107. Barbaro, N.; Pickett, S.M. Mindfully green: Examining the effect of connectedness to nature on the relationship between mindfulness and engagement in pro-environmental behavior. *Pers. Individ. Dif.* **2016**, *93*, 137–142. [CrossRef]
108. Brown, G.; Raymond, C. The relationship between place attachment and landscape values: Toward mapping place attachment. *Appl. Geogr.* **2007**, *27*, 89–111. [CrossRef]
109. Langemeyer, J.; Madrid-Lopez, C.; Mendoza Beltran, A.; Villalba Mendez, G. Urban agriculture—A necessary pathway towards urban resilience and global sustainability? *Landsc. Urban Plan.* **2021**, *210*, 104055. [CrossRef]
110. Pérez-Ramírez, I.; García-Llorente, M.; de la Portilla, C.S.; Benito, A.; Castro, A.J. Participatory collective farming as a leverage point for fostering human-nature connectedness. *Ecosyst. People* **2021**, *17*, 222–234. [CrossRef]
111. Gosling, E.; Williams, K.J.H. Connectedness to nature, place attachment and conservation behaviour: Testing connectedness theory among farmers. *J. Environ. Psychol.* **2010**, *30*, 298–304. [CrossRef]
112. Sarkar, S.F.; Poon, J.S.; Lepage, E.; Bilecki, L.; Girard, B. Enabling a sustainable and prosperous future through science and innovation in the bioeconomy at Agriculture and Agri-Food Canada. *N. Biotechnol.* **2018**, *40*, 70–75. [CrossRef]
113. Han, H. Consumer behavior and environmental sustainability in tourism and hospitality: A review of theories, concepts, and latest research. *J. Sustain. Tour.* **2021**, *29*, 1021–1042. [CrossRef]
114. Bratman, G.N.; Anderson, C.B.; Berman, M.G.; Cochran, B.; de Vries, S.; Flanders, J.; Folke, C.; Frumkin, H.; Gross, J.J.; Hartig, T.; et al. Nature and mental health: An ecosystem service perspective. *Sci. Adv.* **2019**, *5*, eaax0903. [CrossRef] [PubMed]
115. Westley, F.; Olsson, P.; Folke, C.; Homer-Dixon, T.; Vredenburg, H.; Loorbach, D.; Thompson, J.; Nilsson, M.; Lambin, E.; Sendzimir, J.; et al. Tipping toward sustainability: Emerging pathways of transformation. *Ambio* **2011**, *40*, 762–780. [CrossRef]
116. Schaltegger, S.; Lüdeke-Freund, F.; Hansen, E.G. Business Models for Sustainability: A Co-Evolutionary Analysis of Sustainable Entrepreneurship, Innovation, and Transformation. *Organ. Environ.* **2016**, *29*, 264–289. [CrossRef]
117. Van Der Brugge, R.; Rotmans, J.; Loorbach, D. The Transition in Dutch Water Management The transition in Dutch water management. *Reg. Environ. Chang.* **2005**, *5*, 164–176. [CrossRef]
118. Geels, F.W. Technological transitions as evolutionary reconfiguration processes: A multi-level perspective and a case-study. *Res. Policy* **2002**, *31*, 1257–1274. [CrossRef]
119. Garcia, C.A.; Savilaakso, S.; Verburg, R.W.; Gutierrez, V.; Wilson, S.J.; Krug, C.B.; Sassen, M.; Robinson, B.E.; Moersberger, H.; Naimi, B.; et al. The Global Forest Transition as a Human Affair. *One Earth* **2020**, *2*, 417–428. [CrossRef]
120. Darnhofer, I.; Lindenthal, T.; Bartel-Kratochvil, R.; Zollitsch, W. Conventionalisation of organic farming practices: From structural criteria towards an assessment based on organic principles. *Sustain. Agric.* **2009**, *2*, 331–349. [CrossRef]
121. Loorbach, D.; Frantzeskaki, N.; Avelino, F. Sustainability Transitions Research: Transforming Science and Practice for Societal Change. *Annu. Rev. Environ. Resour.* **2017**, *42*, 599–626. [CrossRef]
122. El Bilali, H. The multi-level perspective in research on sustainability transitions in agriculture and food systems: A systematic review. *Agriculture* **2019**, *9*, 74. [CrossRef]
123. Melchior, I.C.; Newig, J. Governing transitions towards sustainable agriculture—Taking stock of an emerging field of research. *Sustainability* **2021**, *13*, 528. [CrossRef]
124. Khanagha, S.; Volberda, H.; Oshri, I. Business model renewal and ambidexterity: Structural alteration and strategy formation process during transition to a Cloud business model. *RD Manag.* **2014**, *44*, 322–340. [CrossRef]
125. Fonte, M. Food consumption as social practice: Solidarity Purchasing Groups in Rome, Italy. *J. Rural Stud.* **2013**, *32*, 230–239. [CrossRef]
126. Dedeurwaerdere, T.; De Schutter, O.; Hudon, M.; Mathijs, E.; Annaert, B.; Avermaete, T.; Bleeckx, T.; de Callataÿ, C.; De Snijder, P.; Fernández-Wulff, P.; et al. The Governance Features of Social Enterprise and Social Network Activities of Collective Food Buying Groups. *Ecol. Econ.* **2017**, *140*, 123–135. [CrossRef]
127. Pérez-Mesa, J.C.; Piedra-Muñoz, L.; Galdeano-Gómez, E.; Giagnocavo, C. Management Strategies and Collaborative Relationships for Sustainability in the Agrifood Supply Chain. *Sustainability* **2021**, *13*, 749. [CrossRef]
128. Lagane, J. When students run AMAPs: Towards a French model of CSA. *Agric. Human Values* **2015**, *32*, 133–141. [CrossRef]
129. Matzembacher, D.E. Sustainability as business strategy in community supported agriculture for producers and consumers. *Br. Food J.* **2019**, *121*, 616–632. [CrossRef]
130. Morgan, K.; Sonnino, R. The urban foodscape: World cities and the new food equation. *Camb. J. Reg. Econ. Soc.* **2010**, *3*, 209–224. [CrossRef]
131. Giagnocavo, C. The Development of the Cooperative Movement and Civil Society in Almeria, Spain: Something from Nothing? *Sustainability* **2020**, *12*, 9820. [CrossRef]
132. Paciarotti, C.; Torregiani, F. The logistics of the short food supply chain: A literature review. *Sustain. Prod. Consum.* **2021**, *26*, 428–442. [CrossRef]

133. Marsden, T.; Banks, J.; Bristow, G. Food Supply Chain Approaches: Exploring their Role in Rural Development. *Sociol. Rural.* **2000**, *40*, 424–438. [CrossRef]
134. He, Z. Analysis of the Agricultural E-Commerce in Rural China. In Proceedings of the 2021 12th International Conference on E-Education, E-Business, E-Management, and E-Learning, Tokyo, Japan, 10–13 January 2021; ACM: New York, NY, USA, 2021; pp. 360–363. [CrossRef]
135. Turner, B. Embodied connections: Sustainability, food systems and community gardens. *Local Environ.* **2011**, *16*, 509–522. [CrossRef]
136. Crowdfarming. Available online: www.crowdfarming.com (accessed on 12 October 2021).
137. Cavaliere, A.; Ventura, V. Mismatch between food sustainability and consumer acceptance toward innovation technologies among Millennial students: The case of Shelf Life Extension. *J. Clean. Prod.* **2018**, *175*, 641–650. [CrossRef]
138. IFOAM Participatory Guarantee Systems (PGS). Available online: https://ifoam.bio/our-work/how/standards-certification/participatory-guarantee-systems (accessed on 5 October 2021).
139. Kaufmann, S.; Hruschka, N.; Vogl, C.R. Bridging the literature gap: A framework for assessing actor participation in participatory guarantee systems (PGS). *Sustainability* **2020**, *12*, 8100. [CrossRef]
140. Bennett, E.M.; Solan, M.; Biggs, R.; McPhearson, T.; Norström, A.V.; Olsson, P.; Pereira, L.; Peterson, G.D.; Raudsepp-Hearne, C.; Biermann, F.; et al. Bright spots: Seeds of a good Anthropocene. *Front. Ecol. Environ.* **2016**, *14*, 441–448. [CrossRef]
141. Giagnocavo, C.; de Cara-García, M.; González, M.; Juan, M.; Marín-Guirao, J.I.; Mehrabi, S.; Rodríguez, E.; van der Blom, J.; Crisol-Martínez, E. Reconnecting Farmers with Nature through Agroecological Transitions: Interacting Niches and Experimentation and the Role of Agricultural Knowledge and Innovation Systems. *Agriculture* **2022**, *12*, 137. [CrossRef]

agriculture

MDPI

Review

Reconnecting with Nature through Good Governance: Inclusive Policy across Scales

Johanna Wilkes

Balsillie School of International Affairs, Wilfrid Laurier University, Waterloo, ON N2H 6T8, Canada; jwilkes@balsillieschool.ca

Abstract: We are disconnected from nature, surpassing planetary boundaries at a time when our climate and social crises converge. Even prior to the emergence of COVID-19, the United Nations and its member states were already off track to achieve the Sustainable Development Goals (SDGs) and fulfil climate commitments made under the Paris Agreement. While agricultural expansion and intensification have supported increases in food production, this model has also fostered an unsustainable industry of overproduction, waste, and the consumption of larger quantities of carbon-intensive and ultra-processed foods. By addressing the tension that exists between our current food system and all that is exploited by it, different scales of governance can serve as spaces of transformation towards more equitable, sustainable outcomes. This review looks at how good governance can reconnect people with nature through inclusive structures across scales. Using four examples that focus on place-based and rights-based approaches—such as inclusive multilateralism, agroecology, and co-governance—the author hopes to highlight the ways that policy processes are already supporting healthy communities and resilient ecosystems.

Keywords: food systems governance; public policy; inclusive multilateralism; agroecology; rights-based approach; biodiversity; climate change; nature

Citation: Wilkes, J. Reconnecting with Nature through Good Governance: Inclusive Policy across Scales. *Agriculture* **2022**, *12*, 382. https://doi.org/10.3390/agriculture12030382

Academic Editors: José Luis Vicente-Vicente, Cristina Quintas-Soriano and María D. López-Rodríguez

Received: 22 February 2022
Accepted: 3 March 2022
Published: 9 March 2022

Publisher's Note: MDPI stays neutral with regard to jurisdictional claims in published maps and institutional affiliations.

1. Introduction

The world is in desperate need of transformation. The COVID-19 pandemic has exacerbated and brought to light much of the structural inequity that already existed. Inequities were enlarged [1] and the number of individuals in poverty is expected to rise [2]. While some countries were structurally prevented from borrowing to address the impacts of the pandemic [3], others were able to launch large-scale social programming responses. Food systems, the focus of this review, have also experienced uneven impacts. It is expected that import-dependent countries may be vulnerable to food price increases due to a multitude of pandemic-related factors, including supply chain disruption and currency depreciation [4].

At the same time, locally led initiatives have emerged in response to urgent and pressing food access challenges at the community scale [5,6]. Local and global policy arenas are not substitutes, but rather interconnected complements and powerful tools for reform. As we move forward through the COVID-19 crisis, the world continues to face a host of collective systemic crises: declining biodiversity [7], the continued surpassing of our planetary boundaries [8], and growing economic inequality [9]. To meet this critical moment in our shared histories, a reorientation is required of our policy environments to prioritize and protect the rights of both individuals and nature across all scales of governance.

Public governance can be a way of reconnecting people with nature rather than commodifying it. This review paper shows that good governance is possible, and that inclusivity and accountability are the preconditions for equitable outcomes. To situate the importance of good governance and the need for change, this paper outlines the ongoing challenges within the current globalized food system (Section 2), the contested landscape across food systems governance (Section 3), and examples in the successful implementation of multi-scalar inclusive governance. Focused on four different examples, this review then explores how multi-scalar solutions create policy pathways that support the pursuit of more equitable, resilient food systems. By focusing on global (the 'most affected' model of the Committee on World Food Security), national (the integrated approaches found within the UNESCO Biosphere Reserve system in Canada), sub-national (the commitment to Zero-Budget Natural Farming (ZBNF) in Andhra Pradesh, India), and local (the growing role of food policy councils and local food systems strategies), this review helps tease out the shared and unique conditions for success across scales and explores why each scale within public governance matters for a better, more sustainable future.

The first and foremost condition of success is good governance, meaning that processes are inclusive, accountable, and place-based. Each of the examples highlights the importance and need for centering inclusivity and a place-based approach that builds agency for those most affected by the food systems governance model under consideration. By increasing the agency of people most connected to the local ecosystems and environments, good governance creates space for nature to thrive.

2. The Race to Zero: Ailing Food Governance in a Globalized Economy

Our industrial food systems are failing both communities and natural environments [10]. In 2020, 2.37 billion people—or roughly one in every three individuals—could not access adequate food for a healthy, nutritious diet [11]. Additionally, even with enough calories produced to feed the current population [12], there remains an alarming trend of increasing food insecurity [11]. Our food systems are also fueling the devastation from climate change. Despite the strong link between food systems and carbon emissions, global industrial agricultural—and the broader agri-food industry—remains one of the largest contributors to the world's greenhouse gas (GHG) emissions, accounting for roughly 31 percent of total GHGs [13]. Agricultural production remains largely outside of nearly all carbon pricing mechanisms—the leading tool within carbon regulatory schemes—and, until recently, a majority of the policy focus has been on adaptation rather than systemic mitigation. To further exacerbate these policy challenges, a significant tool currently used within food policy comprises agricultural subsidies. Intimately tied to questions of trade, these agricultural subsidies can contribute to the encouragement of production decisions that are harmful to human health and the environment [14,15]. The IPCC special report entitled *Climate Change and Land: Summary for Policy Makers* [16], notes: "Expansion of areas under agriculture and forestry, including commercial production, and enhanced agriculture and forestry productivity have supported consumption and food availability for a growing population (*high confidence*). With large regional variation, these changes have contributed to increasing net GHG emissions (*very high confidence*), loss of natural ecosystems (e.g., forests, savannahs, natural grasslands and wetlands) and declining biodiversity (*high confidence*)" (p. 7).

Beyond carbon emissions, there is a need to support Sustainable Development Goal (SDG) 12 (sustainable production and consumption patterns) through better aligning diets and food environments with nutritional requirements. Globally, over a billion tons of food is wasted annually [17]. While some of this waste is due to limited infrastructure, e.g., cold storage, waste happens across the entirety of the value chain—from farm fields to households [18]. The Commissioner of Environmental Cooperation estimates that more than 150 million tons of food—fit for human consumption—is wasted each year in North America. In addition, the report points to the greenhouse gas (GHG) emission implications of loss and waste, equating to 22.1 million hectares of cropland and 3.94 million tons of fertilizer used to create the wasted products [19]. In his recent book, *Eating Tomorrow*, Wise characterized the changes as the borrowing from tomorrow's capacity of soils and environments for the wasting of food today [20].

In response to the shared global challenges facing food systems and nature, the Committee on World Food Security's (CFS) recent High Level Panel of Experts (HLPE) report emphasizes that achieving food security goes far beyond levels of production. The

Food Security and Nutrition: Building a Global Narrative Towards 2030 report highlights the need to reconfigure food security to incorporate two additional dimensions: agency and sustainability [21]. The adoption of this updated definition would bring the concept into line with more recent interpretations of food security that center on the right to food as well as the realities of those individuals experiencing food insecurity directly. Agency refers to both the definition proposed by Sen (1985), which focused on the freedom of pursuit, and "the ability of people to take actions that help improve their own wellbeing, as well as their ability to engage in society in ways that influence the broader context included in their exercise of voice in shaping policies" [21] (p. 8). The authors from within the HLPE argue that agency and sustainability go hand in hand with the established four dimensions: access, availability, utilization, and stability. Sustainability is central to the ability to access food over the long term for current and future generations, while agency plays a central role in the right of each individual to determine how they interact with food systems. Without agency or sustainability, there is no true food security [22]. This updated concept is also much closer to the advocated move towards food sovereignty as a way to recognize the interconnected nature of food systems and the right of individuals, communities, and nations to chart their own food futures.

While the number of people experiencing food insecurity has grown substantially throughout the pandemic, statistics show that it has been on the rise since 2014 and is not experienced evenly [11]. Women and marginalized groups bear a disproportionate burden with rates 10 percent higher than their male counterparts [11]. The HLPE report also notes that deep transformation is required for food systems to support the achievement of SDG 2 (zero hunger) [21]. This urgent call comes at a time when the IPCC reports, with high confidence, that humans are unequivocally affecting the world's climate [23]. There is no question that more individuals are expected to be pushed into poverty due to climate displacement and disruptions [23], concentrated largely within regions already impacted by historical traumas such as (neo)colonialism, exploitative trade practices, and uneven development. It is estimated that small-holders—defined by a recent FAO study as those with farms under 2 hectares—feed roughly one-third of the world [24]. They have only contributed a minor amount to the aggregate carbon output, but will be the most affected by the changing climate [25].

In addition to small-holders, Indigenous communities and other marginalized individuals are some of the most food insecure. In what is now commonly known as Canada, where the author is located, 4.4 million individuals—or 12.7 percent of Canadians—experience food insecurity [26]. Black and Indigenous households experience the most severe levels of food insecurity with rates above 28 percent [26]. This equates to roughly a rate 3.5 times higher than white households [27]. Writing of his experience on a mission to Canada, former Special Rapporteur on the Right to Food, Olivier de Schutter, highlighted the impacts and consequences on food security from climate change in relation to Indigenous communities and access to traditional food systems [28]. Nearly 10 years later, communities across Canada—part of the area known to many Indigenous nations as Turtle Island, which encompasses the region of North America [29]—continue to struggle with the growing and potential effects of climate change [30], as well as the impacts of settler economic systems, environmental disruption, and other pressures, on accessing traditional foods [31].

Reconnecting with nature means more than simply reducing carbon by applying new techniques or adopting additional technology. Narrowly focused solutions, such as Climate-Smart Agriculture (CSA) or Sustainable Intensification (SI), taken independently of more holistic reform, are only likely to embed the by-products of our broken globalized food system even further [32]. Through approaches such as agroecology, food systems, and the communities who depend on them, can thrive as part of nature rather than separate from it. By acknowledging carbon emissions as a symptom or by-product rather than an independent problem allows for more systemic solutions to emerge that incorporate biodiversity, climate, soil health, food insecurity, workers' well-being, and other factors. Unlike the current trends globally, a systemic review of research from over twenty years indicates that initiatives that take a systemic approach focused on the right to food and food sovereignty show a positive impact on food security and nutrition [33]. To aid in this endeavor, public policy provides an accessible tool funded by public resources with clear accountability and jurisdictions. It also plays an end-of-the-line role for governments in state-led decision-making forums, whether it is the United Nations or a local municipality. Public governance more generally—the rules, norms, and choices that govern public institutions—also play an important role in the rethinking of food systems. Pulled between past priorities and future needs, public governance plays a pivotal role in reconfiguring the center of power in decision making. Many recent Intergovernmental Organization (IGO) reports include calls for democratic institutional processes that value the knowledge of youth, Indigenous peoples, and local communities [11,21,34,35]. In addition, these reports highlight the importance of coordinated multi-scalar action to accomplish this transformational change.

3. Multistakeholder, Multilateral, and the Messy Middle

So why, in the context of such urgency, do food systems governance spaces remain deeply entrenched in the current global models of intensification? The answer is both complex and simple at the same time. To begin, not all global engagements are created equally. Decades of path-dependent policies and deep-rooted narratives build the foundations of our modern-day agricultural systems as well as many of our institutionalized understandings (e.g., productivism). Good governance needs to be supported by policy innovations that navigate the space between inclusion and accountability.

Good governance can be thought of as a set of clear rules of engagement that consider who holds power, how that has shaped past policies, and the ways of rebalancing these vested histories. Defined as "the practice of coordinating national policies in groups of three or more states, through ad hoc arrangements or by means of institutions", multilateralism is by nature led by states [36]. Multistakeholderism, on the other hand, sees states as one player within a wide arena of influential actors [37]. While these definitions are considered unique governance arrangements, there is a spectrum of options that exist in implementation. Good governance can, and does, exists across this spectrum. However, the concentration of power and movements towards a more corporate-centered form of multistakeholder engagement, referred to in this review as hyper-multistakeholderism, can reinforce systemic lock-in effects. This form of governance leaves actors focused on symptom-based solutions that drive profits for larger companies who have the resources to invest versus solutions that may be derived through nature—such as agroecology—and provide less financial risk for the producer. The shift towards forms of hyper-multistakeholderism could jeopardize the effectiveness of governance institutions. Multilateralism or multistakeholderism, by their nature, are not necessarily good or bad frameworks. Rather, it is the implementation of these models in the absence of political context that becomes problematic. Without addressing power imbalances and the reorientation of agency throughout the policy process, changes are likely to be superficial, short-term, or in name only. While both incremental and systemic change are essential, the former tends to be far more powerful, generously funded, implemented, and researched than the latter [38,39].

3.1. *Hyper-Multistakeholderism: The Case of the Governance Structures within the UNFSS*

Outcomes and resources derived from multistakeholderism forums that do not recognize these imbalances can continue to reinforce old agricultural models that have negatively impacted biodiversity, climate, and community. A recent example is the United Nations Food Systems Summit of 2021 (the Summit) [40–42]. Called for by the Secretary General of the United Nations in 2019 [43] and delivered in the middle of a global health pandemic, the Summit generated significant attention from a broad audience. Meant to invigorate action towards the achievement of the SDGs, the Summit adopted new language towards global food policy, most significant of which was the 'food systems' lens [44].

While many were enthusiastic about the inclusion of the new language and the high profile of the event, there was concern about the Summit's leadership and the shift towards a form of hyper-multistakeholderism. Disrupting the accountability mechanism of the UN as a member state-based organization, the Summit centered on states as information takers rather than decision makers. As Canfield, Duncan, and Claeys note:

> Whereas the multilateral framework through which global food governance has long located authority in the nation-state and hinged legitimacy on states' fulfillment of their duties and obligations under human rights, in blurring the boundaries between states, corporations, and civil society, the Summit reconstituted the terms through which authority and legitimacy are constituted in global food governance [45] (p. 5).

The framework of the Summit provided private sector representation without adequate safeguards for conflicts of interest [45–50]. In objection, civil society from around the world raised their concerns to the Secretary General and the Summit Secretariat but never received a formal response [49]. Others who decided to engage in the Summit process retracted their involvement. Included in those who stepped down was the International Panel of Experts on Sustainable Food Systems (IPES). To cite the withdrawal letter [51]:

> ... the Summit's rules of engagement were determined by a small set of actors. The private sector, organizations serving the private sector (notably the World Economic Forum), and a handful of scientific experts kick-started the process and framed the agenda.

The Summit process circumvented more accepted United Nations governance norms, such as country-led processes, instead putting broad-based engagement in the driver's seat with little considerations of the political economy of the subject matter. The alignment and implementation of the Strategic Partnership Agreement of the United Nations and the World Economic Forum [52] cemented the early stages of the Summit without adequate consultation [45,47,53]. In addition, processes were guided by the 'all affected' principle rather than the more inclusive 'most affected' approach [45]. Throughout the process, there was a lack of resources provided for civil society participation. Even if members of marginalized communities sought to participate, not all engagement materials or discussions were delivered with translation services—essential for equitable participation across all official UN languages. Criticism of the Summit is wide ranging, but the implication of these governance shifts towards hyper-multistakeholderism are still yet to be fully known [45]. As Guttal notes [53]:

> Using the language of participation and inclusivity, MSIs (multistakeholder initiatives) blur the lines between rights-holders (people), duty-bearers (states) and other stakeholders, while keeping intact power asymmetries and erasing mechanisms of legal accountability and justice. (p. 13)

3.2. *Multilateralism and the Messy Middle: Innovative Processes with Accountability*

By contrast, inclusive forms of multilateralism can leave member states, as governments across scales, as the end-of-the-line decision makers with clear through lines of accountability. This model ensures that the public—rather than shareholders—remain central to the ongoing legitimacy of food systems forum(s). By integrating concepts and processes that acknowledge power imbalances and create structural spaces of agency for those 'most affected' by food systems policy, inclusive multilateralism could create more resilient, transformative outcomes [53]. The critical messy middle, inclusive multilateralism as it is situated between multilateralism and multistakeholderism, could ensure adequate and fair representation without the delegation of accountability away from those who are ultimately held responsible. Inclusive multilateralism does not exclude private sector involvement; rather, it makes participation more transparent through clear rules of engagement and accountability structures. Such inclusive multilateralism will ensure that those protecting essential ecosystems and defending nature are given fair representation in critical policy development processes.

While wading through the messiness, there are examples from different scales of governance that can provide insight into how to balance inclusion with accountability while creating more sustainable outcomes. Governance models that promote feedback loops can effectively integrate connectivity between the local and the global (or vice versa). In essence, inclusive multilateralism is, by nature, multi-scalar through local knowledge, regional networks, and transnational advocacy coalition networks. Examples include the role of La Via Campesina and food sovereignty networks in the 2009 reform of the Committee on World Food Security [45] or the emerging role for cities within global spaces, such as the Conference of the Parties for Climate Change [54], or adapting the Sustainable Development Goals according to localized contexts [55]. Cities have also taken part in leading efforts, such as the Glasgow Declaration [56,57]. In addition, collaborative networks—such as the C40—help band together local actors to advocate for space throughout international forums.

4. Bridging the Local and the Global: Multi-Scalar Pathways in Food Systems

Below are four examples from different scales that are reconnecting people and nature through inclusive, resilient food systems. Each of the examples, presented in Table 1, contributes on at least two levels—at their place of origin/implementation as well as at the global level and/or in the sub-jurisdictional areas that are included—to foster change.

Table 1. Initiatives Across Scales.

Scale	Initiative
Global	The Committee on World Food Security
National	UNESCO Biosphere Sites
State	Zero Budget Natural Farming
Local	Food Policy Groups

4.1. Committee on World Food Security: Principles of Engagement and Fractured Power

While the Summit has been the newly promoted forum of engagement for food systems, by direction of the United Nations General Assembly, global food governance has convened for years in Rome under the mandate of the Committee on World Food Security (CFS). Branding itself "the foremost inclusive international and intergovernmental platform for all stakeholders to work together to ensure food security and nutrition for all", the CFS provides a unique model of global governance built on inclusive principles of participation that incorporates those most affected by food insecurity and food systems policy [58]. Beginning in 1974, the CFS has been an established space of debate on food security and nutrition, but reform in the wake of the 2008/09 food crises fundamentally reshaped the forum. The work by civil society networks to promote the right to food and food sovereignty translated into hard-won gains in the international fora through the introduction of the Civil Society Mechanism at the CFS [45,53,59]. By introducing the 'most affected' principle and resourcing—albeit not sufficiently enough [60]—the CFS has facilitated a way to create structured space for civil society to engage.

Through formal participation throughout the policy process and at meetings, the Civil Society and Indigenous Peoples Mechanism for the Committee on World Food Security (CSM) has made an important contribution to the global food policy space. Expanded in 2018 to represent their membership more appropriately, the civil society mechanism updated their name to include Indigenous Peoples as an important step in representation [61]. Whether it is defending the right to genetic diversity or the inclusion of traditional knowledge, the CSM has pushed for an integrated approach to food systems transformation that reconnects people to nature [62]. In their push to promote agroecology, the 2018 CSM report notes [63]:

> While both environmental degradation and poor nutritional outcomes are results of the dominant industrial food system, the promotion of agroecology and the consumption of diverse diets of locally and agroecologically produced food, can lead to sustainable diets that realize the right to food through improved environmental and nutrition outcomes. (p. 39)

The CSM plays an important role within the development of CFS guidelines and policy products through their continued fight to center governance on a rights-based approach. The guidelines have been built as tools to be implemented at a variety of scales, including within local communities [64]. Through the inclusion of CSM, transnational advocacy networks have found a space to bring the local to the global, keep critical issues on the agenda and hold national governments to account. While there is much work to be completed in realizing the full inclusion of those most affected at the CFS, the CSM is an important mechanism that keeps issues impacting peasant farmers, fisherfolk, and the ecosystems they depend on central to global agendas.

4.2. UNESCO and Breaking a Fortress Model: Biodiversity through Collaboration—International to the National

Encouraged by the work of the Educational, Scientific, and Cultural Organization of the United Nations (UNESCO), the Biosphere program is a unique blend of international advocacy, enabling frameworks and resources through federal involvement, sub-jurisdictional support, and community leadership. Describing the difference between more traditional forms of conservation and the biosphere approach, the Canadian Biosphere Reserves Association [65] notes that:

> Biosphere reserves, . . . , occur wherever an area has conservation value and the surrounding community has pledged to protect biodiversity, cultural heritage and uphold the principles of sustainable development.

As a unique partnership between community and nature, UNESCO Biosphere reserves are located across the world. In Canada, eighteen such partnerships exist across the country's diverse landscapes [66]. Located in different climactic regions of the country, biosphere governance reflects the community through "community-based and cross-representational" membership [67]. The goal for each of these arrangements is to act as a collaboration between nature and community. The considerations included within the planning incorporate: the social and economic needs of the population, the unique characteristics of each of the biosphere's ecosystems, and the cultural heritage of the region. Departing from the fortress model of conservation that separates humans from the environment, biospheres allow for the autonomy of a thriving community to protect, or even restore, parts of the surrounding ecosystem. Models that integrate higher levels of integration between food systems and nature can bolster greater aggregate biodiversity [67,68] and other co-benefits [68]. As Gavin et al. note:

> Effective conservation partnerships are based on mutual respect for the rights, knowledge, practices, and responsibilities of stakeholders. (p. 6)

The UNESCO model fosters relationships, including the use of biospheres to support reconciliation efforts between Indigenous and non-Indigenous people. The UNESCO Biosphere framework allows for international advocacy to enable place-based preservation by encouraging work at the human–nature interface as a critical part of a thriving community.

4.3. Zero Budget Farming and the Potential of the Sub-Jurisdictions in Fostering Change

Emerging from the shadows of the Green Revolution, Zero Budget Natural Farming (ZBNF) has become popular in several Indian states and has even become institutionalized to some degree. In particular, ZBNF, also known as Natural Farming, has blossomed in the state of Andhra Pradesh. This approach to food systems was, in part, a response to the negative impact input-dependent farming was having on communities [69]. Ranjan and Sow note that [70]:

> ZBNF reduces the need of taking loans for farming purpose as it completely depends on the use of internal or naturally available inputs.

With farmers experiencing below-average GDP and high levels of indebtedness in Andhra Pradesh, taking an agroecological approach to food systems allows participants to remove financial pressure from the purchase of external inputs while maximizing the use of local solutions and traditional knowledge. Veluguri et al. found that institutional opportunity and an influential advocate with access to resources were both key in the adoption of ZBNF as a state policy pathway [69]. In addition, early work on farmer field schools through the Food and Agriculture Organization of the United Nations (FAO) allowed for the advancement of farmer-to-farmer knowledge sharing and local resource management [69,71]. A marriage of enabling frameworks, community-led partnerships, well-positioned advocates, and farmer-to-farmer learning has contributed to the success of what is recognized by some as the largest agroecological initiative in the world [69]. Recent research also indicates that ZBNF may not face yield penalties to the same extent as those often associated with production method changes [72]. Agroecology is particularly adept at multiscale impacts since it is both a localized practice and a globalized political social movement. In addition, agroecological transitions have been shown to reconnect farmers with nature in a way that is powerful and regenerative [73]. Utilizing contributions from global and local efforts, ZBNF is projected to reach six million farmers and provide decent livelihoods, prosperous communities, and resilient environments to many more. In addition, the ZBNF model adheres to several of the key recommendations for food systems by the IPBES: promoting agroecological production, integrated landscapes, and the localization of economies [7].

4.4. Groundswell Networks: Food Policy Groups

Johns Hopkins Center for a Livable Future (the Center)'s Food Policy Networks project helps with both tracking and collaborating on food policy group work. Their database of food policy groups across North America allows for researchers and advocates alike to search for initiatives in their region. These networks center on social justice and action-based models. The Center [74] notes that:

> Food policy groups share similar overall goals to make the food system more equitable, sustainable and resilient, but vary in their organizational structure, relationships with government and funding sources.

Food policy groups can be an example of multistakeholder governance that feeds into, or are part of, state decision-making processes. This model allows for accountability to remain within the public domain while incorporating the views and needs of the community. During COVID-19, food policy groups were able to advocate for those most impacted by the pandemic and activate local efforts to support access to food. A recent report by the Center found that over 75 percent of councils advocated for at least one change to improve food access [75]. The report also found that older councils were more likely to engage in policy advocacy versus policy groups that were less than two years old, with 84 percent of established councils engaging in policy advocacy at the "local, state, tribal, or federal related to food systems concerns due to COVID-19" [75] (p. 14).

Food policy groups support systems thinking, the inclusion of a wide range of actors, youth engagement, and feed through lived experiences from cities into global arenas, such as CFS and the Conference of the Parties on Climate Change (COP). The footprint of food policy groups is growing. In Canada, the establishment of a National Food Policy Council will provide advice to the Minister of Agriculture and Agri-Food Canada. European advocates have called a joint food policy council to support an integrated policy approach [76].

5. Shared Strengths and Opportunities

Each of the examples highlights the unique opportunity and successes that come from multi-scalar initiatives. While each effort exhibits unique qualities and contexts, the four examples also share several conditions of success.

By building out a dedicated role for civil society, the CFS gives voice to peasant farmers, fisherfolk, and Indigenous peoples all over the world [77]. Those represented by the CSM are working hard to feed communities while defending the land, ecosystems, and biodiversity we so badly depend on. Whenever possible, delegating the implementation and interpretation of policies to the most granular level—while ensuring the provision of adequate, predictable resources—can help promote community agency. In addition, implementing the right to food as central to all governance practices allows for the realization of our collective fundamental human rights. A rights-based approach moves governance institutions from a narrow axis centered on profit-driven food systems to one centered on people and nature. This can be seen in part through the interventions of the CSMs on the floor of the Plenary [78] of the CFS, as well as the long fight for the recognition of agroecology [79].

The recentering of priorities can lead to robust, resilient environments. UNESCO Biosphere Reserves create co-governance models that connect global institutions with local communities that are supported by the national and provincial governments. By putting co-habitation, rather than isolation, with nature as central to their biospheres model, UNESCO looks to foster sustainable ways to live within critical ecosystems. The farmers and institutions of Andhra Pradesh are also showing how rebuilding our connection with nature is implicit within farmer-led learning and the reduction of input-dependent farming. By creating enabling conditions, providing the right supports, and ensuring communities lead the transition, ZBNF is estimated to create benefits including: carbon emission reductions, bolstering the health of soil and water, empowering women, increasing biodiversity, and creating healthy environments [80].

Finally, adequate resourcing of advocacy networks, such as food policy groups, is essential to sharing knowledge both upstream to global forums and downstream back to communities. Egal and Forster highlight the current gap in resourcing for connecting biodiversity, food systems, and the dense network of activity at the local level [81]. The authors point to the strengthening of rural–urban linkages as an important part of strengthened governance. An increased interest in multi-scalar bridging through concepts such as territorial governance could help build out the critical junctures and support accountability across governance scales. Blay-Palmer et al. argue that frameworks such as the City Region Food System can help build more resilient systems and respond to shocks [82]. There are often strong local or regional networks, but they do not always effectively connect with the global to impact governance outcomes. Local, regional, or even national efforts to promote transformational change are happening in communities across the world, but a lack of resources and attention on how these feed into the global continues to limit the role that these networks could play.

6. Discussion

Good governance in public policy is a precondition for transformation, but it is complemented by actions at the consumer [83], farmer [84], and network [85] level that build trust and shared values [86]. Multi-scalar initiatives, such as those highlighted in this review, help reconnect people with nature through inclusive multilateral frameworks that encourage co-governance and place-based learning. Resourcing and implementation are critical for the success of any initiative. In the different examples in Section 4, governments from across scales have come together to ensure that resources are available for those most affected. This review has shown that good governance can, and has, worked to create more equitable environments that reconnect people with nature. I argue that rights-based policy is not substitutable for other innovative solutions; rather, it is a necessary precursor. By making it the cornerstone of governance, the political context becomes a central consideration rather than an afterthought. Reforming governance back towards public institutions allows for clear accountability. However, good governance extends beyond public policy and institutions. The lessons from the four examples provided can also support good governance as it relates to research through the inclusion of critical questions, such as: Whose knowledge has been included and who benefits from this work? Has the project included the views of those most impacted by this research? Is there a systemic challenge that needs to be named and considered? Does this work support more equitable outcomes? These questions are not divorced from the subjects of technology, efficiency, or economics; rather, they are central to them. By asking these questions, reflection is provided to pause the immediate need to respond and open space for evaluating the more systemic nature of the work across agricultural and food systems science.

7. Conclusions

Food insecurity and environmental degradation (e.g., of soils, atmosphere, and biological diversity) go hand in hand as the festering symptoms of an agricultural system that maximizes output over equity. As Bittman wrote in his recent book, *Animal, Vegetable, Junk: A History of Food, from Sustainable to Suicidal*, the policy of food systems is set with the output function calibrated to the command "of agriculture not food for people, but goods for market" [87] (p. 54).

We are disconnected from nature, surpassing planetary boundaries at a time when our climate and social crises converge. The global community is expected to fall short of achieving the SDGs by 2030 but, according to the State of the World Food Security report of 2021, there will be more individuals suffering from food insecurity than ever before. These stark realities converge with our climate crises, leaving food producers behind and eroding resilience in communities across the world. While the COVID-19 pandemic has shed a light on some of the issues within global food systems, many of the conditions (e.g., corporate concentration, limited access to resources, and eroding workplace safety) were present long before November 2019.

For place-based initiatives and multi-scalar initiatives to thrive, global governance institutions need to be enabling processes centered on people and nature through a rights-based approach. This can be achieved by prioritizing those most affected by food system policies. Inclusive multilateralism, as suggested by critical food scholars and advocates alike, can provide this framework, but to be successful, member states and the institutions who answer to them—such as the United Nations—need to be held to account. Within global public governance, the HLPE has tried to tackle these transformational questions across the work of several reports.

A reconfiguration that is built on the right to food approach and, as the HLPE argues, the addition of agency and sustainability is essential for creating equitable food systems that foster true food security. To implement this call, there is a need to move from ad hoc programming and pilot projects to systems centered at the heart of agriculture and food systems funding.

There is an urgent need to leverage existing resources and use regulations as well as governance processes to reconstitute power in a way that works for all. Integrated and expanding efforts, such as the four outlined in this paper, based on co-governance and place-based principles, will enable more communities to thrive and food systems to support the achievement of the SDGs. By localizing and democratizing governance across scales, we can use public governance and policy to reconnect people with nature to create a prosperous future for generations to come.

Funding: Johanna Wilkes is supported in part by funding from the Social Sciences and Humanities Research Council.

Acknowledgments: This paper was written on the traditional territory of the Anishnawbe, Haudenosaunee, and Neutral peoples, on which I am an uninvited guest. I am grateful for the work of food producers, scholars, global governance experts, advocates, and practitioners across scales who continue to envision a better way forward towards a more inclusive, just, and sustainable future. Thank you for the generosity of insights Mary Caesar, Lucy Hinton, and my tireless advisor while reading earlier iterations of this review. Your comments and suggestions were invaluable throughout the process.

Conflicts of Interest: The author declares no conflict of interest.

References

1. Fisher, M.; Bubola, E. As Coronavirus Deepens Inequality, Inequality Worsens Its Spread. New York Times: The Interpreter. 2020. Available online: https://www.nytimes.com/2020/03/15/world/europe/coronavirus-inequality.html (accessed on 2 March 2022).
2. World Bank. COVID-19 to Add as Many as 150 Million Extreme Poor by 2020. Press Release. 2020. Available online: https://www.worldbank.org/en/news/press-release/2020/10/07/covid-19-to-add-as-many-as-150-million-extreme-poor-by-2021 (accessed on 2 March 2022).
3. Vijaya, R. COVID-19 Further Exposes Inequalities in the Global Financial System. The Conversation: Business + Economy. 2020. Available online: https://theconversation.com/covid-19-further-exposes-inequalities-in-the-global-financial-system-150792 (accessed on 2 March 2022).
4. HLPE. *Impacts of COVID-19 on Food Security and Nutrition: Developing Effective Policy Responses to Address the Hunger and Malnutrition Pandemic*; Committee on World Food Security: Rome, Italy, 2020; Available online: https://www.fao.org/3/cb1000en/cb1000en.pdf (accessed on 2 March 2022).
5. Stahlbrand, L.; Robets, W. Local Food System Responses to COVID-19: Toronto and its City Region. City Region Food Systems Program of the Food and Agriculture Orgaization of the United Nations. 2020. Available online: https://www.fao.org/in-action/food-for-cities-programme/news/detail/en/c/1275076/ (accessed on 2 March 2022).
6. Zollet, S.; Colombo, L.; De Meo, P.; Marino, D.; McGreevy, S.R.; McKeon, N.; Tarra, S. Towards Territorially Embedded, Equitable and Resilient Food Systems? Insights from Grassroots Responses to COVID-19 in Italy and the City Region of Rome. *Sustainability* **2021**, *13*, 2425. [CrossRef]
7. IPBES. *Summary for Policymakers of the Global Assessment Report on Biodiversity and Ecosystem Services of the Intergovernmental Science-Policy Platform on Biodiversity and Ecosystem Services*; Díaz, S., Settele, J., Brondízio, E.S., Ngo, H.T., Guèze, M., Agard, J., Arneth, A., Balvanera, P., Brauman, K.A., Butchart, S.H.M., et al., Eds.; IPBES Secretariat: Bonn, Germany, 2019; p. 56. [CrossRef]
8. Steffen, W.; Richardson, K.; Rockström, J.; Cornell, S.E.; Fetzer, I.; Bennett, E.M.; Biggs, R.; Carpenter, S.R.; de Vries, W.; Wit, C.A.D.E.; et al. Planetary boundaries: Guiding human development on a changing planet. *Science* **2015**, *347*, 1259855. [CrossRef] [PubMed]
9. United Nations. World Social Report 2020: Inequality in a Rapidly Changing World. Department of Economic and Social Affairs. 2020. Available online: https://www.un.org/development/desa/dspd/wp-content/uploads/sites/22/2020/02/World-Social-Report2020-FullReport.pdf (accessed on 2 March 2022).
10. UNEP. 9 Ways Food Systems Are Failing Humanity. Story: Cities and Lifestyles. 2021. Available online: https://www.unep.org/news-and-stories/story/9-ways-food-systems-are-failing-humanity (accessed on 2 March 2022).
11. FAO; IFAD; UNICEF; WFP; WHO. *The State of Food Security and Nutrition in the World 2021. Transforming Food Systems for Food Security, Improved Nutrition and Affordable Healthy Diets for All*; FAO: Rome, Italy, 2021. [CrossRef]
12. United Nations. Can We Feed the World and Ensure No One Goes Hungry? UN News. 2019. Available online: https://news.un.org/en/story/2019/10/1048452 (accessed on 2 March 2022).
13. FAO. New FAO Analysis Reveals Carbon Footprint of Agri-Food Supply Chain. UN News: Climate and the Environment. 2021. Available online: https://news.un.org/en/story/2021/11/1105172 (accessed on 2 March 2022).
14. Franck, C.; Grandi, S.; Eisenberg, M.J. Agricultural Subsidies and the American Obesity Epidemic. *Am. J. Prev. Med.* **2013**, *45*, 327–333. [CrossRef] [PubMed]
15. Laborde, D.; Mamun, A.; Martin, W.; Piñeiro, V.; Vos, R. Agricultural subsidies and global greenhouse gas emissions. *Nat. Commun.* **2021**, *12*, 1–9. [CrossRef] [PubMed]
16. Shukla, P.R.; Skea, J.; Buendia, E.C.; Masson-Delmotte, V.; Pörtner, H.-O.; Roberts, D.C.; Zhai, P.; Slade, R.; Connors, S.; van Diemen, R.; et al. (Eds.) Summary for Policymakers. In *Climate Change and Land: An IPCC Special Report on Climate Change, Desertification, Land Degradation, Sustainable Land Management, Food Security, and Greenhouse Gas Fluxes in Terrestrial Ecosystems*; The Intergovernmental Panel on Climate Change: Geneva, Switzerland, 2019; *in press*.
17. Grosso, M.; Falasconi, L. Addressing food wastage in the framework of the UN Sustainable Development Goals. Waste Manag. *Res. J. Sustain. Circ. Econ.* **2018**, *36*, 97–98.
18. Babbitt, C.W.; Neff, R.A.; Roe, B.E.; Siddiqui, S.; Chavis, C.; Trabold, T.A. Transforming wasted food will require systemic and sustainable infrastructure innovations. *Curr. Opin. Environ. Sustain.* **2022**, *54*, 101151. [CrossRef]
19. CEC. *Characterization and Management of Food Loss and Waste in North America*; Commission for Environmental Cooperation: Montreal, QC, Canada, 2017; p. 48.
20. Wise, T.A. *Eating Tomorrow: Agribusiness, Family Farmers, and the Battle for the Future of Food*; The New Press: New York, NY, USA, 2019.
21. HLPE. Food Security and Nutrition: Building a Global Narrative towards 2030. 2020. Available online: https://www.fao.org/3/ca9731en/ca9731en.pdf (accessed on 2 March 2022).

22. Clapp, J.; Moseley, W.G.; Burlingame, B.; Termine, P. Viewpoint: The case for a six-dimensional food security framework. *Food Policy* **2021**, *106*, 102164. [CrossRef]
23. IPCC. Summary for Policymakers. In *Climate Change 2021: The Physical Science Basis. Contribution of Working Group I to the Sixth Assessment Report of the Intergovernmental Panel on Climate Change*; Masson-Delmotte, V., Zhai, P., Pirani, A., Connors, S.L., Péan, C., Berger, S., Caud, N., Chen, Y., Goldfarb, L., Gomis, M.I., et al., Eds.; Cambridge University Press: Cambridge, UK, 2021; *in press*.
24. Lowder, S.K.; Sánchez, M.V.; Bertini, R. Which farms feed the world and has farmland become more concentrated? *World Dev.* **2021**, *142*, 105455. [CrossRef]
25. Ghimire, D.; Panday, D. Interconnection of climate change, agriculture and climate justice: Complexities for feeding the world under changing climate. *Development* **2016**, *59*, 270–273. [CrossRef]
26. Tarasuk, V.; Mitchell, A. Household Food Insecurity in Canada, 2017–2018. Toronto: Research to Identify Policy Options to Reduce Food Insecurity (PROOF). 2020. Available online: https://proof.utoronto.ca/ (accessed on 2 March 2022).
27. Dhunna, S.; Tarasuk, V. Black–white racial disparities in household food insecurity from 2005 to 2014, Canada. *Can. J. Public Health* **2021**, *112*, 888–902. [CrossRef]
28. United Nations. Report of the Special Rapporteur on the Right to Food, Olivier De Schutter. 2012. Available online: http://www.srfood.org/images/stories/pdf/officialreports/20121224_canadafinal_en.pdf (accessed on 2 March 2022).
29. Robison, A. Turtle Island. The Canadian Encyclopedia. 2021. Available online: https://www.thecanadianencyclopedia.ca/en/article/turtle-island (accessed on 2 March 2022).
30. Panikkar, B.; Lemmond, B. Being on land and sea in troubled times: Climate change and food sovereignty in Nunavut. *Land* **2020**, *9*, 508. [CrossRef]
31. Naylor, A.W.; Ford, J.D.; Pearce, T.; Fawcett, D.; Clark, D.; van Alstine, J. Monitoring the dynamic vulnerability of an Arctic subsistence food system to climate change: The case of Ulukhaktok, NT. *PLoS ONE* **2021**, *16*, e0258048. [CrossRef] [PubMed]
32. Hellin, J.; Fisher, E. The Achilles heel of climate-smart agriculture. *Nat. Clim. Chang.* **2019**, *9*, 493–494. [CrossRef]
33. Sampson, D.; Cely-Santos, M.; Gemmill-Herren, B.; Babin, N.; Bernhart, A.; Bezner Kerr, R.; Blesh, J.; Bowness, E.; Feldman, M.; Gonçalves, A.L.; et al. Food sovereignty and rights-based approaches strengthen food security and nutrition across the globe: A systematic review. *Front. Sustain. Food Syst.* **2021**, *288*, 686492. [CrossRef]
34. HLPE. *Agroecological and Other Innovative Approaches for Sustainable Agriculture and Food Systems That Enhance Food Security and Nutrition*; A report by the High Level Panel of Experts on Food Security and Nutrition of the Committee on World Food Security; HLPE: Rome, Italy, 2019.
35. HLPE. *Promoting Youth Engagement and Employment in Agriculture and Food Systems*; A report by the High Level Panel of Experts on Food Security and Nutrition of the Committee on World Food Security; HLPE: Rome, Italy, 2021.
36. Keohane, R.O. Multilateralism: An agenda for research. *Int. J.* **1990**, *45*, 731–764. [CrossRef]
37. Raymond, M.; DeNardis, L. Multistakeholderism: Anatomy of an inchoate global institution. *Int. Theory* **2015**, *7*, 572–616. [CrossRef]
38. Pimbert, M.P.; Moeller, N.I. Absent agroecology aid: On UK agricultural development assistance since 2010. *Sustainability* **2018**, *10*, 505. [CrossRef]
39. Alexandra, L.; Friedmann, H.; Wittman, H. Can public universities play a role in fostering seed sovereignty? *Elem. Sci. Anthr.* **2021**, *9*, 89.
40. Anderson, M. Corporate Takeover of the Narrative of Food System Solutions: The Case of the UNFSS. 2021. Available online: https://www.researchgate.net/profile/Kursat-Demiryurek/publication/357485868_Organik_ve_Genetigi_Degistirilmis_Organizmali_Urunlere_Karsi_Ogrencilerin_Davranislari_Ondokuz_Mayis_Universitesi_Ornegi/links/61d088c3b8305f7c4b17c316/Organik-ve-Genetigi-Degistirilmis-Organizmali-Ueruenlere-Karsi-Oegrencilerin-Davranislari-Ondokuz-Mayis-Ueniversitesi-Oernegi.pdf#page=97 (accessed on 2 March 2022).
41. McMichael, P. Shock and Awe in the UNFSS. *Development* **2021**, *64*, 162–171. [CrossRef]
42. Clapp, J.; Noyes, I.; Grant, Z. The Food Systems Summit's Failure to Address Corporate Power. *Development* **2021**, *64*, 192–198. [CrossRef]
43. UNFSS. UN Secretary-General António Guterres Announces UN Food Systems Summit | 16 October 2019. Youtube Channel: UN Food Systems Summit. 2019. Available online: https://www.youtube.com/watch?v=6i7x_V6WVaw (accessed on 2 March 2022).
44. Von Braun, J.; Afsana, K.; Fresco, L.; Hassan, M.; Torero, M. Food Systems–Definition, Concept and Application for the UN Food Systems Summit. Available online: https://www.un.org/sites/un2.un.org/files/scgroup_food_systems_paper_march-5-2021.pdf (accessed on 2 March 2022).
45. Canfield, M.C.; Duncan, J.; Claeys, P. Reconfiguring food systems governance: The UNFSS and the battle over authority and legitimacy. *Development* **2021**, *64*, 181–191. [CrossRef] [PubMed]
46. Chandrasekaran, K.; Guttal, S.; Kumar, M.; Langner, L.; Manahan, M.A. Exposing Corporate Capture of the UNFSS Through Multistakeholderism; 2021. Available online: https://www.foodsystems4people.org/wp-content/uploads/2021/09/UNFSSreport2021.pdf (accessed on 2 March 2022).
47. Canfield, M.; Anderson, M.D.; McMichael, P. UN Food Systems Summit 2021: Dismantling Democracy and Resetting Corporate Control of Food Systems. *Front. Sustain. Food Syst.* **2021**, *5*, 103. [CrossRef]
48. CSM. CSM Letter to the CFS Chair on the UN Food Systems Summit. News: Civil Society and Indigenous Peoples Mechanism. 2020. Available online: https://www.csm4cfs.org/letter-csm-coordination-committee-cfs-chair/ (accessed on 2 March 2022).

49. Nisbett, N.; Hoey, L.; Collins, J. Strengthening Governance of the United Nations Food Systems Summit (UNFSS). 2021. Available online: https://www.iatp.org/sites/default/files/2021-07/UN%20FSS%20Governance_Letter_FINAL_7-8-21.pdf (accessed on 2 March 2022).
50. FIAN. Annual Report 2020. 2021. Available online: https://www.fian.org/files/fian-international-annual-report-2020pdf.pdf (accessed on 2 March 2022).
51. IPES-Food. Withdrawal from the UN Food Systems Summit Memo from the IPES-Food Panel, 26 July 2021. 2021. Available online: https://www.ipes-food.org/_img/upload/files/UNFSS%20Withdrawal%20Statement(1).pdf (accessed on 2 March 2022).
52. WEF. The United Nations-World Economic Forum Strategic Partnership Framework for the 2030 Agenda. 2019. Available online: https://weforum.ent.box.com/s/rdlgipawkjxi2vdaidw8npbtyach2qbt (accessed on 2 March 2022).
53. Guttal, S. Re-imagining the UN Committee on World Food Security. *Development* **2021**, *64*, 227–235. [CrossRef] [PubMed]
54. C40 Cities. A Global Network of Mayors Taking Urgent Action to Confront the Climate Crisis and Create a Future Where Everyone Can Thrive. C40 Website, Landing Page. 2021. Available online: https://www.c40.org (accessed on 2 March 2022).
55. Deininger, N.; Lu, Y.; Griess, J.; Santamaria, R. *Cities Taking the Lead on the Sustainable Development Goals: A Voluntary Local Review Handbook for Cities*; Carnegie Mellon University for the Brookings Institute: Pittsburgh, PA, USA, 2019; Available online: https://www.brookings.edu/wp-content/uploads/2019/07/VLR_Handbook_7.7.19.pdf (accessed on 2 March 2022).
56. IPES-Food. Food Systems and Climate Action at COP26. Statement. 2020. Available online: http://www.ipes-food.org/pages/GlasgowDeclaration (accessed on 2 March 2022).
57. Glasgow Food and Climate Declaration. Glasgow Food and Climate Declaration. 2021. Available online: https://www.glasgowdeclaration.org/the-glasgow-declaration (accessed on 2 March 2022).
58. CFS. Committee on World Food Security, about CFS. 2021. Available online: https://www.fao.org/cfs/en/ (accessed on 2 March 2022).
59. McKeon, N. *Food Security Governance: Empowering Communities, Regulating Corporations*; Routledge: Oxford, UK, 2014.
60. Claeys, P.; Duncan, J. *Evaluation of the Civil Society Mechanism (CSM) for relations with the UN Committee on World Food Security (CFS)*. Full Report. 2018. Available online: https://www.csm4cfs.org/wp-content/uploads/2018/02/CSM-Evaluation-Report-2018-ilovepdf-compressed.pdf (accessed on 2 March 2022).
61. CSM. A Mechanism of Civil Society and Indigenous Peoples. Events, News. 2018. Available online: https://www.csm4cfs.org/un-mecanismo-de-la-sociedad-civil-y-de-pueblos-indigenas/ (accessed on 2 March 2022).
62. CSM. CSM Video Interventions at CFS 45! Intervention by Tui Shortland from the International Indigenous Treaty Council, Interventions from CSM Members. 2018. Available online: https://www.csm4cfs.org/csm-video-interventions-cfs-45/ (accessed on 2 March 2022).
63. CSM. Civil Society Report on the United Nations Committee on World Food Security 2018. 2018. Available online: https://www.csm4cfs.org/wp-content/uploads/2018/10/EN-CSM-LR-2018-compressed.pdf (accessed on 2 March 2022).
64. CFS. Implementation and Monitoring. Committee on World Food Security, Policy Products. 2021. Available online: https://www.fao.org/cfs/vgfsn/implementation-monitoring (accessed on 2 March 2022).
65. Canadian Biosphere Reserves Association. FAQ. 2021. Available online: https://www.biospherecanada.ca/faq (accessed on 2 March 2022).
66. Canadian Biosphere Reserves Association. UNESCO Biosphere Reserves Inspire a Positive Future by Connecting People and Nature Today. 2021. Available online: https://www.biospherecanada.ca/biosphere-reserves (accessed on 2 March 2022).
67. Karp, D.S.; Rominger, A.J.; Zook, J.; Ranganathan, J.; Ehrlich, P.R.; Daily, G.C. Intensive agriculture erodes β-diversity at large scales. *Ecol. Lett.* **2012**, *15*, 963–970. [CrossRef]
68. Gavin, M.C.; McCarter, J.; Berkes, F.; Mead AT, P.; Sterling, E.J.; Tang, R.; Turner, N.J. Effective biodiversity conservation requires dynamic, pluralistic, partnership-based approaches. *Sustainability* **2018**, *10*, 1846. [CrossRef]
69. Veluguri, D.; Bump, J.B.; Venkateshmurthy, N.S.; Mohan, S.; Pulugurtha, K.T.; Jaacks, L.M. Political analysis of the adoption of the Zero-Budget natural farming program in Andhra Pradesh, India. *Agroecol. Sustain. Food Syst.* **2021**, *45*, 907–930. [CrossRef]
70. Ranjan, S.; Sow, S. A Way Towards Sustainable Agriculture through Zero Budget Natural Farming. Food and Scientific Reports. 2021. Available online: https://www.researchgate.net/profile/Shivani-Ranjan-2/publication/350545691_A_way_towards_sustainable_agriculture_through_zero_budget_natural_farming/links/606592e3a6fdccad3f628783/A-way-towards-sustainable-agriculture-through-zero-budget-natural-farming.pdf (accessed on 2 March 2022).
71. Khadse, A.; Rosset, P.M. Zero Budget Natural Farming in India–from inception to institutionalization. *Agroecol. Sustain. Food Syst.* **2019**, *43*, 848–871. [CrossRef]
72. Duddigan, S.; Collins, C.D.; Hussain, Z.; Osbahr, H.; Shaw, L.J.; Sinclair, F.; Sizmur, T.; Thallam, V.; Winowiecki, L.A. Impact of Zero Budget Natural Farming on Crop Yields in Andhra Pradesh, SE India. *Sustainability* **2022**, *14*, 1689. [CrossRef]
73. Giagnocavo, C.; de Cara-García, M.; González, M.; Juan, M.; Marín-Guirao, J.I.; Mehrabi, S.; Rodríguez, E.; van der Blom, J.; Crisol-Martínez, E. Reconnecting Farmers with Nature through Agroecological Transitions: Interacting Niches and Experimentation and the Role of Agricultural Knowledge and Innovation Systems. *Agriculture* **2022**, *12*, 137. [CrossRef]
74. Johns Hopkins Center for a Livable Future. Food Policy Groups Around the World. Food Policy Networks. 2021. Available online: https://www.foodpolicynetworks.org/councils/fpg-worldwide/ (accessed on 2 March 2022).
75. Santos, R.; Misiaszek, C.; Bassarab, K.; Harris, D.; Palmer, A. Pivoting Policy, Programs, and Partnership: Food Policy Councils Responses to the Crises of 2020. Report: Food Policy Networks as part of the Johns Hopkins Centre for a Livable Future. 2021. Available online: https://assets.jhsph.edu/clf/mod_clfResource/doc/FPC%202020%20Census%20Report_2021-04-05_FINAL.pdf (accessed on 2 March 2022).

76. IPES-Food. Towards a Common Food Policy for the EU. Report. 2019. Available online: http://www.ipes-food.org/pages/CommonFoodPolicy (accessed on 2 March 2022).
77. CSM. What is the CSM? Civil Society and Indigenous Peoples' Mechanism for Relations with the UN Committee on World Food Security. 2021. Available online: https://www.csm4cfs.org/what-is-the-csm/ (accessed on 2 March 2022).
78. CSMechanism. CSM's Final Statement on the Voluntary Guidelines on Food Systems and Nutrition #VGFSyN. CSMechanism Youtube Channel. 2021. Available online: https://www.youtube.com/watch?v=DeEWG0fKFis (accessed on 2 March 2022).
79. CSM. Civil Society Groups and Indigenous Peoples Demand More Ambitious CFS Policies on Agroecology. CSM, News. 2021. Available online: https://www.csm4cfs.org/civil-society-groups-and-indigenous-peoples-demand-more-ambitious-cfs-policies-on-agroecology/ (accessed on 2 March 2022).
80. UNEP. Zero Budget Natural Farming as a Nature-Based Solution for Climate Action. Working Document. 2020. Available online: https://wedocs.unep.org/bitstream/handle/20.500.11822/28895/Zero_budget.pdf?sequence=1&isAllowed=y (accessed on 2 March 2022).
81. Egal, F.; Forster, T. *Biodiversity, Food Systems, and Urban-Rural Linkages. Biodiversity, Food, and Nutrition: A New Agenda for Sustainable Food Systems*; Routledge: Oxford, UK; New York, NY, USA, 2020; Available online: http://www.b4fn.org/fileadmin/templates/b4fn.org/upload/documents/The_Books/BFN_full_2020_compressed.pdf#page=222 (accessed on 2 March 2022).
82. Blay-Palmer, A.; Santini, G.; Halliday, J.; Malec, R.; Carey, J.; Keller, L.; Ni, J.; Taguchi, M.; van Veenhuizen, R. City Region Food Systems: Building Resilience to COVID-19 and Other Shocks. *Sustainability* **2021**, *13*, 1325. [CrossRef]
83. Mehrabi, S.; Perez-Mesa, J.C.; Giagnocavo, C. The Role of Consumer-Citizens and Connectedness to Nature in the Sustainable Transition to Agroecological Food Systems: The Mediation of Innovative Business Models and a Multi-Level Perspective. *Agriculture* **2022**, *12*, 203. [CrossRef]
84. Chen, Z.; Sarkar, A.; Hasan, A.K.; Li, X.; Xia, X. Evaluation of Farmers' Ecological Cognition in Responses to Specialty Orchard Fruit Planting Behavior: Evidence in Shaanxi and Ningxia, China. *Agriculture* **2021**, *11*, 1056. [CrossRef]
85. González-Rosado, M.; Parras-Alcántara, L.; Aguilera-Huertas, J.; Lozano-García, B. Building an Agroecological Process towards Agricultural Sustainability: A Case Study from Southern Spain. *Agriculture* **2021**, *11*, 1024. [CrossRef]
86. Gugerell, C.; Sato, T.; Hvitsand, C.; Toriyama, D.; Suzuki, N.; Penker, M. Know the Farmer That Feeds You: A Cross-Country Analysis of Spatial-Relational Proximities and the Attractiveness of Community Supported Agriculture. *Agriculture* **2021**, *11*, 1006. [CrossRef]
87. Bittman, M. *Animal, Vegetable, Junk. A History of Food, from Sustainable to Suicidal*; Houghton Mifflin Harcourt: New York, NY, USA, 2021.

agriculture

MDPI

Article

Self-Sufficiency Assessment: Defining the Foodshed Spatial Signature of Supply Chains for Beef in Avignon, France

Michel Mouléry [1,*], Esther Sanz Sanz [1,2], Marta Debolini [1], Claude Napoléone [1], Didier Josselin [3], Luc Mabire [3] and José Luis Vicente-Vicente [2]

[1] French National Institute for Research on Agriculture, Food and Environment (INRAE), UR-Ecodéveloppement, 228 route de l'aérodrome CS 40509, CEDEX 9, 84914 Avignon, France; esther.sanz-sanz@inrae.fr (E.S.S.); marta.debolini@inrae.fr (M.D.); claude.napoleone@inrae.fr (C.N.)

[2] Leibniz Centre for Agricultural Landscape Research (ZALF), Eberswalder Str. 84, 15374 Müncheberg, Germany; vicente@zalf.de

[3] UMR ESPACE 7300, French National Centre for Scientific Research (CNRS), Campus Hannah Arendt, Avignon University, 74 Rue Louis Pasteur, 84029 Avignon, France; didier.josselin@univ-avignon.fr (D.J.); mabire.luc@gmail.com (L.M.)

* Correspondence: michel.moulery@inrae.fr

Abstract: Foodshed approaches allow for the assessment of the theoretical food self-sufficiency capacity of a specific region based on biophysical conditions. Recent analyses show that the focus needs to be shifted from foodshed size portrayed as an isotropic circle to a commodity–group-specific spatial configuration of the foodshed that takes into account the socio-economic and biophysical conditions essential to the development of local food supply chains. We focused on a specific animal product (beef) and used an innovative modeling approach based on spatial analysis to detect the areas of the foodshed dedicated to beef feeding (forage, pasture, and grassland), considering the foodshed as a complex of complementary areas called an archipelago. We used available statistical data including a census to address the city-region of Avignon, France covering a 100 km radius. Our results showed that the factors driving the use of short supply chains for beef feeding areas are the foodshed archipelago's number of patches, the connectivity between them, and the rugosity of the boundaries. In addition, our beef self-sufficiency assessment results differ depending on geographical context. For instance, being located within the perimeters of a nature park seems to help orient beef production toward short supply chains. We discuss possible leverage for public action to reconnect beef production areas to consumption areas (the city) via short supply chains (e.g., green, home-grown school food programs) to increase local food security through increased local food self-sufficiency.

Keywords: foodshed archipelago; proximity food supply chains; spatial signature; city-region; food self-sufficiency; regional food security; agricultural diversification; food planning; regional food system; food policy

Citation: Mouléry, M.; Sanz Sanz, E.; Debolini, M.; Napoléone, C.; Josselin, D.; Mabire, L.; Vicente-Vicente, J.L. Self-Sufficiency Assessment: Defining the Foodshed Spatial Signature of Supply Chains for Beef in Avignon, France. *Agriculture* **2022**, *12*, 419. https://doi.org/10.3390/agriculture12030419

Academic Editor: Christian Schader

Received: 14 February 2022
Accepted: 11 March 2022
Published: 16 March 2022

1. Introduction

Food supply chains are vulnerable to social and economic risks and to natural hazards, as illustrated by the COVID-19 crisis. This is especially true in urban areas, which largely rely on food imported from the global market [1]. In fact, the population of urban areas has exceeded that of rural areas since 2008, and this proportion is expected to increase to 66% by 2050 [2]. Each disruption in global food supply chains has become a social and political issue, prompting a focus on the relocalization of food supply and regionalization of food systems [3]. However, peri-urban agricultural areas are not homogenous, and not every farming system is able to respond to local food demand in terms of foodstuff diversity or quantities [4–6]. In addition, geo-physical spatial heterogeneity means that soils differ in their suitability for agriculture. However, there are few tools available to inform public policies aimed at supporting the regionalization of food systems, particularly to identify the farmland areas where farmers can best respond to incentives.

Current research is highlighting foodshed approaches as a way of identifying the farmland areas functionally linked to cities that could be involved in new short food supply chains [7]. In this paper, we first briefly sketch the state-of-the-art concerning the notion of foodshed. Then, we describe a study case in southern France to which we applied a foodshed approach to analyze beef supply chains using a new methodology. Our method was based on spatial metrics grounded in theories of landscape ecology and on proximity relations in regional development processes. Finally, we present our results and discuss their implications for further research on the regionalization of food systems.

1.1. Foodshed Approaches

The notion of "foodshed" was first used by W. Hedden in 1929 in the book *How Great Cities are Fed* to describe "the geographic area from which food arrives in a community including the rural and urban farmlands, processing and distribution facilities, transportation systems, wholesalers, and retailers that make up a region's food system" [8]. In October 1921, a planned nationwide railway strike threatened New York with the danger of interrupted food supplies to a large city dependent on distant food sources and the loss of nearby farmland to the suburbs. This prompted Walter P. Hedden, Head of the Port Authority of New York's Bureau of Commerce, to write a comprehensive assessment of the city's food supply. Hedden mapped food flows from different locations in the United States, looked at criteria such as seasonality or the origin of food, and studied the logistical infrastructure (rail lines, cooling and storage facilities, distribution centers, and food shops). In 1996, the term 'foodshed analysis' was proposed to inform policy decisions on local food sufficiency or insufficiency [9]. Foodshed analysis can be seen as a comprehensive approach to improving the sustainability of regional food systems [10]. For instance, by determining the potential and risks for agricultural production capacity from the analysis of bioclimatic variables (climate, soil type resources) [11], by assessing the environmental impact and vulnerability of local food systems depending on food origin [12], or by examining whether shortening food supply chains can help maintain agriculture close to urban areas [13].

In this paper, we defined the term "foodshed" as the geographical area in which food is grown to satisfy the food needs of a population from its own domestic production. The foodshed approaches vary depending on the scale of the analysis and the objective: to assess whether total local food demand can be met by local production capacity [14,15], or to assess the production capacity required to meet local food demand [16]. The foodshed approach has also been used to estimate the size of the foodshed required to meet a given rate of food self-sufficiency, taking into account different food system scenarios in terms of food groups, food production systems (conventional versus organic), diets, and levels of food loss and waste (e.g., the Metropolitan Foodshed and Self-sufficiency Scenario: MFSS; [17]). Thus, in addition to food production capacity based on biophysical conditions, our recent work considered socioeconomic features driving the flows and distribution networks of locally-grown food [7]. Our findings showed that analysis needs to be shifted from size assessment of the foodshed represented as an isotropic circle around the city to commodity–group-specific spatial configuration of the foodshed [7].

The aim of the present study was to explore foodshed assessment as a complex of complementary entities (i.e., the "foodshed archipelago"). To this end, we developed a framework grounded in landscape ecology, namely the island biogeography theory, the continent-island model theory, and the connectivity theory. In other words, we assumed that the foodshed is a set of specific food production areas containing patches of different sizes, spatially interconnected or not (i.e., the archipelago). The purpose of this foodshed is to distribute food through short supply chains to feed the local population. In this study, we focused on the landscape structure of the entities composing the foodshed archipelago. To test our hypothesis, we analyzed the beef foodshed of a study case located in Avignon, France. We intentionally focused on animal origin food products, which are little present in Mediterranean city surroundings such as Avignon, due to the prevailing pedo-climatic conditions (water and grassland scarcity). Our objective was to determine whether this beef foodshed archipelago had a specific spatial signature, different to that of beef production areas serving long supply chains. By "spatial signature", we mean particular spatial structures whose arrangement is identifiable in space, resulting in a set of common characteristics such as crop plot shape, location of farmstead, border relationship between farming and urban zones, etc. [5,6,18].

1.2. Foodshed Analysis Based on Landscape Ecology and Proximity Theories

Our analytical framework is grounded in theories of landscape ecology and based on the proximity relations pertaining to regional development processes. The first landscape theory behind our work is the continent-island model theory, which maintains that a local habitat, called a "source," provides individuals to other local habitats, called a "sink" [19]. Within this interplay of colonization and extinction, any habitat can be both "source" and "sink" [20]. Following this approach, we considered a "sink" located in Avignon, which can be portrayed as a large patch requiring resources (beef supply). The "sources" are the other patches, an assembly of pastoral and grassland areas within a radius of 100 km around Avignon. The assembled sources configure the foodshed archipelago.

Second, our analysis of the way they are assembled was based on the connectivity theory. Here, pastoral areas and grasslands are considered as spatial objects or "patches" that are heterogeneous in terms of size and shape. Their actual geographical distribution (i.e., density) is not homogeneous. In general, neighboring patches or adjacent plots are more likely to be connected to each other than an isolated plot. Landscape connectivity is thus defined as the degree to which the landscape facilitates or impedes movement between resource patches [21]. This definition highlights the impacts that the type, quantity, and arrangement of habitat or land use have on movement, and ultimately population dynamics and community structure. Landscape connectivity therefore describes both the physical structure of the landscape and the response of an organism to that structure [22].

From the perspective of the economic geography theory of proximity relations, short food supply chains can only be structured if the three dimensions of proximity are respected, namely, geographical proximity (i.e., distance), organized proximity (i.e., the different ways of being close to other stakeholders, referring to the arranged nature of human activities), and institutional proximity (i.e., the political dimension or adherence to a space that is defined by common rules of action, representations, thought patterns) [23,24]. Our aim here is to define the spatial signature of the beef production areas serving short supply chains in terms of the three dimensions of proximity. We analyzed an empirical case study in the Avignon foodshed using a 4-step methodology. Geographical proximity is measured by distance to the slaughterhouses (cf. 2.4.1.). Organized proximity is considered through geographical proximity and under an analytical framework inspired by landscape ecology, measuring the rugosity of the contours of beef production areas (cf. 2.4.2.) and dominance (cf. 2.4.3.) to account for the territorial embeddedness of these farms [25–27]. Finally, institutional proximity is considered according to whether or not the beef production areas lie within the perimeter of a regional or national nature park (cf. 2.4.4.).

2. Materials and Methods

2.1. Study Area

Following previous work to assess the link between agricultural diversification and self-sufficiency on the Avignon foodshed [7], we defined a radius of 100 km around Avignon, which is a medium-sized city located in south-eastern France. The selected area incorporates three different administrative regions and ten different provinces: Bouches-du-Rhône, Vaucluse, Var, Hautes Alpes in the Provence-Alpes-Côte d'Azur region, Gard, Hérault, Lozère in the Occitanie region, and Ardèche and Drome in the Auvergne-Rhone-Alpes region. It numbers 1358 communes including 738 municipalities containing at least one beef farm selling part of its production in short supply chains. The Avignon peri-urban area is a fertile plain that has historically benefited from irrigation and transport infrastructure, fostering market gardening, fruit growing, and viticulture. More recently, part of Avignon's agriculture has also turned to large-scale cash crops (cereal and lavender). There is a predominance of municipalities specialized in wine-growing in our study area, seeming to form a structured arc along the Rhône, in the Vaucluse, Gard, and Bouches-du-Rhône. In fact, viticulture is a strongly supported and structured sector [28]. Only by moving away from the Avignon conurbation and the arc formed by the wine-growing communes can substantial areas potentially suitable for the grassland and pastoralism linked to beef production be found. They are overwhelmingly concentrated in the north of our study area on the Plateau de Coiron (Ardèche), Lozère, and the Monts de Vaucluse (Vaucluse and Alpes de Haute-Provence).

Beef production in this region of France is based uniquely on extensive grazing systems. The basis of the animals' food is the valorization of spontaneous grass (mountain pastures) and cultivated grass (permanent or temporary). There are no intensive production systems based on importing feeder cattle into feedlots and confining animals in a stall system. Concerning the commercial orientation of the production, farms can either serve short supply chains (SSC) or long supply chains (LSC) (Figure 1). Indeed, many farms supply local slaughterhouses that have existed for a long time and 30% of them sell directly to the consumer. On the other hand, some farms sell live cattle to a livestock trader who will sell beef abroad (e.g., to supermarkets) or export live young cattle to Italian finishers/feedlots where the animals do not return to France [29,30].

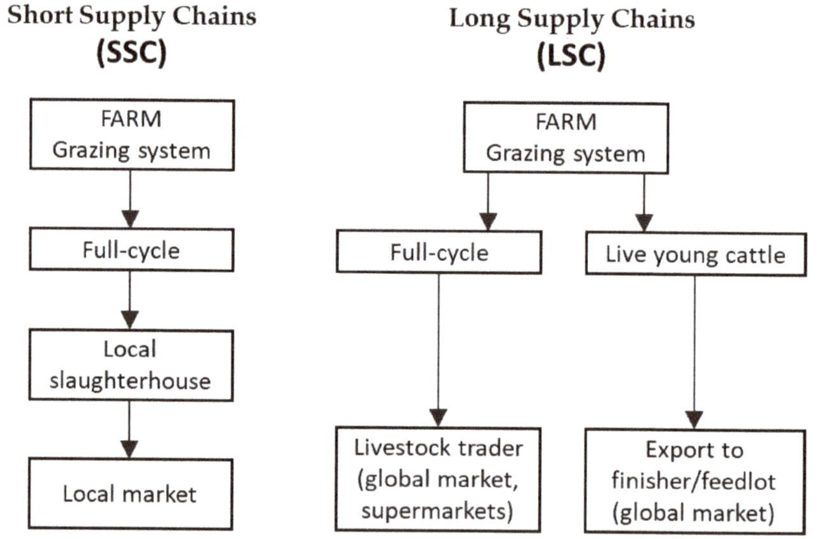

Figure 1. Schema of the type of farms serving short supply chains and long supply chains in the study area around Avignon, France.

Finally, there is also extensive cattle breeding in the municipalities of Camargue and Crau located in the south of the study area. However, these communes specialize in rearing herds of Camargue races for recreational purposes (e.g., bullfighting festivals). They did not focus on food production, so their analysis was discussed (cf. 4.1.) with regard to this specific context.

2.2. *Materials Used to Identify Beef Production Areas*

To spatially identify grasslands and pastoral areas, we used the 2018 plot identification system (LPIS) graphically represented in the French Registre Parcellaire Graphique, which geolocates and informs on areas under different EU Common Agricultural Policy (CAP) aid schemes. This is a very accurate vectoral data source that relies on the farmers' own hand-drawn outlines of their cultivated plots submitted when applying for CAP subsidies. We selected three categories of land used for beef feeding: mountain pastures and moors, permanent grasslands, and temporary grasslands (Table 1).

Table 1. Selected RPG categories of land use for beef feeding.

Category	Description	RPG Code
Mountain pastures, moors	Wood-pasture	BOP
	Pastoral area—predominantly grass and fodder resources. Woody resources present	SPH
	Pastoral area—predominantly woody fodder resources	SPL
Permanent grassland	Permanent grassland—predominantly grass (fodder resources; woody resources absent or little present)	PPH
	grassland in long rotation (6 years or more)	PRL
Temporary grassland	Other temporary grassland 5 years old or less	PTR

After aggregating these land-use categories, we considered patches as potentially serving short supply chains if they fell within the administrative boundary of municipalities with at least one beef farm partly selling through short supply chains, according to the 2010 general agricultural census (source https://www.agreste.agriculture.gouv.fr, accessed on 14 March 2022) at the municipal level. We analyzed patch connection using the "dilation/erosion" method described below. Other sources of complementary data used were the location of slaughterhouses [29,31,32] and the environmental protection perimeters of national and regional nature parks.

2.3. The Dilation/Erosion Methodology

The dilation/erosion methodology is grounded in landscape ecology and widely used for research in different disciplines including medicine and urban planning to analyze the morphology of geometric structures. Applications include the creation of a dilated envelope around built-up areas [33,34], and analysis of the distances between two natural areas to highlight the most direct paths to connect them in a "green and blue grid" [35]. This is based on algebra, topology, and probability concepts.

Here, we connected the vectorized plots of the three selected RPG categories of land use for beef feeding (mountain pastures—moors, permanent grassland, temporary grassland) at a minimum distance of 20 m, taken as the average rough width of roads and paths. The plots were grouped using the dilation method. Then, erosion was generated to refine the contours of the aggregated plots and create patches (Figure 2). Isolated plots more than 20 m away from their nearest neighbor were considered as patches in landscape ecology terms. Technically, a procedure was created using the spatial functions of the UrbanSimul project programmed in postgis, a plugin for PostgreSQL object-relational database used for executing spatial queries [36].

We mapped two sets of data: potential beef feeding areas oriented and not oriented toward short supply chains (Figure 3). The geographical entities of substantial size that are considered patches oriented toward beef short supply chains are shown in green. The largest such patches were located in Ardèche, Drome, Alpes de Haute Provence, areas producing beef breeds such as "Limousine" and "Charolais." Another large patch in the Bouches du Rhône hosts Camargue herds raised for recreational purposes.

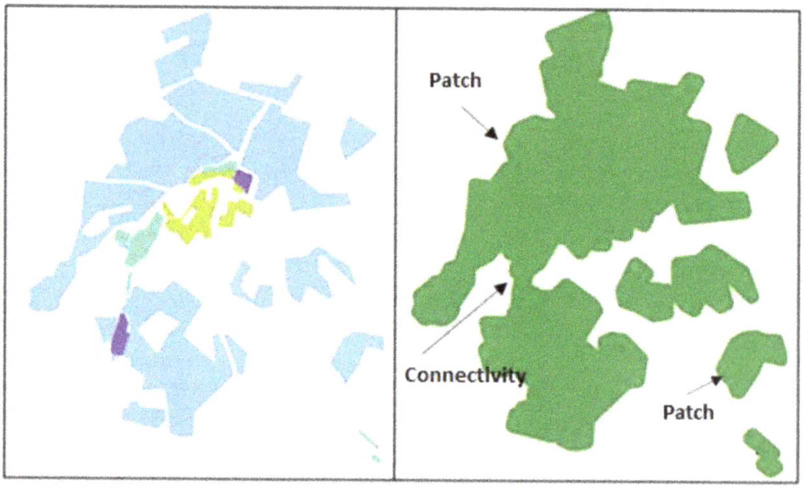

Figure 2. Map on left shows unconnected patches, while map on the right groups patches together using the dilation/erosion method.

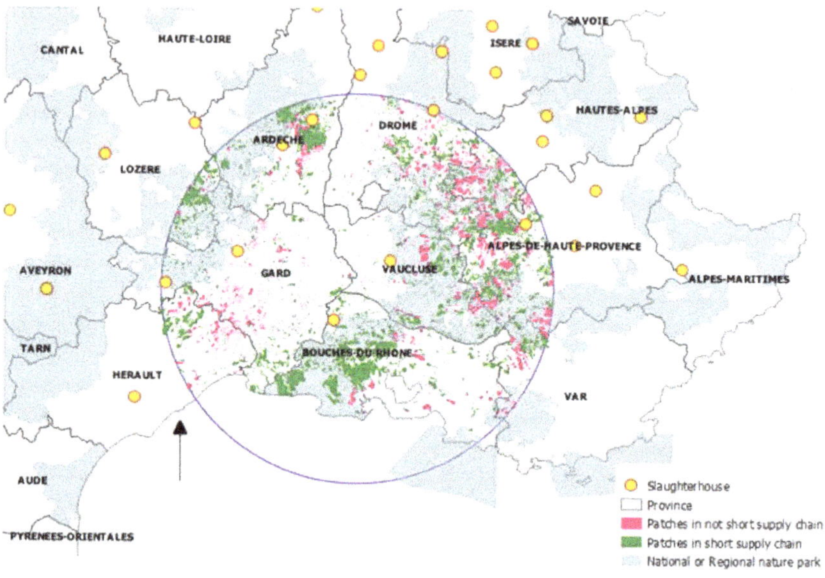

Figure 3. Location of the beef feeding patches within a 100 km radius around the city of Avignon.

2.4. Method Used to Spatially Characterize Beef Feeding Patches in the Archipelago Oriented and Not Oriented toward Short Supply Chains

This section examines whether the spatial signature of areas oriented and not oriented toward short supply chains can be distinguished from each other according to spatial analysis indicators. The goal was to identify areas (patches) more likely to respond to institutional incentives to increase food security by enhancing/promoting short supply chains. We relied on simple tools used in spatial analysis to assess: (1) the effects of distance from nearest slaughterhouse; (2) rugosity, defined as the complexity of the contours of patches; (3) dominance, according to density, number of patches oriented toward short circuits, and total surface area; and (4) the effects of being situated within the perimeters of a regional or a national nature park. The methods used to analyze/assess/calculate the four indicators are described below.

2.4.1. Distance from Nearest Slaughterhouse

The slaughterhouses in the study area are geolocated by a yellow dot (Figure 3). A geographic information system was used to calculate the minimum distance from each centroid of the beef feeding areas (patches) to the nearest slaughterhouse, discriminating between areas oriented and not oriented toward short supply chains. The underlying hypothesis is that production areas used by farmers selling beef in short supply chains are closer to a slaughterhouse than those used by farmers not selling beef in short supply chains. In fact, local slaughterhouses are small-scale structures providing less than 2000 tons of meat per year by slaughtering on behalf of farmers who then sell their meat at markets, to artisan butchers, or in shops specializing in local and organic products. These farmers and butchers rely on short supply chains to reduce intermediaries and add value to their products [37].

2.4.2. Rugosity

The rugosity indicator is based on the complexity of the contours of the patches. This indicator was constructed on the basis of ecology research on the rugosity of coral reefs, showing that the greater the structural complexity of ecological habitats, the greater the diversity of species [38]. This concept of rugosity was taken up by [39] for the urban system, under the hypothesis that high complexity of the contours of the urban–agricultural fringe increases the functional connections between urban and agricultural land uses. It was concluded that increased rugosity is associated with large populations and significant historic peri-urban farm holdings involved in direct marketing. In this paper, we widen this hypothesis beyond direct marketing, seeking to determine whether the rugosity of the farming areas fosters the orientation of beef production toward short supply chains. In other words, whether former cattle production areas oriented toward long supply chains generate more homogeneous limits than newcomers in short supply chains that are more randomly located. Thus, we measured the rugosity of beef feeding areas (patches) both oriented and not oriented toward short supply chains. We used the Gravelius index (K), that is, the ratio of the perimeter of the patch to the circumference of a circle of the same area surrounding it [40]. We applied the formula: $K = \text{perimeter}/2^* \sqrt{(\pi/\text{area})}$. The farther K is from 1, the more complex the contours are. The results are presented in Section 3.

2.4.3. Dominance

After applying the dilation/erosion method, we assessed dominance according to three indicators: (1) the density of the selected RPG categories of land use in the archipelago, discriminating between areas oriented and those not oriented toward short circuits; (2) the number of plots aggregated in the patch (as a proxy of productive crop–plot fragmentation) within the archipelago; and (3) the total area of patches discriminating between those oriented toward short or long supply chains. It was assumed that the larger the patches, the more likely they are to generate a foodshed capable of feeding the city compared to small, scattered pastoral plots. We defined density as the relationship between the surface areas registered in the RGA census (i.e., the selected RPG categories of land for beef feeding, see Table 1) and the surface areas of the patches they lie in, as determined by dilation/erosion, discriminating between those oriented and those not oriented toward short supply chains. The density indicators were calculated over a range of thresholds defined between 50 and 1000 hectares (total average of RPG areas related to the vectorized contours of the patches).

2.4.4. Location within the Perimeters of a Regional or a National Nature Park

The national parks were created in 1973 to ensure the protection of natural areas, both terrestrial and maritime. Pastoral practices are allowed on areas of great biological richness and landscape interest: high mountain pastures and estives (summer pastures), inter-seasonal rangelands, mown natural meadows, etc. In summer, the mountain pastures and estives also host numerous transhumant herds, sometimes coming from distant provinces of southern France (http://www.parcsnationaux.fr/fr/des-connaissances/agriculture-et-pastoralisme, accessed on 14 March 2022). The regional nature parks were created under French regional planning policy in 1967 as an original way of promoting sustainable development strategies based on regional agricultural and agri-food resources [41]. Their participatory approaches contribute to the economic, environmental, and social balance of the territories under a contractual charter signed by the stakeholders. Regional parks generally promote the quality of the landscape and protect small farms, who can add value to their food products through the regional nature park quality label. Therefore, the hypothesis underlying our study is that beef feeding areas located within the perimeters of a regional or a national nature park are more likely to be oriented toward short supply chains.

2.4.5. Beef Self-Sufficiency Ratio

We estimated the quantity of beef produced in the 100 km-radius foodshed. Extracting from RGA 2010 the number of beef per municipality, we multiplied this by an average load of one bovine livestock unit (LSU) per hectare. The LSU is a reference unit for aggregating livestock of different species and ages using specific coefficients initially established on the basis of the nutritional or feed requirements of each type of animal (source Eurostat). To calculate the kg carcass equivalent, we applied a yield of 0.74 tons per hectare [11]. The yearly consumption of bovine meat was estimated by multiplying the number of inhabitants (INSEE 2014) by the kg carcass equivalent of the 2018 bovine meat consumption per capita (http://www.fao.org/faostat/en/#data/FBS, accessed on 14 March 2022). The beef self-sufficiency ratio is therefore the ratio of estimated beef production to estimated consumption in the 100 km-radius foodshed, calculated by province (similar to NUTS-3 level), as follows:

$$\text{Beef self} - \text{sufficiency ratio} = \frac{\text{beef feeding area xyield}}{\text{number of inhabitants} \times \text{average consumption per capita/year}}$$

with beef feeding area = number of bovine livestock × (1 hectare/livestock unit).

3. Results

3.1. Short Supply Chains' Higher Contour Rugosity Than Long Supply Chains

We investigated whether the rugosity of the contours defined by the Gravelius indicator K is a consequential variable distinguishing between short and long supply chains. When this indicator was calculated in the 100 km-radius foodshed, the average K value of entities (isolated patches and archipelago) in short supply chains (SSC) was slightly higher than in long supply chains (LSC) (Table 2). We conclude that contour rugosity is informative on whether beef feeding areas are functionally connected to nearby beef consumption areas, thereby confirming the hypothesis defined in Section 2.4.2.

Table 2. Average rugosity of entities constituting beef feeding areas.

Surface (ha)	K (SSC)	K (LSC)
<50	1.22	1.21
>50	2.19	2.10
>100	2.45	2.32
>300	3.02	2.82
>500	3.36	3.01
>700	3.59	3.29
>1000	3.36	3.01

SSC: short supply chains; LSC: long supply chains.

3.2. Short Supply Chains' Stronger Dominance Compared to Long Supply Chains

Dominance assessment showed twice as many patches oriented toward short supply chains (10458 patches) relative to long supply chains (5296). Both kinds of entities have similar median surface areas devoted to beef feeding (4.85 ha for SSC vs. 4.58 ha for LSC). Average density of beef feeding areas was almost identical for SSC and LSC (Table 3). However, the patches in short supply chains had larger surface areas on average (36 ha for SSC and 28 ha for LSC), which may indicate strong connectivity between the beef feeding areas selling their production in SSC because they are located close to each other. This connectivity may be accentuated by a neighborhood effect, with breeders in short supply chains creating social links and exchanging best practices [30]. Nevertheless, these results should be considered as an overall trend and verified against expert opinion, given that the SSC variable in our study was estimated using census data at the municipal level due to the lack of available data at a finer scale.

Table 3. Density of patches in SSC and LSC.

Area (Hectares)	Density SSC (Percentage)	Density LSC (Percentage)	Area SSC (Hectares)	Area LSC (Hectares)
<50	31.15	31.04	3.33	3.26
>50	68.71	69.48	246.25	174.43
>100	73.32	74.63	410.33	278.06
>300	77.93	79.93	975.01	602.63
>500	78.86	82.67	1389.48	801.39
>700	77.87	80.91	1791.04	973.05
>1000	77.71	82.67	2289.74	1208.73

3.3. Short Supply Chains' Spatial Characteristics

The proportion of areas in SSC to those in LSC was greater within the perimeters of nature parks (regional and national) than in the whole of the study area (Table 4). Furthermore, beef feeding areas selling their production in short supply chains were closer to slaughterhouses than those selling in long supply chains. Pastoral areas operating in short supply chains were on average 24.888 km from the nearest slaughterhouse, whereas those operating in long supply chains were on average 27.294 km from the nearest slaughterhouse.

Table 4. Proportion of beef feeding areas in SSC and LSC for areas located within a national or regional nature park of the study area.

	SSC Beef Feeding Areas (Hectares)	LSC Beef Feeding Areas (Hectares)	Ratio SSC/LSC Areas
Within a nature park in the study area	119,160	38,236	3.11
Total study area	264,953	107,058	2.5

3.4. Beef Self-Sufficiency Ratio Estimates

Finally, we estimated the quantity of beef produced in the 100 km-radius foodshed (Table 5). In the municipalities located in a province with a strong beef production tradition (Drôme and Ardèche), production capacity largely exceeds local consumption, and therefore there is a very high ratio of beef self-sufficiency (1352% and 139%, respectively). Other provinces with less of a beef production tradition (e.g., Var) are dependent on external beef supply. It should be noted that these results confirm those of previous studies [11]. Table 6 summarizes the geographical factors that characterize the beef feeding areas oriented toward short supply chains.

Table 5. Estimation of beef self-sufficiency ratio.

Province	Beef Area (Hectares)	Estimated Beef Production (Tons) [a]	Estimated Beef Consumption, by Year (Tons) [a]	Population	Estimated Beef Self-Sufficiency [b]
Alpes de Haute Provence	714	528	2286	110,466	23%
Hautes Alpes	306	226	381	18,437	59%
Ardèche	6892	5100	3674	177,552	139%
Bouche du Rhone	12,161	8999	40,768	1,970,436	22%
Drôme	4385	3245	4343	209,911	75%
Gard	4581	3390	15,163	732,863	22%
Hérault	642	475	13,085	632,437	444%
Lozère	2733	2022	150	7231	1352%
Var	0	0	1258	60,793	0%
Vaucluse	282	209	11,470	554,393	2%
Total	32,696	24,194	82,255	4,474,519	

[a] The production and consumption of beef was estimated in carcass weight. [b] The beef self-sufficiency ratio was the ratio of estimated beef production to estimated consumption in the 100 km-radius foodshed, calculated by province.

Table 6. Factors that characterize the beef feeding areas oriented toward short supply chains. The soundness of every factor is scored ("+++" high impact to "+" very low impact).

IMPACT	FACTOR
++	Rugosity
+++	Dominance
++	Location within Nature Park
+	Distance from Slaughterhouse

4. Discussion

4.1. Determinants of the Spatial Signature of Beef Feeding Areas Oriented toward Short Supply Chains

Our results confirm the initial hypotheses that the rugosity of beef feeding areas is informative on the orientation of food production toward short supply chains [39,42]. On the other hand, there were more patches in short supply chains (SSC) than in long supply chains (LSC). One explanation may be increasing urban demand for food grown "close to home", interacting with the processes of rural restructuring to foster small-scale farming and its direct food linkages to cities [43]. In addition, the surface areas of the patches in SSC were larger. This result could be an artifact caused by the dilation/erosion methodology used to aggregate the patches: those of the farms in SSC are close to each other and those in LSC are more isolated. It would be interesting in further work to use other spatial analysis methodologies such as Moran's I or LISA [44] to compare the results obtained here.

In addition, another research avenue would be to analyze the neighborhood effects over small surface areas to better understand how landscape pattern, and in particular, fragmentation (in the sense of landscape ecology), impacts the functioning of the landscape (i.e., agricultural activities on farms). Small surfaces areas with high rugosity can be used for extensive feeding (mountain pastures), a good use for areas that are far from the farm or that are too fragmented [45]. Moreover, neighborhood effects should take into account social relations between producers and between producers and urban demand, since an important driver of the archipelago structure is the supply chain.

This work was limited by a lack of sources of statistics on short supply chains at a finer scale than the municipality. As a result, we may be overestimating the beef feeding areas in SSC. Indeed, we assigned to SSC all the cattle feeding areas of any municipality that had at least one beef livestock farmer who declared marketing via SSC (RGA 2010). Our research perspectives included working at a finer scale based on quantitative field surveys and expert opinion, coupled with data from the upcoming RGA 2020, which will be available in 2022. Analyzing this database would also enable us to identify the part of the land in the Bouches du Rhone that is used for raising bulls for bullfighting, currently included in the "cattle" section of the RGA in the same way as beef cows. In addition, a field survey would make it possible to specify the type of beef cattle breeding (Charolais, Limousine) and refine production estimates (e.g., yield/carcass).

Regarding the effect of public policies on the orientation of land use toward SSC, we analyzed the effect of being located within a regional or national nature park. Our results showed that there were 3.11 times more areas under SSC than under LSC inside parks (see Table 4). This may be due to both the parks' actions in support of SSC (e.g., supplying public school canteens with local food products, promoting food quality labels, organizing farmers' markets) and to the territorial dynamics of proximity that the parks promote [4,24].

Finally, distance from slaughterhouses seems to be a factor explaining orientation toward SSC. This should be further addressed by research examining the typology of slaughterhouses (small versus large) and the differences in slaughtering costs. In addition, the possible introduction of mobile slaughterhouses currently being discussed by stakeholders (chamber of agriculture, livestock associations) would likely impact orientation toward short supply chains, attracting small farms and isolated cattle farms in particular. An interesting future extension would therefore be to compare our approach with the stakeholders' expertise by means of a participatory process. It should be noted that the effect of distance from consumption points was not analyzed here, since for beef and for the study area, the average distance in short supply chains is 200 km (expert opinion).

4.2. *What Role and Leverage for Public Action?*

Our findings revealed that the analysis of local food self-sufficiency must be shifted from foodshed size assessment to a commodity–group-specific spatial configuration of a complex of complementary components, the so-called foodshed archipelago (Figure 4). These results specifically question the theoretical agricultural land use model by von Thünen, where the type of agriculture is determined by the distance to the city center [46] (based on circumcentric rings). Biophysical features, for instance, soil fertility, are very often not distributed around the urban area in a gradient [7]. Furthermore, the spatial distribution of agricultural production also responds to socioeconomic features resulting from the particular history of each place in terms of its urbanization, development of the agricultural sector, organization of activities, and environmental protection, as we have shown in previous work [5,47]. Actually, the multilevel aspect of food systems remains a remarkable scientific challenge to integrate the stakeholders' local vision and global statistical data and thus tailor regional food security-oriented policies [48]. One promising research avenue to help fill this gap is the partial least squares-path modeling methodological approach (PLS-PM), which makes it possible to analyze complex relationships between the socio-economic structure of farms, demography, landscape structure, and landscape management and function, in order to characterize the spatial signature of peri-urban agriculture [6]. Beyond the methodological contributions, the findings can be used to inform public decisions.

What leverage is there for public action to reconnect beef production areas to consumption areas (the city) through short supply chains? The obvious direction is using development initiatives to increase the connectivity of beef feeding areas (e.g., land acquisition to install new breeders) and rugosity (e.g., protection of small pastoral areas on the outskirts of the city). In addition, public action can play a decisive role in fostering short supply chains through nature parks, as we have seen above. Moreover, public procurement (e.g., local food public procurement for school canteens) can promote local food supply chains by encouraging producer groups, developing partnerships with intermediaries (e.g., butchers for custom cutting), and securing outlets under contract for part of the production [30,49].

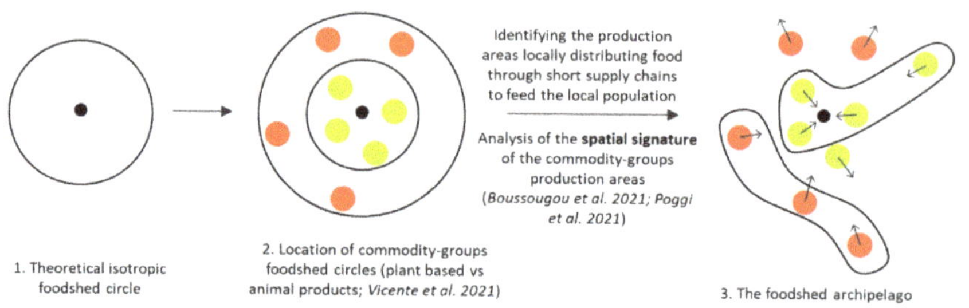

Figure 4. Three-step methodological approach used to shift the focus of a foodshed analysis from an assessment of its size (isotropic circle) to its assessment as a complex of complementary pieces (foodshed archipelago), Vicente et al. 2021 [7]; Boussougou et al. 2021 [6]; Poggi et al. [18].

Furthermore, from a regional food security perspective, even if all the arable land oriented toward the production of food sold in short supply chains (see Table 5) were used for beef production, none of the provinces in our study area, except for Lozère and Ardèche, would be self-sufficient. Would it be possible (and desirable) to encourage farmers to redirect certain pastoral areas (e.g., those used for leisure activities involving horses) to beef feeding, in order to produce beef to feed the city? In the end, our results show that the spatial arrangement of areas is also an important consideration, to be added to the biophysical and agro-climatic conditions such as altitude, hygrometry, and soil characteristics. It would be interesting to explore whether the foodshed approach—based on the concept of sustainable city-region food systems—could be integrated more intensively into food policies to sustainably increase food self-sufficiency at the regional level [50].

5. Conclusions

The paper attempts to delimit and characterize the foodshed using concepts from landscape ecology (rugosity, connectivity, patches, and the archipelago) as applied to beef supply chains. We discriminated between beef feeding areas oriented toward short supply chains (SSC) and those oriented toward long supply chains (LSC) using available statistical data to confirm or challenge our hypothesis of a particular spatial signature of agricultural areas oriented toward SSC. Our results show that the beef feeding areas in SSC have a particular spatial arrangement: they are small patches very closely situated (<20 m) and connected to each other, forming large areas with high-rugosity contours. This confirms the hypothesis of a spatial signature of areas in SSC. In other words, the functioning and management of the landscape are translated in space into particular spatial structures whose arrangement is identifiable, as our previous work has shown [4–6,18].

These areas composed of small, connected patches contribute more to city food supply than isolated patches due to their functional connection in short supply chains. These results, although they do not call into question the productive capacity of isolated farms, are relevant in terms of food security at a regional level from a food planning perspective. By revealing the positive impact of nature parks on the existence of short supply chains, we have shown the decisive role that public action can play.

Author Contributions: Conceptualization, M.M. and E.S.S.; Validation, M.M., E.S.S., C.N. and J.L.V.-V.; Formal analysis, D.J., J.L.V.-V., M.M. and E.S.S.; Resources, M.D.; Writing—original draft preparation, M.M, E.S.S. and L.M.; Writing—review and editing, M.M. and E.S.S.; Supervision, M.M., E.S.S., C.N. and D.J. All authors have read and agreed to the published version of the manuscript.

Funding: This work was carried out as part of the ongoing project FOODSHIFT2030, funded from the European Union's Horizon 2020 Research and Innovation Program under grant agreement number 862716. The project (Grant Agreement 862716, H2020-SFS-2018-2020/H2020-SFS-2019-1) on transforming the European food system towards a low carbon circular future applies a string of mechanisms for maturing, combining, upscaling and multiplying sustainable food system innovations. The project has received EUR 7.5 million from the EU Horizon 2020 programme.

Institutional Review Board Statement: Not applicable.

Informed Consent Statement: Not applicable.

Acknowledgments: The authors wish to thank the anonymous reviewers for their diligent work and their helpful advice. Our thanks to Avignon University, UMR 7300, Laboratoire ESPACE for providing the resources of the ESPACE Laboratory and Avignon University as well as to Guillaume Ollivier (INRAE—UR Ecodéveloppement) for his remarks. We are grateful to the Urbansimul team for their contribution on dilation/erosion (computer code of the Urbansimul 2016 project), hal-01604422. We thank Catherine Brinkley (Department of human ecology, University of California, Davis), Jacques Lasseur (INRAE, UMR SELMET), Noé Guiraud, (Urban School of Lyon), Albert Alameldine, REGAL network project manager, and Hugues Fortuna, manager of the Avignon public schools central kitchen. We thank Marjorie Sweetko for the English language editing.

Conflicts of Interest: The authors declare no conflict of interest.

References

1. IPES-Food. *COVID-19 and the Crisis in Food Systems: Symptoms, Causes, and Potential Solutions*; International Panel of Experts on Sustainable Food Systems: Brussels, Belgium, 2020; p. 11.
2. United Nations Organisation (UNO). *World Urbanization Prospects: The 2014 Revision*; UNO. Department of Economic and Social Affairs, Population Division: New York, NY, USA, 2014; p. 517.
3. Moragues-Faus, A. Distributive Food Systems to Build Just and Liveable Futures. *Agric. Hum. Values* **2020**, *37*, 583–584. [CrossRef] [PubMed]
4. Sanz, E.S. *Planification Urbaine et Agriculture. Méthodologie Systémique de Caractérisation de L'agriculture Périurbaine à Partir d'une Recherche Empirique En France et En Espagne. Thèse Pour L'obtention du Titre de Docteur en Géographie (UAM) et en Etudes Urbaines (EHESS)*; EHESS/UAM: Paris, France, 2016.
5. Sanz, E.S.; Martinetti, D.; Napoléone, C. Operational Modelling of Peri-Urban Farmland for Public Action in Mediterranean Context. *Land Use Policy* **2018**, *75*, 757–771. [CrossRef]
6. Boussougou, G.B.; Sanz, E.S.; Napoléone, C.; Martinetti, D. Identifying Agricultural Areas with Potential for City Connections: A Regional-Scale Methodology for Urban Planning. *Land Use Policy* **2021**, *103*, 105321. [CrossRef]
7. Vicente-Vicente, J.L.; Sanz-Sanz, E.; Napoléone, C.; Moulery, M.; Piorr, A. Foodshed, Agricultural Diversification and Self-Sufficiency Assessment: Beyond the Isotropic Circle Foodshed—A Case Study from Avignon (France). *Agriculture* **2021**, *11*, 143. [CrossRef]
8. Hedden, W.P. *How Great Cities Are Fed*; D. C. Heath & Cp.: Boston, MA, USA, 1929.
9. Kloppenburg, J.; Hendrickson, J.; Stevenson, G.W. Coming in to the Foodshed. *Agric. Hum. Values* **1996**, *13*, 33–42. [CrossRef]
10. Butler, M. *Analyzing the Foodshed: Toward a More Comprehensive Foodshed Analysis*; Portland State University: Portland, OR, USA, 2000.
11. Feagan, R. The Place of Food: Mapping out the "local" in Local Food Systems. *Prog. Hum. Geogr.* **2007**, *31*, 23–42. [CrossRef]
12. Peters, C.J.; Bills, N.L.; Wilkins, J.L.; Fick, G.W. Foodshed Analysis and Its Relevance to Sustainability. *Renew. Agric. Food Syst.* **2009**, *24*, 1–7. [CrossRef]
13. Aubry, C.; Kebir, L. Shortening Food Supply Chains: A Means for Maintaining Agriculture Close to Urban Areas? The Case of the French Metropolitan Area of Paris. *Food Policy* **2013**, *41*, 85–93. [CrossRef]
14. Griffin, T.; Conrad, Z.; Peters, C.; Ridberg, R.; Tyler, E.P. Regional Self-Reliance of the Northeast Food System. *Renew. Agric. Food Syst.* **2015**, *30*, 349–363. [CrossRef]
15. Tichenor, N.E.; van Zanten, H.H.E.; de Boer, I.J.M.; Peters, C.J.; McCarthy, A.C.; Griffin, T.S. Land Use Efficiency of Beef Systems in the Northeastern USA from a Food Supply Perspective. *Agric. Syst.* **2017**, *156*, 34–42. [CrossRef]
16. Li, K.; Jin, X.; Ma, D.; Jiang, P. Evaluation of Resource and Environmental Carrying Capacity of China's Rapid-Urbanization Areas—A Case Study of Xinbei District, Changzhou. *Land* **2019**, *8*, 69. [CrossRef]
17. Zasada, I.; Schmutz, U.; Wascher, D.; Kneafsey, M.; Corsi, S.; Mazzocchi, C.; Monaco, F.; Boyce, P.; Doernberg, A.; Sali, G.; et al. Food beyond the City—Analysing Foodsheds and Self-Sufficiency for Different Food System Scenarios in European Metropolitan Regions. *City Cult. Soc.* **2019**, *16*, 25–35. [CrossRef]
18. Poggi, S.; Vinatier, F.; Hannachi, M.; Sanz Sanz, E.; Rudi, G.; Zamberletti, P.; Tixier, P.; Papaïx, J. How Can Models Foster the Transition towards Future Agricultural Landscapes? *Adv. Ecol. Res.* **2021**, *64*, 305–368. [CrossRef]
19. Pulliam, H.R. Sources, Sinks, and Population Regulation. *Am. Nat.* **1988**, *132*, 652–661. [CrossRef]
20. Décamps, H.; Décamps, O. Organisation de l'espace et processus écologiques. *Economierurale* **2007**, *297*, 41–54. [CrossRef]
21. Taylor, P.D.; Fahrig, L.; Henein, K.; Merriam, G. Connectivity Is a Vital Element of Landscape Structure. *Oikos* **1993**, *68*, 571. [CrossRef]
22. Taylor, P.D.; Fahrig, L.; With, K.A. Landscape connectivity: A return to the basics. In *Connectivity Conservation*; Crooks, K.R., Sanjayan, M., Eds.; Cambridge University Press: Cambridge, UK, 2006; pp. 29–43. ISBN 978-0-511-75482-1.

23. Torre, A.; Wallet, F. Introduction: The role of proximity relations in regional and territorial development processes. In *Regional Development and Proximity Relations*; Edward Elgar Publishing: Cheltenham, UK, 2014; pp. 1–44. ISBN 978-1-78100-289-6.
24. Angeon, V.; Bertrand, N. Les Dispositifs Français de Développement Rural: Quelles Proximités Mobilisées ? *Géographie Économie Société* **2009**, *11*, 93–114. [CrossRef]
25. Hinrichs, C.C. Embeddedness and Local Food Systems: Notes on Two Types of Direct Agricultural Market. *J. Rural Stud.* **2000**, *16*, 295–303. [CrossRef]
26. Brinkley, C. Visualizing the Social and Geographical Embeddedness of Local Food Systems. *J. Rural Stud.* **2017**, *54*, 314–325. [CrossRef]
27. Chiffoleau, Y. From Politics to Co-Operation: The Dynamics of Embeddedness in Alternative Food Supply Chains. *Sociol. Rural.* **2009**, *49*, 218–235. [CrossRef]
28. Guiraud, N.; Laperrière, V.; Rouchier, J. Une Géographie Des Circuits Courts En Région Provence-Alpes-Côte d'azur. *L'Espace Géographique* **2014**, *4*, 356–373. [CrossRef]
29. Confédération paysanne Provence-Alpes-Côte-d'Azur. *Confédération paysanne Provence-Alpes-Côte-d'Azur*; Séminaire "L'abattage de Proximité"; Confédération paysanne Provence-Alpes-Côte-d'Azur: Buoux, France, 2019; p. 23.
30. Agence Nationale de la Cohesion des Territoires, Massif des Alpes. *Compte Rendu Du Séminaire "Valoriser Les Viades de Montagne Massif Des Alpes"*; Chanmbre d'Argiruclure à Grenoble: Grenoble, France, 2020.
31. Confédération paysanne Provence-Alpes-Côte-d'Azur. *Région PACA en région Provence-Alpes-Côte-d'Azur Contexte de l'abattage En Région Provence-Alpes-Côte-d'Azur*; Presented at the Séminaire "L'abattage de proximité"; Confédération paysanne Provence-Alpes-Côte-d'Azur: Buoux, France, 2019.
32. Confédération paysanne Provence-Alpes-Côte-d'Azur. *Région Auvergne Rhône-Alpes, Région SUD Présentation Du Projet "Enjeux de Structuration et de Développement Des Filières Viande Du Massif Alliant Valorisation Du Produit Pour l'éleveur et Apport d'un Produit Local de Qualité Au Consommateur"*; Presented at the Séminaire "Valoriser les viandes de montagne"; Confédération paysanne Provence-Alpes-Côte-d'Azur: Grenoble, France, 2020.
33. Girardet, X.; Foltête, J.-C.; Clauzel, C.; Vuidel, G. Restauration de la connectivité écologique: Proposition méthodologique pour une localisation optimisée des passages à faune. *Vertigo* **2016**, *24*, 17337. [CrossRef]
34. Guérois, M. *Les Formes Des Villes Européennes Vues Du Ciel. Une Contribution de l'image CORINE Land Cover à La Comparaison Morphologique Des Grandes Villes d'Europe Occidentale*; Université Panthéon-Sorbonne-Paris I: Paris, France, 2003.
35. Amsallem, J.; Deshayes, M.; Bonnevialle, M. Analyse comparative de méthodes d'élaboration de trames vertes et bleues nationales et régionales. *Sci. Eaux Territ.* **2010**, *3*, 40. [CrossRef]
36. Geniaux, G.; Leroux, B.; Samaké, A.D.; Cezanne-Bert, P.; Moulery, M.; Code Informatique Du Projet Urbansimul 2016. 2016, hal-01604422. Available online: https://hal.archives-ouvertes.fr/hal-01604422 (accessed on 14 March 2022).
37. Du Gerny, S.; Les Abattoirs de Proximité de plus En plus Sollicités Par Les Petits Producteurs. Les Echos 2015. Available online: https://www.lesechos.fr/2015/09/les-abattoirs-de-proximite-de-plus-en-plus-sollicites-par-les-petits-producteurs-253395 (accessed on 10 March 2022).
38. Dustan, P.; Doherty, O.; Pardede, S. Digital Reef Rugosity Estimates Coral Reef Habitat Complexity. *PLoS ONE* **2013**, *8*, e57386. [CrossRef]
39. Brinkley, C. High Rugosity Cities: The Geographic, Economic and Regulatory Pathology of America's Most Non-Concentric Urban Areas. *Land Use Policy* **2018**, *73*, 215–224. [CrossRef]
40. Barké, M.K.; Oussein, I.; Bielders, C.; Ambouta, K.J.M.; Tychon, B. Caractérisation morphologique des cuvettes oasiennes du Centre-Est du Niger. *Physio-Geo* **2017**, *11*, 255–276. [CrossRef]
41. Angeon, V.; Boisvert, V.; Caron, A. La marque « Parc naturel régional »: Un outil au service d'un développement local durable et un modèle pour les pays du Sud ? *Afr. Contemp.* **2007**, *222*, 149. [CrossRef]
42. Forman, R.T.T. *Urban Regions: Ecology and Planning beyond the City*; Harvard University: Cambridge, MA, USA, 2008; ISBN 10.1017/CBO9780511754982.
43. Jarosz, L. The City in the Country: Growing Alternative Food Networks in Metropolitan Areas. *J. Rural Stud.* **2008**, *24*, 231–244. [CrossRef]
44. Anselin, L. Local Indicators of Spatial Association—LISA. *Geogr. Anal.* **1995**, *27*, 93–115. [CrossRef]
45. FiBL. *Engraissement Au Pâturage*; FiBL: Frick, Switzerland, 2001.
46. Von Thünen, J.H. *Von Thunen's Isolated State*; Pergarmon Press: Glasgow, UK, 1826.
47. Sanz, E.S.; Napoléone, C.; Hubert, B. A Systemic Methodology to Characterize Peri-Urban Agriculture for a Better Integration of Agricultural Stakes in Urban Planning. *L'espace Géographique* **2017**, *46*, 174–190.
48. Sanz, E.S.; Napoléone, C.; Mouléry, M. Agroecological Transitions and Local Food Self-Sufficiency Assessment. From the Isotropic Circle to the Archipelago Foodshed. *Agroecol. Transform. Sustain. Food Syst. Insight Fr. CGIAR Res.* **2021**, *98*, 23708.
49. Sanz, E.S. Public procurement as a booster of midscale food supply chains. In *Public Food Procurement for Sustainable Food Systems and Healthy Diets*; Swenson, L., Hunter, D., Schneider, S., Eds.; FAO: Rome, Italy, 2022.
50. Gugerell, C.; Sato, T.; Hvitsand, C.; Toriyama, D.; Suzuki, N.; Penker, M. Know the Farmer That Feeds You: A Cross-Country Analysis of Spatial-Relational Proximities and the Attractiveness of Community Supported Agriculture. *Agriculture* **2021**, *11*, 1006. [CrossRef]

agriculture

MDPI

Article

Nature's Contributions to People Shape Sense of Place in the Coffee Cultural Landscape of Colombia

Beatriz E. Murillo-López [1,*], Antonio J. Castro [2,3] and Alexander Feijoo-Martínez [1]

1 Faculty of Environmental Sciences, Technological University of Pereira, Pereira 660003, Colombia; afeijoo@utp.edu.co
2 Andalusian Centre for the Assessment and Monitoring of Global Change (CAESCG), Department of Biology and Geology, University of Almeria, 04120 Almeria, Spain; acastro@ual.es
3 Department of Biological Sciences, Idaho State University, Pocatello, ID 83209, USA
* Correspondence: betymu@utp.edu.co; Tel.: +57-30-0652-3575

Abstract: Understanding the roots of a sense of place in farmlands is crucial for stopping rural exodus to urban areas. Farmers' experiences related to their way of life, peace and quiet, rootedness, pleasure, and inspiration are fundamental components of a sense of place in farmlands. Here, we used the city of Pereira located in the Coffee Cultural Landscape of Colombia (CCLC) to examine the role of nature's contributions to people (NCP) in forming meanings and attachments that shape their sense of place to this region. This region has experienced intense agricultural lands abandonment due to rapid urbanization over the last decades. To do so, a mixture of qualitative and quantitative methods was used, including semi-structured interviews, observation, and dialogue, to capture farmers' perceptions and emotions associated with farmlands, reasons for remaining, and the diversity of NCPs. Results indicated that farmers recognized farmlands as a quiet and safe space that support family cohesion. Results also showed that the characteristics of the farms (e.g., agricultural practices, distance to cities, and gender) play an important role in articulating a farmer's attachment to farmlands. Finally, farmers identified nonmaterial NCP (e.g., physical and psychological experiences and supportive identities) to be the most important contributions for shaping their sense of place. We call for the need to include robust and transparent deliberative and negotiation mechanisms that are inclusive of all relevant stakeholders, to aim to address unequal power, and to recognize and strengthen communities' mechanisms of action on the CCLC.

Keywords: socioecological systems; local identity; rural abandonment; agroecology; world heritage site

Citation: Murillo-López, B.E.; Castro, A.J.; Feijoo-Martínez, A. Nature's Contributions to People Shape Sense of Place in the Coffee Cultural Landscape of Colombia. *Agriculture* **2022**, *12*, 457. https://doi.org/10.3390/agriculture12040457

Academic Editors: José Luis Vicente-Vicente, Cristina Quintas-Soriano and María D. López-Rodríguez

Received: 9 February 2022
Accepted: 18 March 2022
Published: 24 March 2022

1. Introduction

According to the latest report of the Intergovernmental Science-Policy Platform on Biodiversity and Ecosystem Services (IPBES), the supply of food, energy, and materials to human communities is increasing at the expense of nature's capacity to provide, producing drastic effects on ecosystems that sustain livelihoods [1]. In the processes of human use and modification of nature's resources, relationships between people and lands are formed and evolve over time, shaping cultural roots to the land. Understanding this human–nature relationship requires approaches that capture factors that articulate a sense of place, including meaning, attachment, characteristics of places, the complexity of environmental values, and individual experiences within the landscape [2].

The transformation of the ecosystems in the central Andes of South America has configured in the Colombian coffee-growing region environments in which the cultivation of diverse varieties of coffee has predominated, which have given rise to exports to international markets [3–5]. Traditional coffee crops are accompanied by multiple subsystems that form mosaics and patches between successions of natural vegetation, riparian areas close to bodies of water, *Guadua angustifolia* and the predominance of cultivated plants as companions of the systems, which are friendly to the conservation of the biodiversity of the macrofauna of the soil [6]. However, the intense use and transformation of the traditional farming and natural system (gallery and/or riparian forest and bamboo forest) in favor of urban expansion (discontinued urban fabric) is producing a decline of traditional farmlands systems (traditional coffee and plantain crops) and their biodiversity [7], thus altering the sustainable way of living of rural communities [8].

This context of the Coffee Cultural Landscape of Colombia (CCLC) led to its declaration in 2011 as a World Heritage Site by the United Nations Educational, Scientific and Cultural Organization (UNESCO). The CCLC is considered a landscape that should be prioritized for preservation because of its tangible and intangible significance to the territory, and it is at risk of losing its unique sociocultural roots that rural families have formed with traditional farming systems present there [9]. Among the major risks are urban expansion (e.g., construction of condominiums increases the discontinuous urban fabric) and the intensification of the agriculture (e.g., cattle pastures and plantain and avocado monocultures), which have caused a simplification and homogenization of the landscape, displacing agricultural lands with traditional uses and their communities, leading to the loss of agricultural culture, biodiversity, and sense of place [9]. Together, these land transformations have particularly changed the agricultural practices of the city of Pereira located in the western foothills of the Cordillera Central above the Cauca River valley.

The most dominant farming practices in the CCLC are peasant and semi-industrial styles. The semi-industrial style centralizes labor productivity and growth, mainly based on the mobilization of external resources, which leads to a disconnection between traditional farming and nature, while the peasant style focuses on autonomy, family labor, and self-controlled resources that depend on the sustainable use of ecological capital [10,11]. These farming styles highlight the different ways in which farmers relate to farm resources and production as a business, as well as provide care for families [11,12]. These farming styles also differ in the environmental pressure they place on ecosystems and in the diversity of nature´s contributions (NCP) they provide to people [13].

This investigation adds to the growing body of research addressing the connection between people and nature through the assessment of how NCP shape a sense of place in rural settings (Figure 1). Our intention is to understand the relationship between farming style, sense of place, and NCP, because these concepts are solidly rooted in cultural repertoires. To advance this aim, it is necessary to not only recognize and integrate the characteristics of farmers and farms, but also to explore rootedness, security, and feelings associated with farmlands [14]. By addressing these factors, it will be possible to provide a better and more informed guidance in the future on sustainable land management in these areas [15].

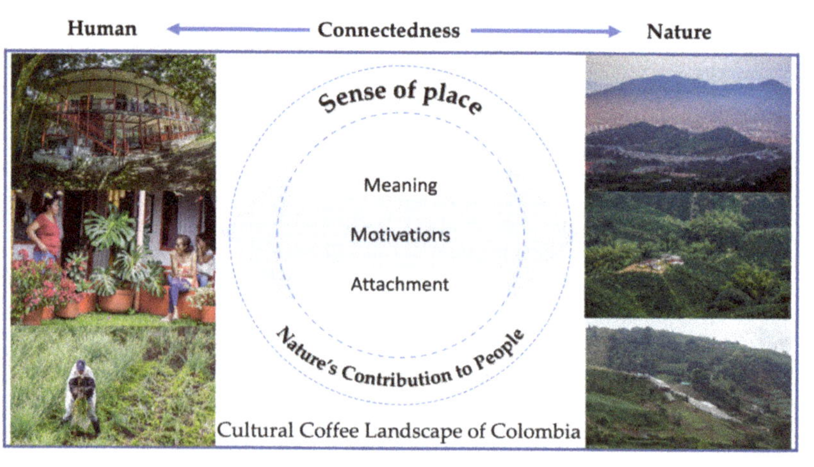

Figure 1. Conceptual framework of a sense of place through the NCP lens.

Sense of place is defined as a motivation for stewardship and actions to care for the environment and use the resources it provides. It is also presented as a cognitive and emotional variable that mediates how people respond to social-ecological change [16,17]. The human–nature relationship is nonlinear and often depends on the formations of relational values, i.e., values that arise from a relationship with nature, encompassing a sense of place, feelings of well-being (mental and physical health), and cultural, community, or personal identities [18–21]. Farmers have a complex relationship with farmlands as they have the ability to read nature and make decisions to protect or use resources. Additionally, farms are multifunctional landscapes (e.g., areas production, conservation, and relaxing zone) that can be related to specific relational values of farmlands [12,22,23]. The CCLC is shaped by mosaics (e.g., patches of interconnected crops and natural areas) and are inhabited by rural families holding beliefs, attitudes, and social norms that create farmland with high cultural value. Sense of place in this region has been described as a wide range of connections between people and places that develop based on the place meanings and attachment a person has for a particular setting [16,24]. We integrated the concept of nature's contributions to people (NCP) framework developed by IPBES to capture a broad range of worldviews, knowledge systems, and stakeholders. The NCP approach recognizes the central and pervasive role that culture plays in defining all links between people and nature [21], and the importance of local knowledge for understanding meanings, motivations, and attachment to agricultural landscapes (Figure 1).

Within this context, this study aims to examine the role of NCP in shaping the sense of place of farmers in the CCLC. Specifically, we focused on examining the role of meanings, attachments, values, and connection associated with nature in shaping the sense of place to this region. To do so, a mixture of qualitative and quantitative methods were used to (i) characterize the diversity of farmers and farms of a case study located in the CCLC; (ii) examine the diversity of emotions associated with farmlands, as well as sociodemographic factors that explain them; (iii) explore the sense of place of local communities through exploring motivations to remaining in the region; (iv) identify the diversity of nature's contributions to people that articulated farmers´ sense of place; (v) to explore the visions of local communities regarding the future of the CCLC.

2. Materials and Methods

2.1. Study Area: The Coffee Cultural Landscape of Colombia

The study was conducted in the rural area of the city of Pereira, Risaralda, Colombia, located between 4°43′4.8″ N and 75°50′38.4″ W and 4°52′15.6″ N and 75°36′18″ W. The farms are located between 1221 and 1922 m.a.s.l. (meters above sea level) (Figure 2). The average temperature is 21.2 °C; the average total annual rainfall was 2301 mm and the relative air humidity ranges yearly between 73 and 79% [25]. Pereira occupies an area of 607 km^2 and the approximate population is 467,269 inhabitants, of which 81,432 (17.4%) are residents of the rural area [26].

2.1.1. Land-Use and Land-Cover Change in the CCLC

Over the last three decades, significant changes in land use and cover have been documented in the CCLC affecting the agricultural production of coffee and other native crops. In 1997 the export in Colombia of agricultural products was 32.5% of the total exported; however, in 2011 it was reduced to 8.2% of the export of agricultural products [27]. Changes in land cover and urban expansion in the city of Pereira begin to show the decrease in lands used for coffee cultivation (from 1997 to 2014 it went from 10,706 ha to 5454 ha). Likewise, permanent crops decreased from 5747 ha to 3646 ha for the same period of analysis and transitory crops decreased by 214 ha [9], which placed more pressure in the rural sector due to the change in the type of agricultural production (i.e., pastures for cattle, industrial avocado cultivation) and livelihood of rural communities (i.e., land for human occupation—gentrification), thus influencing factors that shape the sense of place, identity, and heritage.

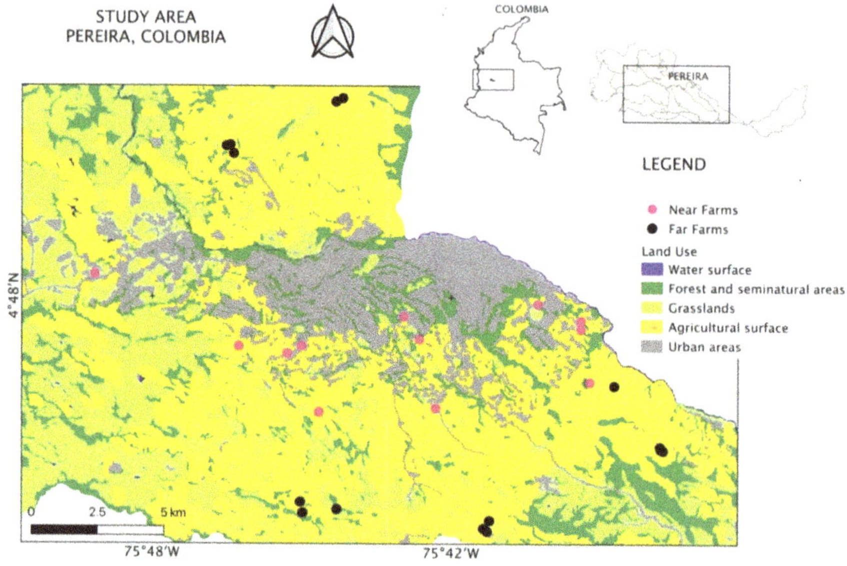

Figure 2. Geographic location of farms in the CCLC and current land cover and land-use type.

2.1.2. Farms Characteristics and Locations in the CCLC

Pereira is a municipality of the CCLC and extends through some of the coffee-producing areas at the foothills of the western and central mountain ranges of the Cordillera de los Andes. The characteristics of the area reflect the process of adaptation of coffee cultivation to the complex conditions imposed by the Colombian Andes [28]. The CCLC represents traditional forms of human settlements with small-to-medium-sized production units (between 0.5–2.6 ha), with steep slopes (15–50% inclination), elevation between 1000–2000 m.a.s.l., precipitation between 1600–2700 mm, and average temperature of 22.2 °C [9,29,30].

The CCLC is a continuously productive landscape that has shaped the cultural connection of rural communities to the land over decades. The coffee-growing families have mainly planted coffee, accompanied by subsistence crops (corn, beans, plantain, fruit trees) and with a low level of mechanization. The cultural practices have been passed down through generations and reflect a knowledge based on experience and understanding of the surroundings [8]. In addition, this small-scale production is distinguished by its family-based workforces, whereby the producer and family all work on the farm. Most families tend to live on the premises and so are able to constantly supervise their coffee plants and other crops. Only when the production cycle is at a peak are workers from outside the family hired—on a temporary basis—to help with harvesting [28,30,31]. The farm work is often built on the family farm by doing, making mistakes, correcting them by repeatedly reperforming the activities, and by observing and hearing experiences of neighboring farmers [8]. Farms are centers of (informal) education for families, mainly about crops, practices, and strategies, making the families and their farms into an expression of coffee culture.

Exploring the sense of place in the CCLC requires methodologies that can reveal meanings, attachments, relational, and historical values to these lands [32]. We selected 27 farms based on their proximity to agricultural areas of Pereira (i.e., no forest and seminatural areas, no artificial surfaces), primary productive activity (i.e., no livestock, no tourism), and farmer willingness to participate in this study (Figure 2).

2.2. Social Sampling Strategy

Farms were selected by the willingness and desire of rural families to provide information on the values and perceptions they hold in relation to farms and rural landscapes. This study conducted a qualitative research method through the use of semistructured interviews, in-person observation, and informal conversation with farmers [33,34] (Figure 3). The strength of these techniques lies in the creation of bonds of trust between farmers and interviewers to obtain information reflecting meanings and attachment to farmers' values related to the agricultural landscape. Since farmers are often heterogeneous in terms of their relationship with the environment, it was crucial to develop a relationship of trust. This method has been previously used to collect information about emotional connections to natural features and rural landscapes [8]. A total of 27 in-person interviews across all selected farms were conducted between August and December 2018.

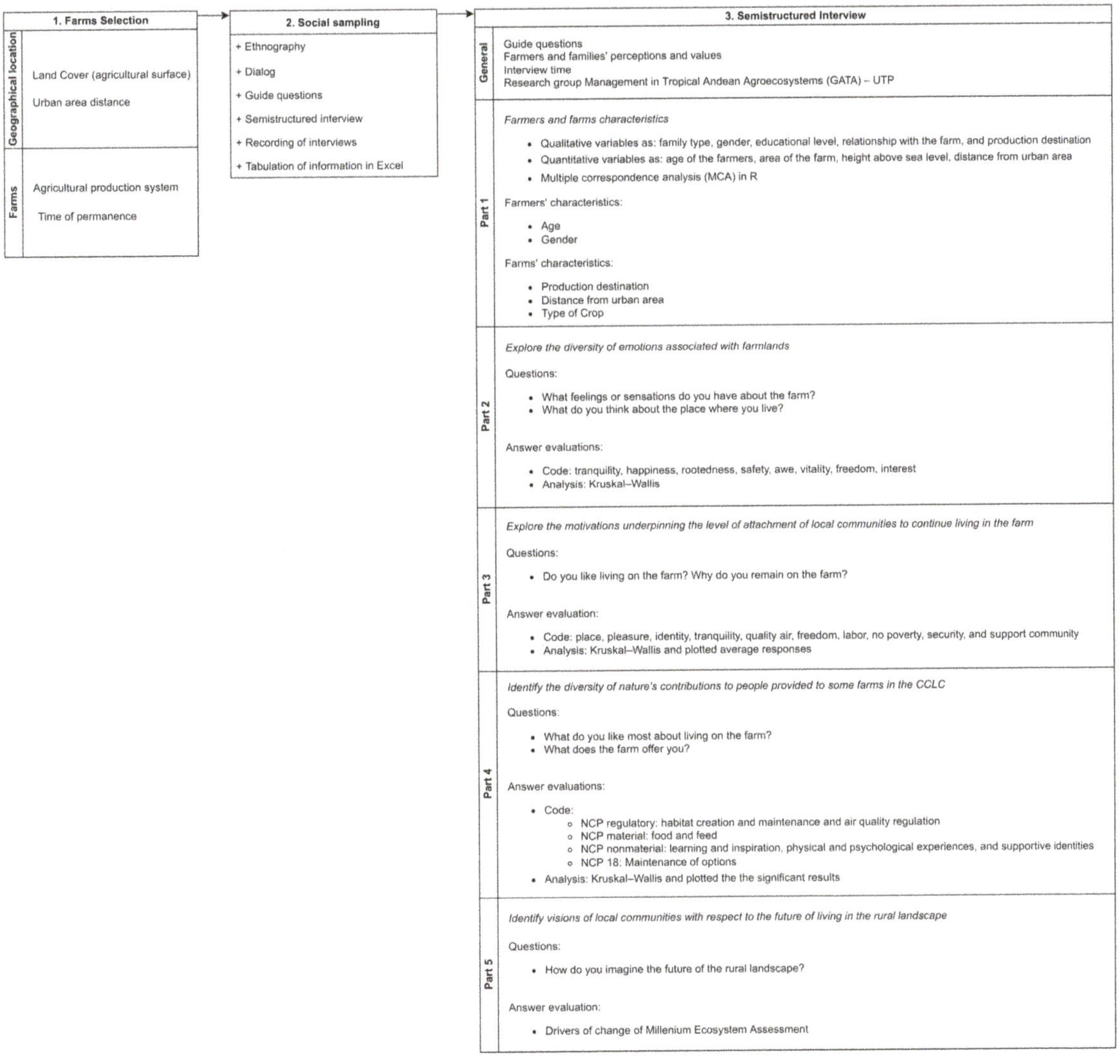

Figure 3. Methodological steps of the research approach.

2.3. Semistructured Questionnaire Design

The semistructured questionnaire was separated into five sections (Figure 3). In-person interviews were on average one and a half hours and were conducted by the research group Management in Tropical Andean Agroecosystems (GATA, Spanish acronym)—Technological University of Pereira (UTP). The semistructured interviews included open questions aiming to explore farmers and farm characteristics, as well as their perceptions and feelings associated to rural landscapes. In addition, farmer's motivations to remain in the farm and the diverse contributions (i.e., NCP) they perceived from the rural landscapes were explored. Once permission was obtained from the interviewees, each interview was recorded to facilitate the information collection of the interviewee's story and keep the details exactly as they were expressed [35].

2.3.1. Farmers and Farms' Characteristics

The questionnaire collected qualitative information such as family type (childless couples, nuclear, extended) [8], gender (female and male), educational level (primary school, high school, technical, technology, university degree), origin, type of relationship with farms (managers, owner-managers, owners, workers) and destination of crop production (sale and self-consumption, sale). Additionally, quantitative information related to the farmers' age, farm surface area, altitude, time of tenure and time spent on the farm were collected [7]. Farmers and the 27 farms' characteristics were classified based on the data provided by the interviewees and farms' information. We chose two ranges for each qualitative variable; the range was calculated by subtracting the minimum value from the maximum value of the data set, and this range was divided by two to classify farmers and farms according to the characteristics of the group (Figure 3).

A farming style is defined as a distinctive way of ordering the many sociomaterial interrelations involved in farming [11]. Each farming style is a description of the way farmers and rural families arrange the available resources (e.g., labor, land, input, and time) for the exploitation and replication of the production system [10,11,22,36–39]. Information collected from each farm was used to classify them as peasant or semi-industrial style. We used variables such as farmer's relationship to the farm and time living in the farm as well as farm surface, crop types, and which crops generate income; information related to the tenure of the farm, hiring personnel, and destination of the production were taken into account.

Farms were classified as near or far from Pereira City. To determine the distance (near or far), a layer of roads of the municipality was assembled [40] and a distance matrix was created. The type of road was taken into account (levels of difficulty according to the conditions of the roads—earthen roads to cement concrete road—where the value ranged from 1 to 7, with 7 being the weight of the road with the greatest difficulty to be traveled by farmers to carry agricultural production to the city). The matrix was generated from the farms to the market place in Pereira. The result was a matrix with the weight of the roads (distance in meters and the value in difficulty of the roads to reach the center of Pereira) (Table A1). To analyze the characteristics of the farmers and farms, a multivariate analysis was performed using the age and gender of farmers and the distance to the urban area, type of crops, production destination, and area. A multiple correspondence analysis (MCA) in R was used to explain the relationship between types of farmers and farms' characteristics.

2.3.2. Farmer's Emotions Associated with Farm Landscapes in the CCLC

We asked farmers about their emotions generated by living on these farm landscapes. We introduced different questions to explore their perceptions and facilitate the dialogue with farmers. The following questions were asked: What feelings or sensations do you have about the farm? What do you think about the place where you live? Responses were coded according to eight emotions associated with living on the farm, including tranquility, happiness, rootedness, safety, awe, vitality, freedom, and interest (see Tables A1 and A2). Several emotions could be associated with one single response. A Kruskal–Wallis analysis was performed to find correlations between farmers' and farms' characteristics and the diversity of emotions (Figure 3).

2.3.3. Sense of Place of Local Communities in the CCLC

The sense of place within farm landscapes was examined by using multiple questions, including Do you like living on the farm? Why do you remain on the farm? Responses were coded according to motivations to continue living on the farm and classified as place, pleasure, identity, tranquility, air quality, freedom, labor, no poverty, security, and support community (see Tables A1 and A2). The Kruskal–Wallis analysis is a nonparametric test for comparing variances of more than two variables and it was used to explore differences between farmers' and farms' characteristics with motivations to continue living in the CCLC (Figure 3).

2.3.4. Diversity of Nature's Contributions to People Provided by Farms in the CCLC

To explore the diversity of NCP associated with farms, the following questions were asked, including What do you like most about living on your farm? What does the farm offer you? Each response was transcribed and classified into the material and nonmaterial NCP proposed by Díaz et al. [21]. Considering the mean of the responses, a Kruskal–Wallis analysis was performed to explore the relationship between farmers' and farms' characteristics and NCP (Figure 3). NCP were grouped into material, nonmaterial, and regulating categories. In these categories NCP18 was not included because this contribution is considered in the three groups (material, nonmaterial and regulating NCP) for Diaz et al. [21]. For this reason, we analyzed it separately (see Tables A1 and A2).

2.3.5. Visions of Local Communities Regarding to the Future of the CCLC

To explore how farmers and their families perceive the future of the rural landscape in the CCLC, we asked how do you imagine the future of the rural landscape? Responses were classified according to three categories: disappearance of rural areas, displacement, and uncertainty due to change. Additionally, we asked farmers to express motivation underpinning their responses, which were classified as both direct and indirect drivers of global change [41,42], including sociopolitical change and land-use change, as well as economic, cultural, and climate change (Figure 3). Two direct drivers were mainly recognized as change promoters in the region, i.e., land-use change and climate change. Additionally, we recognized visions associated with three indirect drivers: economy, political, and culture [43]. The economy was defined as per capita income and the taxes and subsidies provided by the government; the political reasons were defined as the mechanisms for the development of the rural sector; the culture was determined as values, beliefs, and norms that a group of people share.

3. Results

3.1. Farmers and Farms' Characteristics

Farmers interviewed were mainly from Risaralda (15 farmers), Valle del Cauca (5), Caldas (3), Quindío (2), and Antioquia (1). Only one farmer did not express its place of origin. The interviewees were made up of farm owners (48%), owner-managers (26%), managers (19%), and farm workers (7%). The age of the interviewees ranged from 26 to 85 years old, and the time spent in the region ranged from 3 to 69 years.

We found that 33.3% of farmers were female. It was also found that 44.4% of them were between 60 and 85 years of age (elderly). The educational level was heterogeneous, with 33% of farmers with no studies, 22% with elementary school, 19% with high school, 11% with a university degree (15% of responses were not registered). We found that the most common family type was the extended family (i.e., more family members live in the household, such as grandparents, aunts, uncles, cousins, etc.), followed by the nuclear family (parents and children). Regarding the farms' characteristics, we found that 66.7% of the farms showed changes in land use between 1997 and 2014, the most dominant being a land transition from coffee to heterogeneous agriculture practices. We also found that some farms persisted despite being located in urban cover areas. Finally, 55.6% of the farms were located far from the urban area (Table 1).

Table 1. Farmers and farms' characteristics in the CCLC.

	Variable	Category	Range	*n*	Average	Used in MCA
Farmers	Gender	Female		9	33%	√
		Male		18	67%	
	Age (years)	Adult	26–59	15	56%	√
		Elderly	60–85	12	44%	
	Relationship with the farm	Managers		5	19%	
		Owner-managers		7	26%	
		Owners		13	48%	
		Workers		2	7%	
	Educational level	No data		4	15%	
		No study		9	33%	
		Primary school		6	22%	
		High school		3	11%	
		Technical		1	4%	
		Technology		1	4%	
		University degree		3	11%	
	Family type	Childless couples		3	11%	
		Extended		12	44%	
		NA		2	7%	
		Nuclear		10	37%	
Farms	Altitude (m.a.s.l.)	Low	1221–1572	16	59%	
		High	1573–1922	11	41%	
	Area (ha)	<14 ha	0.5–14	23	85%	√
		>14 ha	14–28.8	4	15%	
	Type of crops *	Traditional		20	74%	√
		Innovative		7	26%	
	Type of crops generating income **	Monoculture		7	26%	
		Subsidiary		20	74%	
	Time on the farm (years)	>36	36–69	14	52%	
		<36	26–36	13	48%	
	Hiring personnel	No		11	41%	
		Yes		16	59%	
	Destination of the production	Sale and self-consumption		17	63%	√
		Sale		10	37%	
	Farming style	Peasant		18	67%	
		Semi-industrial		9	33%	
	Land-cover change in 1997–2014	Yes		18	67%	
		No		9	33%	
	Distance	Near	18,957–38,530	12	44%	√
		Far	38,531–58,102	15	56%	

* Type of crops: Traditional, are defined as those crops that have always been cultivated in the area of the farm (coffee, banana and citrus); Innovative, refers to crops that have not been traditional in the area, are new to the area of study (tropical flowers, succulents, vegetables). ** Type of crops generate income: It is related to the type and number of crops that provide the economic income for the farm. Monoculture: one crop; Subsidiary: Several crops contribute to income.

Two farming styles were identified, 66.7% with arrangements tending towards peasant and 33.3% towards semi-industrial farms (Table 1). The peasant style was characterized by farms with an area of less than 14 ha, with traditional crops, crop association, no hired personnel, and production destined for sale and self-consumption. In addition, in the peasant style, the person in charge of the farm's activities and administration was the owner or an administrator who had been on the farm for more than 37 years. On the other hand, farms with a semi-industrial style were represented by farms with more than 14 ha, dominated by novel monocultures, with hired personnel for field work and the production was destined for sale. Additionally, we found that the person in charge of the farm was an administrator or hired worker who had been with the farm for less than 36 years.

The MCA differentiated significant associations between farmers' and farms' characteristics. Dimension 1 identified the relationship between the variables farm with area greater than 14 ha and destination of the production for sale, while in dimension 2, the variables that contributed the most were crop type, monoculture, and distance near and far; in dimension 3, they were female and male genders (Figures 4 and A2–A4). The first three dimensions explained 76.1% of the variance. We found an associated statistical significance in dimension 1 (36.7% of the variance) and in dimension 2 (26.3%). On the X-axis (dimension 1), we found a good separation of farms according to area and production destination. On the Y-axis (dimension 2), the farms were distributed in relation to type of crop and distance (Figures 4 and A1). Farmers older than 60 years old were mainly female and their production was for sale and self-consumption.

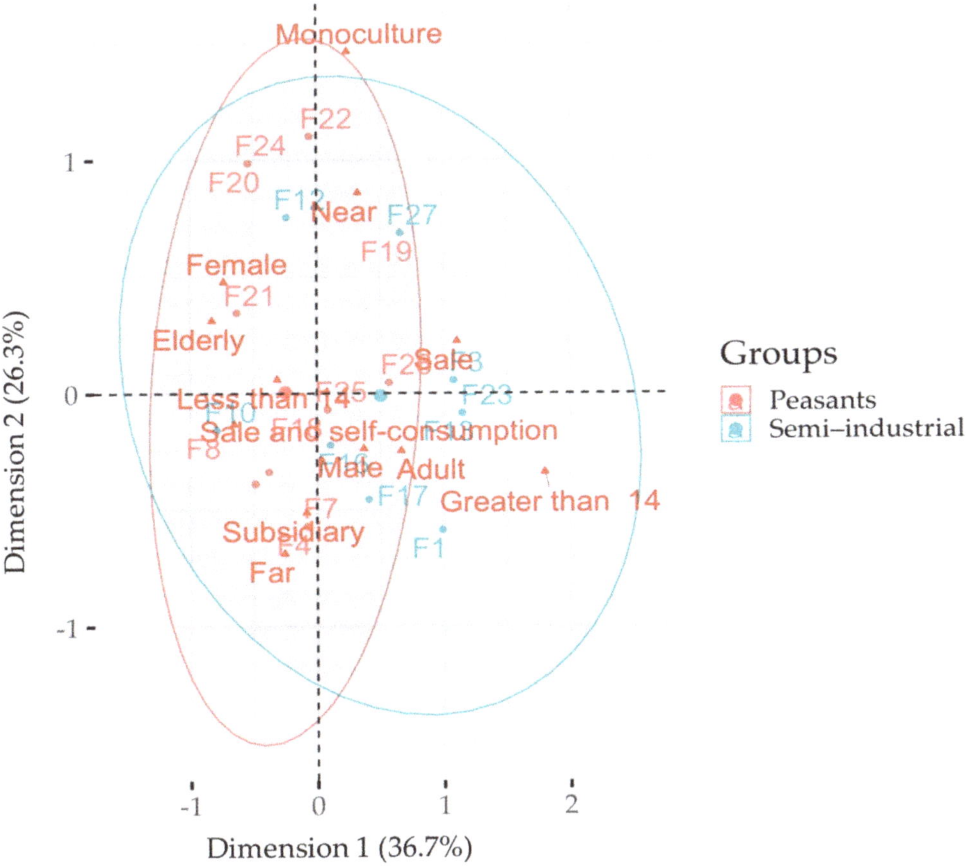

Figure 4. Multiple correspondence analysis of farmers' and farms' characteristics.

3.2. Diversity of Emotions Associated with Farmlands

Results showed that farmers identified multiples emotions associated with living on the farmlands of the CCLC. Examples of these emotions included "the farm is a lot of peace, silence and tranquility" (tranquility); "The farm makes my soul happy" (happiness); "I don't know. I feel nostalgia when I work in the fields because I remember my father, I imagine him working there" (rootedness); "The farm generates security" (safety); "The farm is wonderful" (awe); "The farm is life, I breathe pure and clean air" (vitality); "The farm is freedom" (freedom); and "Through the work on the farm I think and begin to philosophize" (interest). According to the classification of the emotions used, we found that tranquility (69%), happiness (31%), rootedness (27%), and safety (23%) were the most common emotions or feelings associated with farm landscapes, followed by awe and vitality (15%), freedom (12%), and interest (12%) (Figure 5).

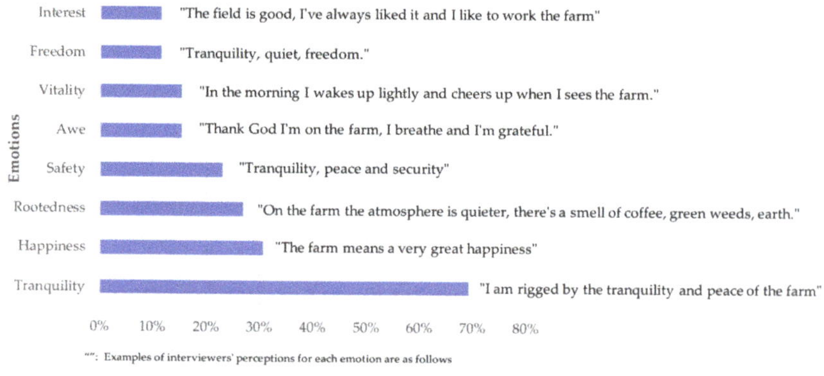

Figure 5. Recognition of emotions generated by living on the farm.

We found that gender and age were significantly related to family rootedness ($p < 0.05$ for gender and $p < 0.10$ for age). Rootedness is understood as the affective bond they have in accordance with the identity to the farm. Additionally, according to the farming style, we found a significant relationship between the contribution of the farm to human safety ($p < 0.05$), tranquility ($p < 0.1$) and happiness, admiration and vitality ($p < 0.15$). Finally, we also found a correlation between the farm distance to urban areas and the perceptions regarding farm rootedness and safety ($p < 0.15$ for both emotions) (Table 2).

Table 2. Variables that influence the different types of senses on the farm.

Variables	Tranquility	Happiness	Rootedness	Safety	Awe	Vitality	Freedom	Interest
Farming style	**	*	***			*	*	
H of Kruskal–Wallis	2.889	2.138	8.357	1.486	0.000	2.261	2.261	1.625
Degree of freedom	1	1	1	1	1	1	1	1
Two-sided p-value	0.089	0.144	0.004	0.223	1.000	0.133	0.133	0.202
Distance			*	*				
H of Kruskal–Wallis	0.000	0.214	2.321	2.684	0.650	0.057	0.057	0.650
Degree of freedom	1	1	1	1	1	1	1	1
Two-sided p-value	1.000	0.644	0.128	0.101	0.420	0.812	0.812	0.420
Gender				***				
H of Kruskal–Wallis	0.722	1.368	0.929	4.550	1.625	0.565	0.141	1.625
Degree of freedom	1	1	1	1	1	1	1	1
Two-sided p-value	0.395	0.242	0.335	0.033	0.202	0.452	0.707	0.202
Age				**				
H of Kruskal–Wallis	0.000	1.445	0.371	3.352	0.163	1.710	0.057	0.163
Degree of freedom	1	1	1	1	1	1	1	1
Two-sided p-value	1.000	0.229	0.542	0.067	0.687	0.191	0.812	0.687

Signification of codes: 0.05, '***'; 0.1, '**'; 0.15, '*'.

3.3. Sense of Place of Local Communities in the CCLC

Regarding the farmers' motivation to remain on these farm landscapes in the near future, we found that 85% of the farmers expressed a positive motivation to remain in the CCLC, while 11% of farmers responded negatively, and 4% felt uncertainty. The most frequent motivations for remaining in this region were associated with the recognition of farms as their place (85%), followed by pleasure and well-being of living there (37%), a collective recognition of the countryside as a home (identity) (33%), tranquility, air quality (clean and no noise), and the freedom of being in open spaces (22%). To a lesser extent, we also found labor (19%), fullness (15%), and farming security (11%) to be important motivations to remain in the region (Figure 6).

Figure 6. Farmer's motivations to remaining in the CCLC. Signification of codes: 0.05, '***'; 0.1, '**'; 0.15, '*'.

Regarding gender, we found that men were strongly connected to farm tasks ($p < 0.1$), while women mainly valued being on the farm the most ($p < 0.05$) and the recognition of the farms as a home ($p < 0.15$). Regarding age, we found that older adults (over 60 years old) were more willing to remain on the farm due to the recognition of the farm as a place to live ($p < 0.05$) (Figure 6).

We also found that farms with a semi-intensive farming style valued tranquility more than farms with a peasant style ($p < 0.05$). However, peasant farms recognized farms as dwelling, providing pleasure and identity ($p < 0.1$) as motivations to remain. Regarding the distance to urban areas, we observed that the farms closer to the urbanized areas showed motivations to remain associated with fullness ($p < 0.05$). On the contrary, farms located farther were more associated with benefits linked to tranquility, air quality, and security ($p < 0.15$) (Figure 6).

3.4. Nature's Contributions to People in the CCLC

Of the eighteen NCPs, farmers identified seven NCPs associated with the farm landscapes of the CCLC (Figure 7). We found that nonmaterial NCP were the most commonly associated with farmlands, including physical and psychological experiences (NCP16, 85%), maintenance of options (NCP18, 74%), and supportive identities (NCP17 56%). We also found regulating NCP such as habitat creation and maintenance (NCP1, 52%) and air quality regulation (15%) to be important contributions in this region.

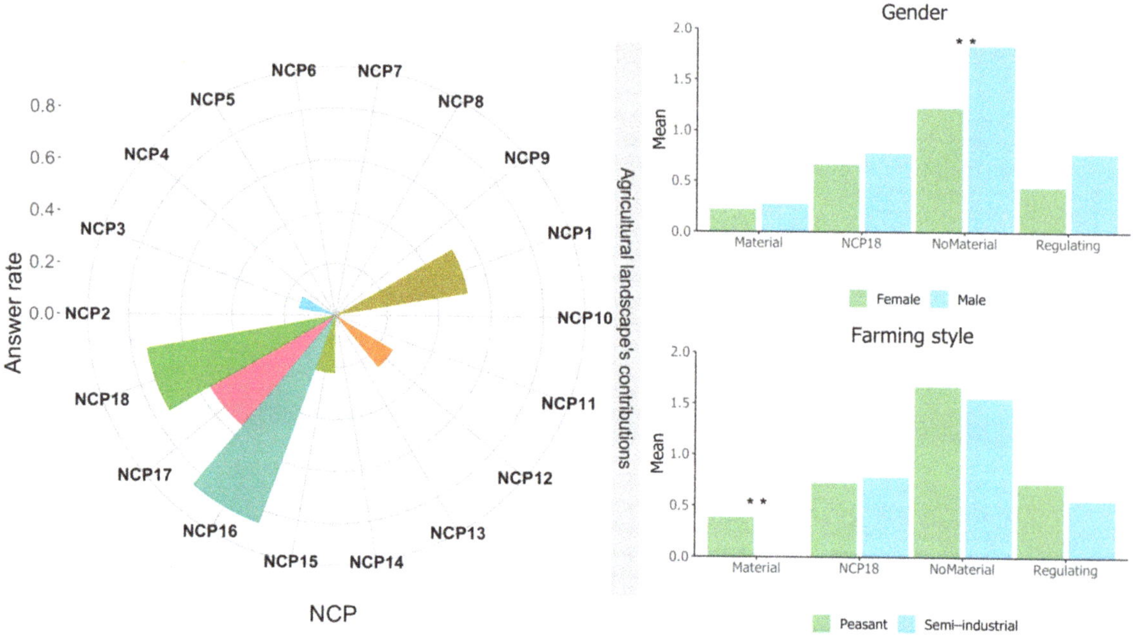

Figure 7. Farmer's perception of NCP provided by farms in the CCLC. NCP1, habitat creation and maintenance; NCP2, pollination and dispersal of seeds and other propagules; NCP3, regulation of air quality; NCP4, regulation of climate; NCP5, regulation of ocean acidification; NCP6 regulation of freshwater quantity, location, and timing; NCP7, regulation of freshwater and coastal water quality; NCP8, formation, protection, and decontamination of soils and sediments; NCP9, regulation of hazards and extreme events; NCP10, regulation of detrimental organisms and biological processes; NCP11, energy; NCP12, food and feed; NCP13, materials, companionship, and labor; NCP14, medicinal, biochemical, and genetic resources; NCP15, learning and inspiration; NCP16, physical and psychological experiences; NCP17, supporting identities; NCP18, maintenance of options. Signification of codes: 0.1, '**'.

We found significant differences in the mean response for nonmaterial and material NCP across gender and farming style. Male identified more nonmaterial NCP than women ($p < 0.1$) (Figure 7). Male recognized farms as spaces where identities are supported, a source of satisfaction and experiences, family rootedness, and agricultural traditions. Among the stories recorded, we found examples such as "Every night there is a longing for the work of the other day" (rural man, 71 years old); "All my life I have lived in the countryside, I have always liked it. And in the area, everything is very quiet, it is safe" (rural man, 72 years old); "The farm gives me tranquility and brings back memories of my childhood, of my tradition. And it is also safe" (rural man, 26 years old).

Regarding farming styles, we found significant differences in relation to the material NCP ($p < 0.05$). In this sense, farms with peasant farming styles identified the importance to secure food for families. An example of stories reflecting this is: "On the farm there is always food within reach and there is no money involved" (rural woman, 37 years old); in the peasant style the production of the farm is destined both for sale and for self-consumption; on the contrary, farms with a semi-intensive style orient all their production for sale and do not recognize these material contributions of the agricultural landscape to the well-being of the rural family (Figure 7).

3.5. Visions of Local Communities Regarding the Future of the CCLC

Diverse visions were found associated with future changes in land use, including "the growth of the city" and "destruction of the natural environment for urban expansion", while the climate change was mostly recognized with visions such as "the change that has occurred in the rural sector has been mainly due to climate change" and "changes in the climate are quite perceived, the rainy and sunny seasons are more intense". We found visions related to "rural work is very hard and poorly paid", "rural people want to go to the city in search of better opportunities", "agricultural production is not profitable" and "the government will not let agricultural production end" are reasons included in the economy category. The political visions found were mainly related to "farmer is unprotected, has no social security", "the government does not support the field for lack of regulation and protection" and "the promotion of sustainable tourism with the people of the area". Moreover, the visions linked to culture were related to arguments such as "young people do not want to continue with the farm and work it" and "there is no one to work the land".

Of all visions found, 41.7% of farmers considered that the rural areas will disappear in the near future, while 33.3% of them expressed uncertainty and 25% believed that displacement to another site was the most likely option. Forty-seven reasons were collected supporting these visions of the future of the CCLC, mostly justified by arguments related to changes in land uses (27.7% of farmers), followed by economic arguments (23.4%), and sociopolitical and cultural arguments (21.3%). Farmers recognized climate change as a lesser force for future changes in rural areas, with 6.4% (Figure 8).

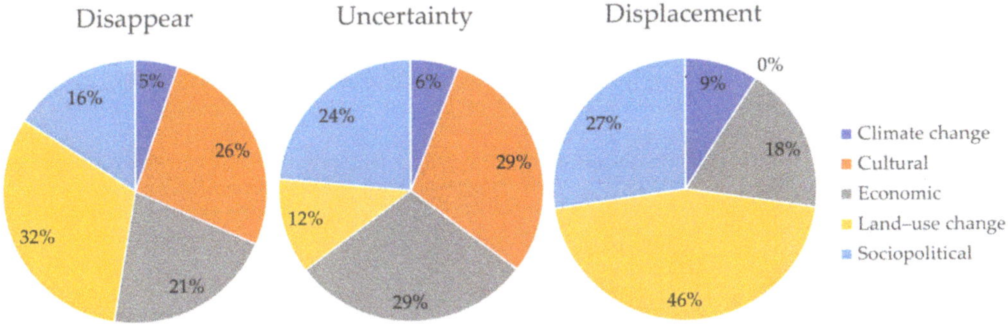

Figure 8. Farmers' perspectives and supporting arguments regarding the future of CCLC.

4. Discussion

4.1. Farmers and Farms' Characteristics in the CCLC

Our results identified nonmaterial NCP (e.g., as physical and psychological experiences maintenance of options and supportive identities) to be the most important contributions shaping the sense of place of farmers in the CCLC. This is consistent with several studies that have shown the long history of how rural families have developed cultural roots and have coevolved with farming landscapes in multiple intangible ways and forms [9,13]. Additionally, we found that farms' characteristics (e.g., farming styles, distance to cities, and gender) may play an important role in articulating farmers' meanings and attachment to these farmlands. In this sense, different farming styles appeared to be associated with the particular meanings and perceptions that farmers hold to the territory. This finding is consistent with several other studies where inhabited places reflected people's values, histories, material, and symbolic practices [16,17,32], thus indicating the importance of farming practices in shaping different levels of human connection to nature and in forming land stewardship [44–47]. Specifically, we found that two farming styles, peasant and semi-industrial, are shown to be influencing the farmers' perception toward particular NCP and emotions associated with farmlands (Figure 7, Table 2), and with motivations to remain in the CCLC (Figures 6 and 7). These results are consistent with findings in the study Heterogeneity reconsidered [11], that showed the importance of comanagement of territory with communities for promoting land transitions that preserve and shape the sense of place within the land.

We found that gender played an important role related to the emergence of "pluriactivity" in farmlands, which in the theories of the new rurality, stands out as the incursion of women to generate income in especially nonagricultural activities. This is a relevant result because it changes the configuration of the sense of place, incorporates into future analyses the perspective of gender equity and the participation of different social actors in development processes and projects. Then, the examination of the role of NCP in the configuration of the farmers' sense of place in the CCLC allowed an inquiry about the new family configuration with increasing participation of women (33%), which assigns new functions to rural spaces in the ways of perceiving material, nonmaterial and regulatory NCP (Figures 4, 6 and 7). A gender-inclusive analysis showed that men and women often value NCP in different ways and may possess diverse knowledge, with implications for the value of places for management priorities [48] and the formulation and implementation of sustainable and equitable policies and interventions [49].

4.2. Sense of Place in the CCLC

Sense of place is defined as the meanings and attachments that people possess in a territory [12,16,17] (Figure 9). Our findings were able to identify specific NCP underpinning the diversity meanings and the attachment of Pereira's farmers to the farmlands in the CCLC.

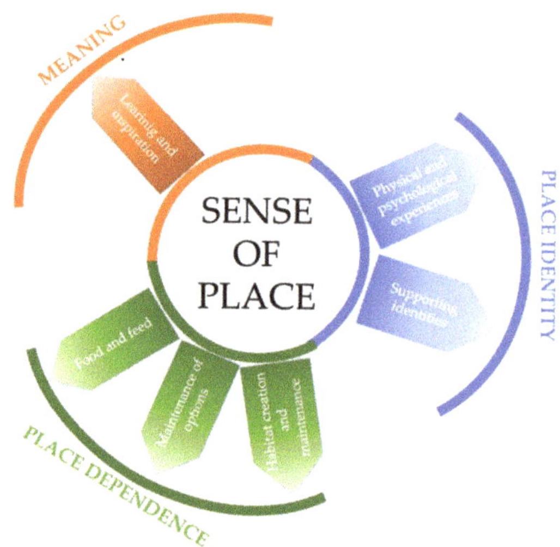

Figure 9. Characterization of the sense of place in the CCLC (adapted from Masterson et al. [16]).

Firstly, the diversity of meanings found were mainly interpreted through the diversity of emotions towards farmlands and the opportunities associated with learning and inspiration (NCP 15). Examples of these emotions included tranquility, happiness, freedom, and interest, and can be interpreted as reflections of farmers' experiences of living in the CCLC. This result is aligned with findings of Rajala and Sorice [12] that showed how landowners' emotions can contribute directly to farmers' emotional health.

Secondly, as defined by [50], attachment to a place is developed through daily and sustained interactions, as well as a strong motivation to maintain the relationship with a place over time. Our findings identified a relationship between farmers and place in the CCLC (i.e., interpreted the place identity and dependence). Place identity defines an individual's personal identity with the physical environment [16,17]. Our study captured the reasonings for remaining in the CCLC such as "I am a peasant and very proud of my roots", which reflects on how a person's identity is linked to a place and depends on specific farmland contributions, such as supporting identity (NCP17) and psychological experiences (NCP16) associated with living on a farm. Here, we argue that this qualitative information must be used for the understanding of identity and attachment along agricultural landscapes [21]. These findings also support recent insights that have shown how the landscapes of the CCLC are intrinsically connected with cultural assets and meanings ascribed to farmlands [19,21,32].

The place dependency to farmlands conveys an instrumental connection between people and place, conceived and measured as the capacity of an environment to facilitate the achievement of goals and satisfy important needs [16]. Our results found that the most important material nature's contribution to people identified in farms of the CCLC was food production (NCP 12), which is crucial to sustain livelihoods of local communities. In addition, one of the strongest reasons given by farmers to remain on the farmlands of the CCLC was the place where they inhabit themselves, which can be interpreted as a way to recognize the capacity of this region to provide security and support tranquility of livelihoods (Figure 6). Another example of place dependency bonds to the land was revealed to be communities' perceptions of farms as a space that maintains the options for a good quality of life (NCP 18) and as a place where they have been able and can continue to develop their livelihoods and persist. This may reflect how place dependence enhances place identity and in turn influences people's responsible behavior [46,50]. Here, we argue that this finding can be interpreted as evidence that farmers do not perceive themselves as separated from their farms in the Pereira CCLC.

Finally, results obtained in this study must be interpreted in the context of some limitations. First, one limitation had to do with the impossibility of sampling a larger number of farms and farmers due to the lack of financial resources and the need for additional fieldwork research assistants. Second, another limitation had to do with the difficulty in building trust with farmers, which influenced our ability to run more extensive interviews and obtain more precise information regarding the institutional aspects of farmland governance in the CCLC. Finally, the lack of security in the Pereira region greatly hindered sampling efforts in the study due to the local communities' distrust of visitors or foreigners.

5. Conclusions

UNESCO recognized in 2011 the CCLC as a world heritage site, which influenced Colombia laws and management plans for its preservation and care. This study provides empirical evidence of the important role that nature's contributions to people play in shaping the sense of place and land heritage in the CCLC. The diverse farms studied in the CCLC showed how the heterogeneity of farming styles are key for preserving biocultural diversity of this region, which demonstrates the strong relationship between sense of place and human behavior and provides evidence that affective attachment to lands can shape behavior towards nature protection. However, progressing on this direction requires time to build trust with farmers and financial and human resources to create collective planning strategies. Future work must address the need for robust and transparent deliberative and negotiation mechanisms that are inclusive of all relevant stakeholders (i.e., their perceptions and cultural differences), aim to address unequal power, and recognize and strengthen communities' governance within the CCLC.

Author Contributions: Conceptualization, B.E.M.-L., A.J.C. and A.F.-M.; methodology, B.E.M.-L. and A.J.C.; formal analysis, B.E.M.-L.; investigation, B.E.M.-L. and A.F.-M.; writing—original draft preparation, B.E.M.-L.; writing—review and editing, A.J.C. and A.F.-M.; visualization, B.E.M.-L.; supervision, A.F.-M. and A.J.C. All authors have read and agreed to the published version of the manuscript.

Funding: The doctoral thesis was supported by the internal funds of the Vice-Rectory of Research of the Technological University of Pereira, Colombia (E2-18-2). The first author was funded by a scholarship of COLCIENCIA-COLFUTURO (647), Colombia and a mobility grant funded by the AUIP.

Institutional Review Board Statement: Not applicable.

Informed Consent Statement: Informed consent was obtained from all subjects involved in the study. The authors were authorized by all interviewees based on Law 1581 of 2012, article 9 and 12, Regulatory Decree 1377 of 2013 article 10 of Colombia.

Data Availability Statement: Not applicable.

Acknowledgments: We thank all the farmers who kindly allowed us to enter their farms and who generously responded to our semistructured interview. This publication is part of the doctoral thesis in environmental sciences of the first author in the Graduate Programme of the Faculty of Environmental Sciences of the Technological University of Pereira. The doctoral thesis was supported by the internal convocation of the Vice-Rectory of Research of the Technological University of Pereira, Colombia (E2-18-2). The first author had a forgivable loan scholarship from COLCIENCIA-COLFUTURO for her doctoral studies (647), Colombia. This publication was funded by UAL. Finally, the authors appreciate the participation and support provided by the research group Socio-Ecological Research Lab at the University of Almeria and the Management in Tropical Andean Agroecosystems (GATA—Spanish acronym).

Conflicts of Interest: The authors declare no conflict of interest.

Appendix A

Table A1. Definition and summary of variables.

Objectives	Variables	Variable Definition. Brief Explanation
Farmers and farms' characteristics	Gender	Gender of the interviewed farmer
	Age (Years)	Age of the interviewed farmer
	Farming style	Comprise ways of organizing and reorganizing the internal and external requirements of the farms and are firmly rooted in a stock of cultural knowledge
	Land-use change	Whether or not there was a change in land cover around the farms between 1997 and 2014
	Distance	Distance variable as near and far from the most central collection center in the city
Human Emotions	Tranquility	Quality or state of being tranquil; calmness; peacefulness; quiet; serenity; free from or unaffected by disturbing emotions; unagitated; serene; placid
	Happiness	State of pleasant spiritual and physical satisfaction
	Rootedness	An affection, a virtue, a use or a habit: to become very firm; to establish oneself permanently in a place, binding oneself to people and things
	Safety	Quality of a site that provides security, certainty, confidence
	Awe	To see, contemplate or consider with special esteem or pleasure something that calls our attention because of qualities judged as extraordinary.
	Vitality	1. f. Quality of having life; 2. f. activity or efficiency of the vital faculties (quality of life)
	Freedom	The natural ability of people to act in one way or another, and not to act, so they are responsible for their actions
	Interest	Inclination or attraction felt towards an object or activity they like; activity that is done habitually and for pleasure in leisure time

Table A1. *Cont.*

Objectives	Variables	Variable Definition. Brief Explanation
Reasons to remain	Place	Farmers define it as the space in which their place is; the farm is more than that space for agricultural production; it is the personal relationship with the territory, where production takes place, where the family lives and is formed
	Pleasure	Pleasure is related to the feeling of well-being generated by staying on the farm and not being in the city
	Identity	Farmers are defined in relation to the farm, the rural life, working in the field and being a farmer; it is related to the roots and tradition
	Tranquility	Quality or state of being tranquil; calmness; peacefulness; quiet; serenity. Free from or unaffected by disturbing emotions; unagitated; serene; placid.
	Air quality	They express that the air on the farm is clean
	Freedom	The natural ability of people to act in one way or another, and not to act, so they are responsible for their actions. Farmers express freedom on the farm as open space, open doors and windows; the possibility of going from one place to another without restrictions in the space itself
	Labor	Related to always having something to do, being busy, and feeling useful
	No poverty	They express that there is never a lack of food on the farm no matter how difficult the situation
	Safety	Quality of a site that provides security, certainty, confidence
	Support community	Strength in the relationship with the community; the neighborhood that exists; the support and care provided to each other
Nature's contributions to people (NCP)	Habitat creation and maintenance	"... conditions necessary or favorable for living beings of direct or indirect importance to humans"
	Regulation of air quality	Perception "... Filtration, fixation, degradation or storage of pollutants that directly affect human health or infrastructure"
	Food and feed	"Production of food from wild managed, or domesticated organisms"
	Learning and inspiration	"Provision, by landscapes, seascapes, habitats or organisms, of opportunities for the development of the capabilities that allow humans to prosper through education, acquisition of knowledge and development of skills for well-being, information, and inspiration for art and technological design"
	Physical and psychological experiences	"Provision, by landscapes, seascapes, habitats or organisms, of opportunities for physically and psychologically beneficial activities, healing, relaxation, recreation, leisure, tourism and aesthetic enjoyment based on the close contact with nature"
	Supporting identities	Landscapes, seascapes, habitats or organisms being the basis for religious, spiritual, and social-cohesion experiences; source of satisfaction derived from knowing that a particular landscape, seascape, habitat or species exists
	Maintenance of options	Capacity of ecosystems, habitats, species or genotypes to keep options open in order to support a good quality of life

Appendix B

Methodological Approach

We investigated the role of the diversity of NCP in shaping the sense of place in the CCLC. We followed the approach of Masterson et al. (2017) where place attachment and place meanings are described as key concept to understand the motivation for stewardship and actions to care for the environment and use the resources.

In this sense, (i) the farmers' emotions associated with farm landscapes are connected with place meanings; (ii) the reasons of local communities for remaining on the farm, approaches to place attachment and (iii) the diversity of nature's contributions to people provided by farm landscapes are used to try to explain both meanings and attachment to place (Table A2).

Table A2. Questions, quotes and codes.

Objective	Questions	Examples of Responses	Interpretation Source	Code
Emotion	What do you like most about living here on your farm? What feelings or sensations do you have about the farm? What do you think about the place where you live?	"I don't know. I feel nostalgic when I work in the fields because I remember my father, I imagine him working over there" "The farm generates a feeling of tranquility, happiness to the soul and security" "I am happy to see this beautiful place. The morning rises lightly and perks up when you see the farm"	Responses were interpreted and classified as different emotions that emerged from the dialogue.	Tranquility Happiness Rootedness Safety Awe Vitality Freedom Interest
Reasons to remain	Do you like living on the farm? Why do you remain on the farm?	"To live in the field because I don't like to live in the city, because in the city there are bad influences. I would not like to live in the city because there is nothing to do there, I would only interfere with the family" "Life in the city is very busy. Here in the field everything is peaceful, I only worry about having my plants well and if I want a banana or a mango, the land itself gives them to me"	Responses were interpreted and classified as different reasons to remain on the farm that emerged from the dialogue.	Place Pleasure Identity Tranquility Quality air Freedom Labor No poverty Security Support community
Nature's contributions to people	What do you like most about living on your farm? What does the farm offer you?	" . . . I always like to be working the field"; "It is easier to educate children in the field, neighbors take care of all children" (NCP code: Learning and inspiration. NCP15); "It is a calm and healthy life"; "The farm gives me tranquility" (NCP code: Physical and psychological experiences. NCP 16); "I consider myself a peasant of -pura cepa- and very proud, it's my rootedness "; "I like the farm because I was born there, it is a matter of tradition" (NCP code: Supporting identities. NCP 17) "You find everything on the farm. I like everything on the farm, living on it because you live in peace" (NCP code: Maintenance of options. NCP 18)	Responses were interpreted and classified as different NCP that emerged from the dialogue. We use the framework developed by Díaz et al. 2018	NCP1, habitat creation and maintenance; NCP3, regulation of air quality; NCP12, food and feed; NCP15, learning and inspiration; NCP16, physical and psychological experiences; NCP17, supporting identities; NCP18, maintenance of options

Appendix C

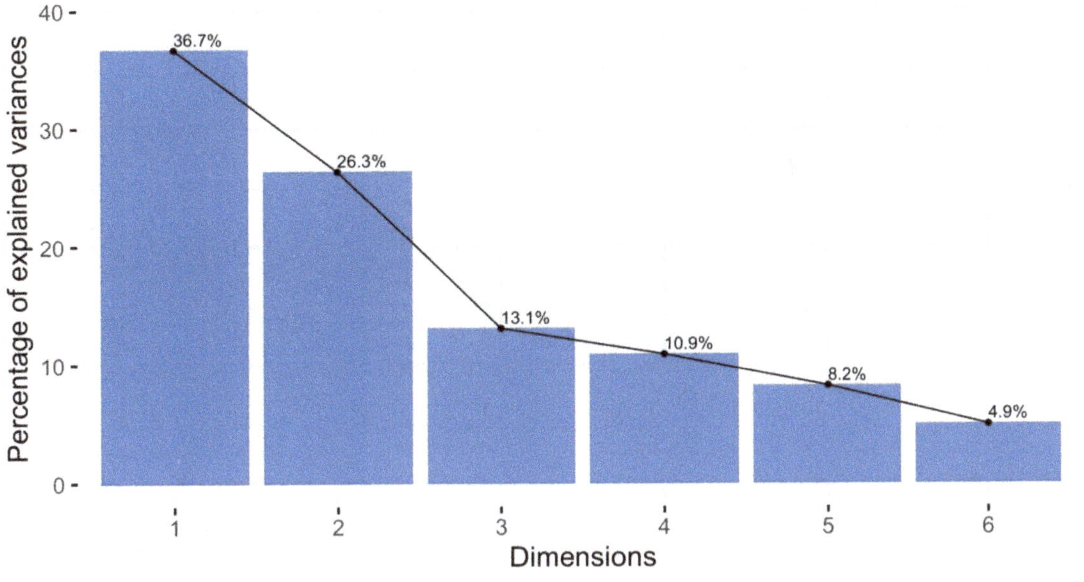

Figure A1. ACM's dimensions and percentage of explained variances.

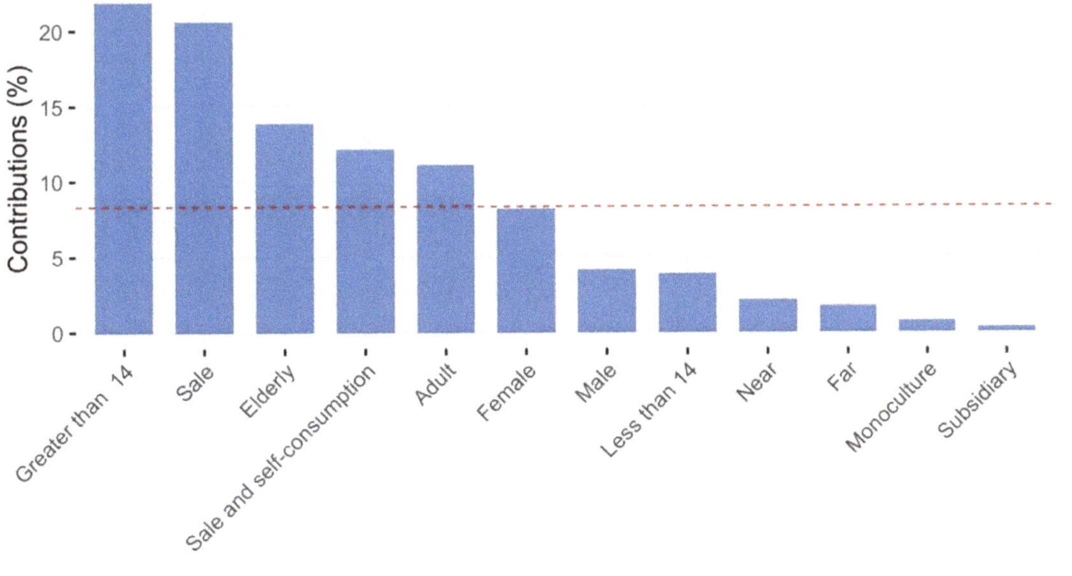

Figure A2. Contribution of variables to dimension 1.

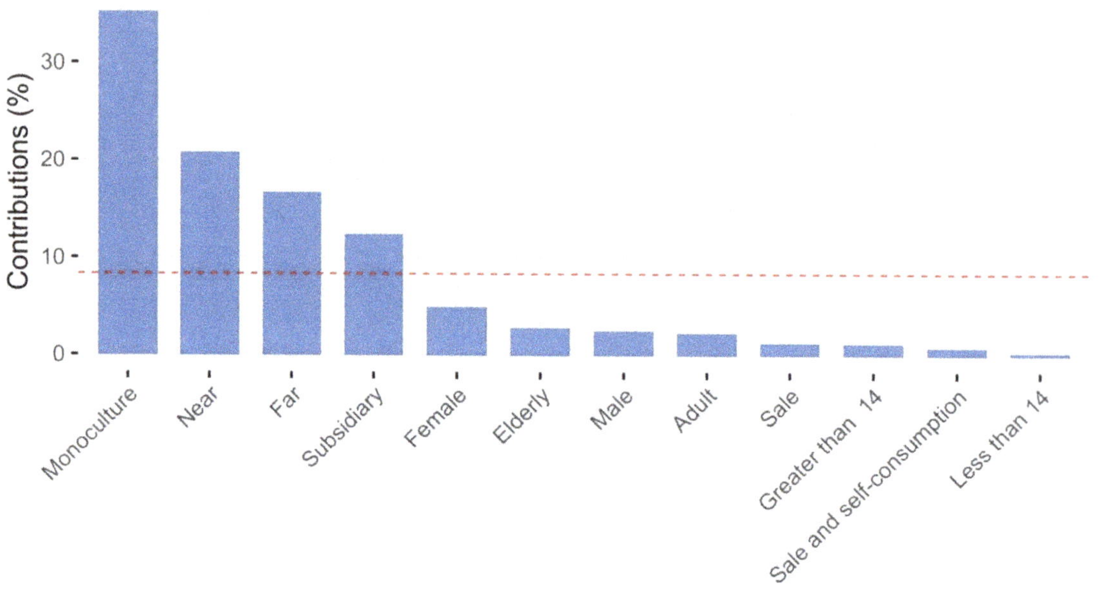

Figure A3. Contribution of variables to dimension 2.

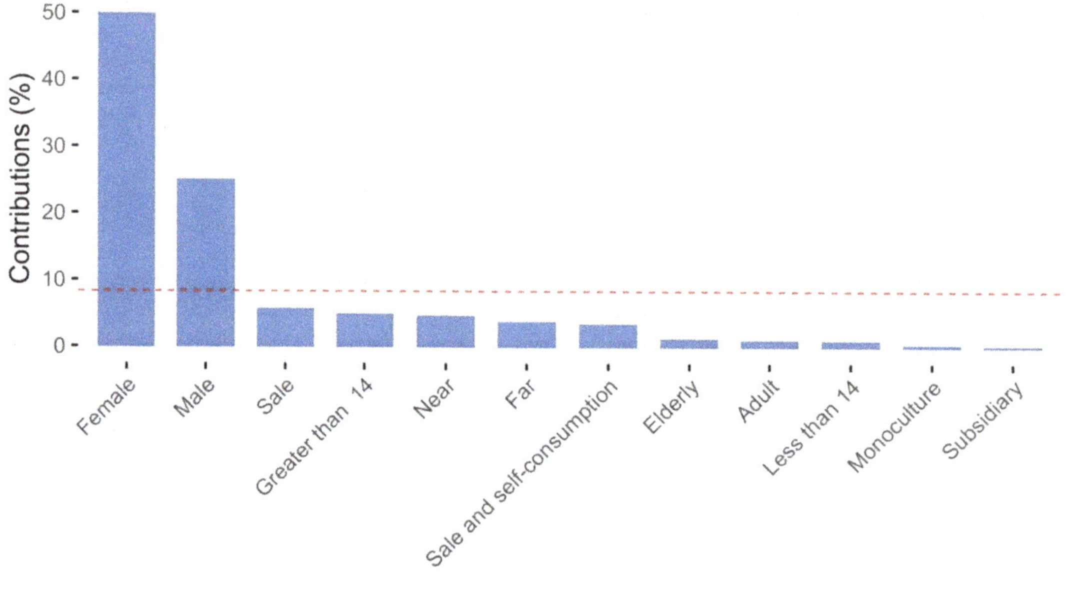

Figure A4. Contribution of variables to dimension 3.

References

1. IPBES. *Global Assessment Report on Biodiversity and Ecosystem Services of the Intergovernmental Science-Policy Platform on Biodiversity and Ecosystem Services*; Brondizio, E.S., Settele, J., Díaz, S., Ngo, H.T., Eds.; IPBES Secretariat: Bonn, Germany, 2019; 1148p. [CrossRef]
2. Khettab, S.; Chabbi-Chemrouk, N. Sense of place in the coastal town of Tipaza in Algeria: Local-community's socio-cognitive representations. *Int. J. Sustain. Built Environ.* **2017**, *6*, 544–554. [CrossRef]

3. Rivera Pabon, J.A. Del periodo precolombino al mito fundacional de Pereira: Cien siglos de historia previa. *Rev. Antropol. Sociol. Virajes* **2014**, *16*, 67–102. Available online: https://revistasojs.ucaldas.edu.co/index.php/virajes/article/view/924 (accessed on 8 February 2022).
4. Toro Zuluaga, G. Eje cafetero Colombiano. *Rev. Cienc. Hum.* **2005**, *35*, 127–149.
5. Rettberg, A. Global markets, local conflict. Violence in the Colombian coffee region after the breakdown of the International Coffee Agreement. *Lat. Am. Perspect.* **2010**, *37*, 111–132. [CrossRef]
6. Feijoo, A.; Carvajal, A.F.; Zúñiga, M.C.; Quintero, H.; Fragoso, C. Diversity and abundance of earthworms in land use systems in central-western Colombia. *Pedobiologia* **2011**, *54*, S69–S75. [CrossRef]
7. Rojas-Cano, A.; Martínez, A.F.; Molina-Rico, L.J.; Zúñiga-Torres, M.C.; Vargas, H.Q. Caracterización de agricultores y estrategias conducentes a políticas públicas en el Eje Cafetero colombiano. *Rev. Colomb. Cienc. Soc.* **2021**, *12*, 165–192. [CrossRef]
8. Zuñiga Torres, M.C. Perspectivas Ambientales para las Juventudes Rurales en Sistemas de Actividades de Fincas, Eje Cafetero Colombiano. Ph.D. Thesis, Universidad Tecnológica de Pereira, Pereira, Colombia, 2019.
9. Murillo-López, B.E.; Feijoo-Martínez, A.; Carvajal, A.F. Land cover changes in coffee cultural landscapes of Pereira (Colombia) between 1997 and 2014. *Cuad. Investig. Geogr.* **2022**. [CrossRef]
10. Van Der Ploeg, J.D. Farming styles research: The state of the art. In Proceedings of the Workshop on 'Historicising Farming Styles', Melk, Austria, 21–23 October 2010; pp. 1–15.
11. Van der Ploeg, J.D.; Ventura, F. Heterogeneity reconsidered. *Curr. Opin. Environ. Sustain.* **2014**, *8*, 23–28. [CrossRef]
12. Rajala, K.; Sorice, M.G. Sense of place on the range: Landowner place meanings, place attachment, and well-being in the Southern Great Plains. *Rangelands* 2021, *in press*. [CrossRef]
13. Hammes, V.; Eggers, M.; Isselstein, J.; Kayser, M. The attitude of grassland farmers towards nature conservation and agri-environment measures—A survey-based analysis. *Land Use Policy* **2016**, *59*, 528–535. [CrossRef]
14. Kosanic, A.; Petzold, J. A systematic review of cultural ecosystem services and human wellbeing. *Ecosyst. Serv.* **2020**, *45*, 101168. [CrossRef]
15. Ramos de Robles, S.L.; Feria Cuevas, Y. La noción de sentido de lugar: Una aproximación por medio de textos narrativos y fotografías. *Innov. Educ.* **2016**, *16*, 83–110.
16. Masterson, V.A.; Stedman, R.C.; Enqvist, J.; Tengö, M.; Giusti, M.; Wahl, D.; Svedin, U. The contribution of sense of place to social-ecological systems research: A review and research agenda. *Ecol. Soc.* **2017**, *22*, 49. [CrossRef]
17. Newell, R.; Canessa, R. From sense of place to visualization of place: Examining people-place relationships for insight on developing geovisualizations. *Heliyon* **2018**, *4*, e00547. [CrossRef]
18. Tengberg, A.; Fredholm, S.; Eliasson, I.; Knez, I.; Saltzman, K.; Wetterberg, O. Cultural ecosystem services provided by landscapes: Assessment of heritage values and identity. *Ecosyst. Serv.* **2012**, *2*, 14–26. [CrossRef]
19. Fish, R.; Church, A.; Winter, M. Conceptualising cultural ecosystem services: A novel framework for research and critical engagement. *Ecosyst. Serv.* **2016**, *21*, 208–217. [CrossRef]
20. Chan, K.M.; Gould, R.K.; Pascual, U. Editorial overview: Relational values: What are they, and what's the fuss about? *Curr. Opin. Environ. Sustain.* **2018**, *35*, A1–A7. [CrossRef]
21. Díaz, S.; Pascual, U.; Stenseke, M.; Martín-López, B.; Watson, R.T.; Molnár, Z.; Hill, R.; Chan, K.M.A.; Baste, I.A.; Brauman, K.A.; et al. Assessing nature's contributions to people. *Science* **2018**, *359*, 270–272. [CrossRef] [PubMed]
22. Murillo-López, B.E. Disponibilidad de Recursos y Tipos de Sistemas de Cultivo de Café y Plátano en la Cuenca del río La Vieja, Colombia. Master's Thesis, Universidad Tecnológica de Pereira, Pereira, Colombia, 2010.
23. Allen, K.E.; Quinn, C.E.; English, C.; Quinn, J.E. Relational values in agroecosystem governance. *Curr. Opin. Environ. Sustain.* **2018**, *35*, 108–115. [CrossRef]
24. Rajala, K.; Sorice, M.G.; Thomas, V.A. The meaning(s) of place: Identifying the structure of sense of place across a social-ecological landscape. *People Nat.* **2020**, *2*, 718–733. [CrossRef]
25. IDEAM (Instituto de Hidrología, Meteorología y Estudios Ambientales). Sistema de Información Hidrológica y Meteorológica —SISDHIM. 2018. Available online: http://www.ideam.gov.co/web/tiempo-y-clima/seguimiento-tiempo (accessed on 8 February 2022).
26. DANE (Departamento Administrativo Nacional de Estadística). *Resultados Censo Nacional de Población y Vivienda 2018*; DANE: Pereira, Colombia, 2019.
27. DANE (Departamento Administrativo Nacional de Estadística). *Colombia, Exportaciones (FOB) por Productos, Según Clasificación Cuentas Nacionales a Precios Corrientes*; DANE: Pereira, Colombia, 2017.
28. UNESCO. Coffee Cultural Landscape of Colombia. 2021. Available online: https://whc.unesco.org/en/list/1121/ (accessed on 8 February 2022).
29. Jaramillo, A. *El Clima de la Caficultura en Colombia*; Federación Nacional de Cafeteros Cenicafé: Bogota, Colombia, 2018; ISBN 978-958-8490-21-2.
30. PCC (Paisaje Cultural Cafetero). Paisaje Cultural Cafetero. 2021. Available online: http://paisajeculturalcafetero.org.co/ (accessed on 8 February 2022).
31. Fernández Retamoso, M.P.; Rincón Jaime, C.; Isaza Londoño, J.L.; Nieto González, C.E.; CRECE (Centro de Estudios Regionales Cafeteros y Empresariales). *Coffee Cultural Landscape. An Exceptional Fusion of Nature, Collective Human Effort and Culture*; Conceptos y Estrategias: Bogotá, Colombia, 2010.

32. Ryfield, F.; Cabana, D.; Brannigan, J.; Crowe, T. Conceptualizing 'sense of place' in cultural ecosystem services: A framework for interdisciplinary research. *Ecosyst. Serv.* **2019**, *36*, 100907. [CrossRef]
33. Zúñiga, M.C.; Feijoo, A.; Quintero, H.; Aldana, N.J.; Carvajal, A.F. Farmers' perceptions of earthworms and their role in soil. *Appl. Soil Ecol.* **2013**, *69*, 61–68. [CrossRef]
34. Tubb, D. Muddy decisions: Gold in the Chocó, Colombia. *Extr. Ind. Soc.* **2015**, *2*, 722–733. [CrossRef]
35. Auer, A.; Maceira, N.; Nahuelhual, L. Agriculturisation and trade-offs between commodity production and cultural ecosystem services: A case study in Balcarce County. *J. Rural Stud.* **2017**, *53*, 88–101. [CrossRef]
36. Van der Ploeg, J.D. Rural sociology and the new agrarian question. *Sociol. Rural* **1993**, *33*, 240–260.
37. Van Averbeke, W.; Mohamed, S.S. Smallholder farming styles and development policy in South Africa: The case of Dzindi Irrigation Scheme, Agrekon. *Agrekon* **2006**, *45*, 136–157. [CrossRef]
38. Vander Vennet, B.; Schneider, S.; Dessein, J. Different farming styles behind the homogenous soy production in southern Brazil. *J. Peasant Stud.* **2016**, *43*, 396–418.
39. Díaz-Porres, M. Producción-Descomposición de Hojarasca y Macroinvertebrados Fragmentadores en Cuatro Agroeco-Sistemas de la Cuenca del Río La. Master's Thesis, Universidad Tecnológica de Pereira, Pereira, Colombia, 2009.
40. IGAC (Instituto Geografico Agustin Codazzi). Geodatabase. 2019. Available online: https://geoportal.igac.gov.co/contenido/datos-abiertos-cartografia-y-geografia (accessed on 8 February 2022).
41. Quintas-Soriano, C.; Castro, A.J.; Castro, H.; García-Llorente, M. Impacts of land use change on ecosystem services and implications for human well-being in Spanish drylands. *Land Use Policy* **2016**, *54*, 534–548. [CrossRef]
42. MEA (Millenium Ecosystems Assessment). MA conceptual framework. In *Ecosystems and Human Well-Being: A Framework for Assessment*; Island Press: Washington, DC, USA, 2005; pp. 1–25.
43. Pörtner, H.O.; Scholes, R.J.; Agard, J.; Archer, E.; Arneth, A.; Bai, X.; Barnes, D.; Burrows, M.; Chan, L.; Cheung, W.L.; et al. *IPBES-IPCC Co-Sponsored Workshop Report on Biodiversity and Climate Change*; Intergovernmental Science-Policy Platform on Biodiversity and Ecosystem Services (IPBES) and Intergovernmental Panel on Climate Change (IPCC): Bonn, Germany, 2021. [CrossRef]
44. Gosling, E.; Williams, K. Connectedness to nature, place attachment and conservation behaviour: Testing connectedness theory among farmers. *J. Environ. Psychol.* **2010**, *30*, 298–304. [CrossRef]
45. Lokocz, E.; Ryan, R.L.; Sadler, A.J. Motivations for land protection and stewardship: Exploring place attachment and rural landscape character in Massachusetts. *Landsc. Urban Plan.* **2011**, *99*, 65–76. [CrossRef]
46. Ellis, E.C. Sharing the land between nature and people. *Science* **2019**, *364*, 1226–1228. [CrossRef] [PubMed]
47. Riechers, M.; Balázsi, Á.; Abson, D.J.; Fischer, J. The influence of landscape change on multiple dimensions of human–nature connectedness. *Ecol. Soc.* **2020**, *25*, 3. [CrossRef]
48. Bravo-Monroy, L. Coffee and potato agroecosystems: Social construction of spaces as a concept to analyse nature's contributions to people. *Front. Ecol. Evol.* **2021**, *9*, 607230. [CrossRef]
49. Fortnam, M.; Brown, K.; Chaigneau, T.; Crona, B.; Daw, T.M.; Gonçalves, D.; Hicks, C.; Revmatas, M.; Sandbrook, C.; Schulte-Herbruggen, B. The gendered nature of ecosystem services. *Ecol. Econ.* **2019**, *159*, 312–325. [CrossRef]
50. Hernández, B. Place attachment: Antecedents and consequences. *PsyEcology* **2021**, *12*, 99–122. [CrossRef]

MDPI
St. Alban-Anlage 66
4052 Basel
Switzerland
Tel. +41 61 683 77 34
Fax +41 61 302 89 18
www.mdpi.com

Agriculture Editorial Office
E-mail: agriculture@mdpi.com
www.mdpi.com/journal/agriculture

www.ingramcontent.com/pod-product-compliance
Lightning Source LLC
Chambersburg PA
CBHW040823250926
51850CB00007BA/1247
* 9 7 8 3 0 3 6 5 3 8 5 5 6 *